NETWORKS OF LEARNING AUTOMATA
Techniques for Online Stochastic Optimization

NETWORKS OF LEARNING AUTOMATA
Techniques for Online Stochastic Optimization

M. A. L. Thathachar

P. S. Sastry

Dept. of Electrical Engineering
Indian Institute of Science
Bangalore, India

KLUWER ACADEMIC PUBLISHERS
Boston / Dordrecht / New York / London

Distributors for North, Central and South America:
Kluwer Academic Publishers
101 Philip Drive
Assinippi Park
Norwell, Massachusetts 02061 USA
Telephone (781) 871-6600
Fax (781) 871-6528
E-Mail: <kluwer@wkap.com>

Distributors for all other countries:
Kluwer Academic Publishers Group
Post Office Box 322
3300 AH Dordrecht, THE NETHERLANDS
Telephone 31 78 6576 000
Fax 31 78 6576 254
E-Mail: <services@wkap.nl>

 Electronic Services <http://www.wkap.nl>

Library of Congress Cataloging-in-Publication Data

Networks of Learning Automata: Techniques for Online Stochastic Optimization
M.A.L. Thathachar and P.S. Sastry
ISBN 1-4020-7691-6

Copyright © 2004 by Kluwer Academic Publishers

All rights reserved. No part of this work may be reproduced, stored in a retrieval system, or transmitted in any form or by any means, electronic, mechanical, photocopying, microfilming, recording, or otherwise, without prior written permission from the Publisher, with the exception of any material supplied specifically for the purpose of being entered and executed on a computer system, for exclusive use by the purchaser of the work.

Permission for books published in Europe: permissions@wkap.nl
Permissions for books published in the United States of America: permissions@wkap.com

Printed on acid-free paper.

Printed in the United States of America

TO OUR PARENTS

Contents

Dedication		v
Preface		xiii
1.	INTRODUCTION	1
	1.1 Machine Intelligence and Learning	1
	1.2 Learning Automata	8
	1.3 The Finite Action Learning Automaton (FALA)	10
	1.3.1 The Automaton	10
	1.3.2 The Random Environment	11
	1.3.3 Operation of FALA	12
	1.4 Some Classical Learning Algorithms	14
	1.4.1 Linear Reward–Inaction (L_{R-I}) Algorithm	14
	1.4.2 Other Linear Algorithms	17
	1.4.3 Estimator Algorithms	18
	1.4.4 Simulation Results	20
	1.5 The Discretized Probability FALA	20
	1.5.1 DL_{R-I} Algorithm	22
	1.5.2 Discretized Pursuit Algorithm	22
	1.6 The Continuous Action Learning Automaton (CALA)	24
	1.6.1 Analysis of the Algorithm	26
	1.6.2 Simulation Results	30
	1.6.3 Another Continuous Action Automaton	31
	1.7 The Generalized Learning Automaton (GLA)	33
	1.7.1 Learning Algorithm	36
	1.7.2 An Example	38
	1.8 The Parameterized Learning Automaton (PLA)	41
	1.8.1 Learning Algorithm	44

	1.9	Multiautomata Systems	46
	1.10	Supplementary Remarks	47
2.	GAMES OF LEARNING AUTOMATA		51
	2.1	Introduction	51
	2.2	A Multiple Payoff Stochastic Game of Automata	54
		2.2.1 The Learning Algorithm	56
	2.3	Analysis of the Automata Game Algorithm	63
		2.3.1 Analysis of the Approximating ODE	65
	2.4	Game with Common Payoff	74
	2.5	Games of FALA	76
		2.5.1 Common Payoff Games of FALA	78
		2.5.2 Pursuit Algorithm for a Team of FALA	79
		2.5.3 Other Types of Games	82
	2.6	Common Payoff Games of CALA	83
		2.6.1 Stochastic Approximation Algorithms and CALA	85
	2.7	Applications	89
		2.7.1 System Identification	89
		2.7.2 Learning Conjunctive Concepts	92
	2.8	Discussion	102
	2.9	Supplementary Remarks	103
3.	FEEDFORWARD NETWORKS		105
	3.1	Introduction	105
	3.2	Networks of FALA	106
	3.3	The Learning Model	107
		3.3.1 G-Environment	107
		3.3.2 The Network	109
		3.3.3 Network Operation	113
	3.4	The Learning Algorithm	116
	3.5	Analysis	116
	3.6	Extensions	127
		3.6.1 Other Network Structures	127
		3.6.2 Other Learning Algorithms	128
	3.7	Convergence to the Global Maximum	129
		3.7.1 The Network	130
		3.7.2 The Global Learning Algorithm	131
		3.7.3 Analysis of the Global Algorithm	132

	3.8	Networks of GLA	136
	3.9	Discussion	137
	3.10	Supplementary Remarks	138
4.		**LEARNING AUTOMATA FOR PATTERN CLASSIFICATION**	139
	4.1	Introduction	139
	4.2	Pattern Recognition	140
	4.3	Common Payoff Game of Automata for PR	146
		4.3.1 Pattern Classification with FALA	148
		4.3.2 Pattern Classification with CALA	150
		4.3.3 Simulations	151
	4.4	Automata Network for Pattern Recognition	154
		4.4.1 Simulations	158
		4.4.2 Network of Automata for Learning Global Maximum	161
	4.5	Decision Tree Classifiers	164
		4.5.1 Learning Decision Trees using GLA and CALA	166
		4.5.2 Learning Piece-wise Linear Functions	170
	4.6	Discussion	173
	4.7	Supplementary Remarks	176
5.		**PARALLEL OPERATION OF LEARNING AUTOMATA**	177
	5.1	Introduction	177
	5.2	Parallel Operation of FALA	178
		5.2.1 Analysis	180
		5.2.2 ϵ-optimality	181
		5.2.3 Speed of Convergence and Module Size	184
		5.2.4 Simulation Studies	185
	5.3	Parallel Operation of CALA	185
	5.4	Parallel Pursuit Algorithm	187
		5.4.1 Simulation Studies	190
	5.5	General Procedure	190
	5.6	Parallel Operation of Games of FALA	192
		5.6.1 Analysis	194
		5.6.2 Common Payoff Game	196
	5.7	Parallel Operation of Networks of FALA	197
		5.7.1 Analysis	199

		5.7.2	Modules of Parameterized Learning Automata (PLA)	199
		5.7.3	Modules of Generalized Learning Automata (GLA)	201
		5.7.4	Pattern Classification Example	202
	5.8	Discussion		203
	5.9	Supplementary Remarks		204
6.	SOME RECENT APPLICATIONS			205
	6.1	Introduction		205
	6.2	Supervised Learning of Perceptual Organization in Computer Vision		206
	6.3	Distributed Control of Broadcast Communication Networks		213
	6.4	Other Applications		218
	6.5	Discussion		221

EPILOGUE				223
Appendices				
A	The ODE Approach to Analysis of Learning Algorithms			227
	A.1	Introduction		227
	A.2	Derivation of the ODE Approximation		229
		A.2.1	Assumptions	230
		A.2.2	Analysis	231
	A.3	Approximating ODEs for Some Automata Algorithms		234
		A.3.1	L_{R-I} Algorithm for a Single Automaton	235
		A.3.2	The CALA Algorithm	236
		A.3.3	Automata Team Algorithms	239
	A.4	Relaxing the Assumptions		239
B	Proofs of Convergence for Pursuit Algorithm			241
	B.1	Proof of Theorem 1.1		241
	B.2	Proof of Theorem 5.7		245
C	Weak Convergence and SDE Approximations			247
	C.1	Introduction		247
	C.2	Weak Convergence		247
	C.3	Convergence to SDE		248
		C.3.1	Application to Global Algorithms	250
	C.4	Convergence to ODE		251

References	253
Index	265

Preface

The idea of machines which provide intelligent service to humans without complaints of fatigue or boredom, has been around for a long time. It moved from the world of dreams to the realm of possibility only in recent times. While there is no rigorous definition of what constitutes the multidimensional concept of intelligence, it is commonly conceded that the intelligence component of machines of various types has been steadily increasing in the past few decades, although it is still nowhere near the level observed in human beings. There are many ongoing efforts in different disciplines, all contributing in the direction of improving the intelligence of machines in several dimensions.

Learning is a crucial aspect of intelligence and machine learning has emerged as a vibrant discipline with the avowed objective of developing machines with learning capabilities. This book represents an effort in the same direction and discusses one approach to the problem of learning from examples.

The book considers synthesis of complex learning structures from simple building blocks. The simplest element is a learning automaton which learns to select the best action by repeated interactions with an unknown random environment. It uses stochastic algorithms for refining probabilities of selecting actions for this purpose. The paradigm is that of learning from a probabilistic teacher and constitutes a reinforcement learning model. Using the single learning automaton as the building block, systems consisting of several learning automata such as games and feedforward networks are constructed. Mathematical analysis of their behavior with suitable learning algorithms is provided and a detailed discussion of how learning automata solutions can be constructed in a variety of applications is presented.

This book may be regarded as a natural successor of the book, K.S.Narendra and M.A.L.Thathachar, 'Learning Automata: An Introduction', Prentice Hall, 1989. However, the present book can be read independently. We provide a fairly comprehensive account of learning automata models with emphasis on multiautomata systems. A good part of the book essentially describes our research in the area subsequent to the publication of the earlier book.

We start with a brief introduction to the concept of a learning automaton and consider a variety of learning automata models (including automata with continuous action sets and automata that accept context vector inputs) in the first chapter. The next two chapters give a detailed treatment of complex structures such as games of automata and networks made of several teams of learning automata. In each case, we describe effective learning algorithms and present an analysis of their asymptotic behavior. We also give some examples to illustrate different models. The remaining three chapters of the book can be read almost independently. In Chapter 4 we provide a detailed account of how the automata algorithms can be employed in an application by considering the pattern recognition problem. The next chapter deals with the issue of parallel operation of these models which considerably improves efficiency of learning. In the final chapter of the book we briefly discuss some applications of these models in computer vision and communication networks, and also point out many other recent applications.

In the first three chapters, we have used ideas from pattern recognition to illustrate and motivate some of the concepts and models. The reader does not need any detailed knowledge of pattern recognition to appreciate such comments; familiarity with the general problem of learning in pattern recognition is sufficient for the purpose. We have provided a brief introduction to pattern recognition at the beginning of Chapter 4. A major theme underlying all the algorithms in this book is that of optimization of a functional based only on some noisy observations of the function values. Such online stochastic optimization techniques are relevant in almost all aspects of adaptive and learning systems. The analyses of various learning algorithms in this book also bring out this aspect. We have illustrated the relevance of such optimization problems in the context of learning in pattern recognition at the beginning of Chapter 4.

One of the strengths of the learning automata approach is that analytical results regarding the long-term behavior of complex systems of automata are available. Hence, we present mathematical analysis of many of the algorithms discussed here. However, we have tried to make the analysis accessible to a large audience. The mathematical background needed to understand it has been kept roughly at the level of a graduate engineering student. The mathematically sophisticated parts of the analysis are summarized in the appendices. Wherever possible, we have tried to keep the analysis simple and thus some of the results are presented under more stringent assumptions than needed.

The book can be used partially as a text book for graduate level courses on Machine Learning, Soft Computing, Reinforcement Learning and Real-time Stochastic Optimization. We hope the book will be of interest to researchers in the above areas as well as many other related areas.

PREFACE

A major part of the book contains research carried out at the Department of Electrical Engineering, Indian Institute of Science, Bangalore, and we would like to thank all our graduate students who have enthusiastically participated in this endeavor. In particular, we would like to thank Vijay V. Phansalkar, G. Santharam, K. Rajaraman, M. T. Arvind and K. Krishna. Vijay Phansalkar gave very enthusiastic support for this book and has provided some crucial inputs. The method of analysis detailed in Appendix A is essentially due to him. We also thank Ambika Prasad, Srivatsan Laxman and H.K.Aditya for their help with some of the simulations and figures.

Investigation in the field of learning automata started at the Indian Institute of Science in the late sixties and a research group has gradually emerged. A major part of this book grew out of the efforts of this group. The liberal academic policies of Indian Institute of Science as well as the stimulating atmosphere in the department created by our colleagues, have all contributed to the flourishing of this research and we record our appreciation for the same.

M.A.L.T. would like to thank All India Council for Technical Education (AICTE) for the award of an Emeritus Fellowship during the course of writing this book. We thank the staff of Kluwer Academic who were involved in the production of this book. In particular, we wish to thank Alexander Greene, Senior Editor, for his support of this project and Melissa Sullivan, Editorial Assistant, for timely help while preparing the manuscript.

Finally, we would like to thank all our family members and in particular, our spouses, Yadu Thathachar and Padmaja Sastry, for their patience and help throughout the course of writing of this book.

M.A.L.Thathachar

P.S.Sastry

August, 2003.

Chapter 1

INTRODUCTION

1.1 Machine Intelligence and Learning

Learning has long been recognized as an important aspect of intelligent behavior. Over the last few decades, the process of learning, which was studied earlier mostly by psychologists, has become a topic of much interest to engineers as well, in view of its role in machine intelligence. Psychologists or biologists who conduct learning experiments on animals or human subjects try to create models of behavior through analysis of experimental data. Engineers are more interested in studying learning behavior for helping them to synthesize intelligent machines. While the above two goals are distinctly different, the two endeavors are nonetheless interrelated, because success in one helps to improve our abilities in the other. Different approaches (with varying levels of relationship to human learning) for incorporating learning ability in machines have been investigated by engineers interested in the general area of machine intelligence.

We can distinguish two broad approaches to machine intelligence. One of these is the conventional AI approach which is based on the assumption that intelligence can be realized through a formal symbol manipulation process. When symbols can represent ideas, thinking or intelligent behavior can be regarded as computing with symbols using some formal rules. This approach has led to many useful applications such as expert systems which mimic the reasoning abilities of human experts in narrow domains of specialization by representing the underlying knowledge as a set of rules used in a formal reasoning process. The second approach is that of connectionist systems or neural networks which try to mimic the architectural structure of human brain. If the brain is viewed as a network of neurons, then it stands to reason that an intelligent machine should be built from a network of units or artificial neurons each of which is a mathematical abstraction of some of the essential properties of the natural neurons. Such artificial neural networks have been studied extensively in recent times and have been applied with considerable success in a variety of areas. There is, however, no clear indication which approach is superior. Depending on the application or the aspect of intelligent behavior that is of interest, one or the other approach is appropriate.

Whatever be the path one takes to machine intelligence, learning appears to be an important question that has to be addressed. In artificial neural networks, learning is often the central issue. Most abilities of an adult human being (including the 'simple' ones such as sitting, standing, walking, talking etc.) are all learnt through practice in infancy and childhood. As per our current understanding of brain, all such learning comes about by changing the nature of connections between neurons in our brain. Since artificial neural networks are inspired by the structure of human brain, it is to be expected that learning plays an important role here. In an artificial neural network, the parameters (often called weights) associated with each neuron have to be learnt (from some examples provided) for achieving proper operation of the network. In the AI approach, the performance of any system is critically dependent on the quantity and quality of knowledge available, which is often represented in the form of rules. It is, in general, difficult to hand-code all such rules. It is often more effective to have the machine learn the best set of rules from examples. Thus, even in the symbolic AI approach, learning is an important problem to be considered.

In psychology, one comes across several types of learning such as classical conditioning, operant conditioning, cognitive learning etc. While these distinctions are useful for analyzing human or animal learning behavior, only some of the basic features of learning as studied in psychology need be carried over to the domain of learning machines. In this context, a useful engineering definition would be the following: *a machine or system is said to learn if it improves its performance through experience gained over a period of time without complete information about the environment in which it operates.* Such self-improvement in performance with or without external guidance could be regarded as the hallmark of learning machines. The experience using which the system is to improve its performance is often in the form of examples of the kind of tasks to be performed by it. The learning algorithm analyzes the performance of the system on these examples and modifies the the system to improve its performance. This book presents a comprehensive account of one approach to machine learning that is useful in a variety of situations involving learning from examples.

The approach to learning followed here may be described as learning by doing. This is the type of learning often associated (in psychology) with acquisition of manual skills like learning to type. Learning a task is accomplished by doing it; that is, by applying a suitable action out of a repertoire of available actions. The effect of the action is measured by a performance index which, in turn, helps to suitably modify the procedure for selecting actions. Often, the measured performance itself may be random and many repeated trials may be necessary to find out the suitability of a particular action. Learning in this manner needs two ingredients: a modifiable procedure for generating actions and a method for evaluating results of an action. Such a procedure has been called generate and test method. The learner generates an action (based on its current knowledge), tests its efficacy and uses the results of the test to modify its knowledge and hence the procedure for generating actions. This process continues till the results of test are satisfactory. This theme of generate and test is closely related to trial and error learning in psychology, conjecture and refutation in philosophy and induction and deduction in logic. In an engineering context, the learner is essentially doing a *hill climbing* to improve the performance.

The performance evaluation needed in the learning cycle as above, often comes through an external agency termed the teacher or the environment. If the teacher can tell the learner the correct action to be chosen in every conceivable situation, then learning is trivial. However, learning with the aid of a teacher is often nontrivial and significant. The teacher provides examples and tells the learner which action is suitable in each situation represented by the examples. However, the learner has to *generalize* this experience to contexts or situations not pointed out by the teacher. This type of learning is called *supervised learning* and it is typically encountered in pattern recognition problems with perfect or noise-free examples.

A somewhat harder learning problem arises when the teacher or the environment does not indicate the correct action (even for the example situations); but provides only a scalar evaluative response to the selection of an action by the learner. Such response is called a *reinforcement* and this structure is called *reinforcement learning* or *learning with a critic*. The analogy here is with a critic who only says how good an artistic creation is and cannot, in general, say how it should have been done. Since the teacher does not say which is the correct action, the learner has to try different actions to know which is better and this type of learning involves more effort than supervised learning. It may be noted that the requirement of generalizing the experience to properly tackle *contexts* not associated with the examples, exists here also.

In many situations, the reinforcement that can be supplied as an evaluation, is itself a random variable. Here, learning takes place with a *probabilistic teacher* or equivalently, in a *random environment*, thus making the problem even harder. Now all actions have to be tried a number of times to evaluate the mean reinforcement associated with them in order to find the best action.

Naturally, the need here is for an effective mechanism of selecting successive actions as the learning proceeds, so that the process is efficient.

The latter type of reinforcement learning is the subject matter of this book. The basic difficulty of the problem is that the characteristics of the teacher, i.e., the probability distributions of the reinforcement signal associated with different actions are unknown. In the context of optimizing the performance, the problem can be viewed as *stochastic hill climbing* or *online stochastic optimization*.

Reinforcement learning in this situation involves a sequence of trials. The learner tries an action and obtains an indication of its performance in the form of a reinforcement. On the basis of this reinforcement, it chooses a new action, obtains a new reinforcement and so on. For successful learning, there should be an appropriate mechanism for selection of actions that can utilize the experience in the form of actions selected and reinforcements obtained so far. However, each action needs to be tried a number of times in view of the randomness of the teacher. Thus, at each trial, the learner has two different goals to balance. The learner should try different actions so that it knows more about the actions. At the same time it has to select the best action based on the knowledge gained so far. How much effort should be put in each of these two directions is a basic issue in learning. These two can be called exploration and exploitation. Such a dilemma exists in different forms in different fields. In statistics it goes under the name of bandit problems or sequential design of experiments. In controlling an unknown system, it is called dual control, as one has to balance the dual requirements of identification and control.

We can gain an intuitive feel for this kind of reinforcement learning by considering the classical problem of controlling an inverted pendulum. The pendulum consists of a pole hinged to a cart at the bottom. The cart can move along a straight line horizontally. The pole may have a weight attached to it at the top. The objective is to keep the pendulum in a nearly vertical position by applying a proper sequence of forces to the cart.

If a good model in the form of appropriate equations of motion and all the parameter values of the cart-pole system is available, one can, in principle, apply one of a variety of methods from control theory to solve the problem. However, when there is lack of knowledge or uncertainty about the system characterization, it becomes a learning control problem. The type of learning systems that we consider in this book do not need any information regarding the system. The idea is to learn by doing or by interacting with the system. Basically, if the pole is inclined away from the front of the cart, a push is needed and if it is inclined toward the front, a pull is needed. If the time instants of force application are sufficiently close, we can balance the pole by a proper sequence of three actions: a push of a constant magnitude, a pull of a constant magnitude or no force. The appropriateness of each action depends on some observable quantities such as the angle from vertical and rate of change of this angle. The learning system can gain experience through a series of trials of applying an action and observing the results. Initially, the learning system could be unsuccessful; that is, the pole could fall down after short intervals. However, if the learning mechanism works properly, the intervals between failures become longer and longer. An advantage of the learning approach is that changes in the system can be handled more easily. For example, if the weight at the top end of the pole changes or if the length of the pole is changed, the same algorithm can balance the pole by going through a few more learning cycles.

This example gives a flavor of the kind of learning discussed in this book. In the methods discussed here, the basic model is the learning automaton. It is a stochastic model operating in the framework of reinforcement learning. An overview of this model is given in the next section. By combining a number of such automata in different ways, we can have methods to tackle many different learning situations.

The field of machine learning, which is now a flourishing research area, has been exploring many different (though related) approaches to understand learning behavior and to develop design methodologies for synthesizing learning machines. Such machine learning techniques are now routinely applied in many areas such as pattern recognition, datamining etc. In fact, in areas such as speech recognition, only learning techniques deliver acceptable levels of performance. There are a variety of machine learning algorithms and techniques which are appropriate in different situations based on the nature of the task and the nature of representation needed for the learnt knowledge.

However, there is a unifying theme underlying all problems of learning from examples. All such learning situations essentially involve solving a special kind of optimization problem. Learning techniques differ in how they tackle this optimization problem and what they assume regarding the problem. Below, we briefly present a mathematical formalism that brings out this common characteristic of learning problems. Here we assume that the reader is familiar with some of the machine learning techniques. We explain this framework again in the specific context of pattern recognition in Chapter 4.[1]

We can define the objective in a learning problem to be that of optimizing a performance index [Tsy71]

$$\mathcal{J}(\Lambda) = \int_{\mathcal{X}} \mathcal{R}(\mathbf{x}, \Lambda) d\mathcal{P} \qquad (1.1.1)$$

where $\mathcal{R}(\mathbf{x}, \Lambda)$ is a function of a *parameter vector* or *model* Λ and an *observation vector* $\mathbf{x}$, and $\mathcal{X}$ is the space of all possible $\mathbf{x}$. The *performance index*

---

[1] The mathematical details in the rest of this section may be skipped without any loss of continuity.

$\mathcal{J}$ is the expectation of $\mathcal{R}$ with respect to the distribution $\mathcal{P}$ which includes the randomness in $\mathbf{x}$ as well as any randomness in the function $\mathcal{R}$ itself. In the context of reinforcement learning discussed above, we can identify Λ with actions and $\mathcal{R}(\mathbf{x}, \Lambda)$ with reinforcement obtained in a specific random context represented by $\mathbf{x}$. The special feature of the learning problem is that the probability distribution $\mathcal{P}$ is unknown. We are required to find the optimizer of $\mathcal{J}$ using the experience with a sequence of $\mathbf{x}$'s drawn according to the unknown distribution $\mathcal{P}$. Many different problems involving learning from examples can be put in this framework by choosing a suitable parameterization Λ and a suitable structure for $\mathcal{R}$. For example, we illustrate this in the context of pattern recognition in Chapter 4.

A complicating feature of this optimization problem is that here we do not have full knowledge of the objective function. Since the distribution $\mathcal{P}$ is unknown, we cannot calculate $\mathcal{J}(\Lambda)$ for any given Λ and thus no standard optimization methods would work. The learning automata methods discussed in this book are all suitable for optimizing $\mathcal{J}$ using only a sequence of values of the random variable $\mathcal{R}$, that is, the sequence of reinforcements obtained in response to actions selected in different contexts or examples. In this respect these methods constitute an online stochastic optimization technique. Another class of methods which can handle the optimization in a similar way is the so called stochastic approximation methods. In the stochastic approximation techniques one assumes that the parameter represented by Λ is a real vector. As we shall see in this book, automata based techniques can handle richer classes of parameterizations.

Another way in which this optimization is handled in most learning algorithms is to approximate $\mathcal{J}$ based on its value on the set of examples. The RHS of (1.1.1) is an expectation of $\mathcal{R}$, but we cannot calculate it because $\mathcal{P}$ is unknown. However we can approximate this expectation integral by the sample average of the observed values of $\mathcal{R}$ on the set of examples. Provided that the examples are drawn according to the distribution $\mathcal{P}$ in an *independent and identically distributed* (iid) fashion and that there are 'sufficient' number of examples, such an approximation should be good. This kind of approximation is called an empirical expectation of $\mathcal{R}$. We illustrate all the details involved here in the context of a general pattern recognition problem in Chapter 4. In approximating $\mathcal{J}$ by this empirical expectation, we are essentially measuring the performance of the learning system *only* on the set of examples provided. On the other hand, the original $\mathcal{J}$ as defined by (1.1.1) measures the performance on *all* possible situations. Thus an important question in statistical learning theory is whether or not the optimizer of this empirical expectation is a good approximation to the optimizer of the original expectation. This is the issue of *generalization ability* of the learning machine, namely, whether the learning machine is able to generalize the experience over the example set

Learning Automata 7

to all the other situations. One can answer this question in a precise manner and the generalization abilities depend not only on having sufficient examples but also on the parameterization chosen for representing the learning system. That is, it depends on the set of all possible Λ over which we search for the optimum. We will not be dealing with this theoretical issue in this book. The reader is referred to [Vap97] for details. However, we note here that the conditions needed for ensuring proper generalization are satisfied for all systems considered in this book.

From the viewpoint of the learning algorithm, the advantage of approximating our objective function by the empirical expectation is that its values can be calculated and one can, in principle, use a standard optimization procedure to find the optimizer of this approximated objective function. The popular back-propagation algorithm used for learning in feedforward neural networks is an example of this approach.

In any way of handling this optimization problem, another important consideration is the specific representation chosen for the learning system which determines the set of all Λ over which we need to search to find the optimum. Different techniques in machine learning differ in the kind of representations that they handle and the amount of uncertainty assumed in the values of the performance index $\mathcal{R}$. In this context, a distinguishing characteristic of learning automata models, which form the subject matter of this book, is that the search for the optimum parameter, Λ, is conducted in the space of probability distributions over the parameter space rather than in the parameter space itself. While this may appear to complicate the problem further, it means that we do not need to assume any specific algebraic structure on the parameter space. This permits a great deal of flexibility in designing learning solutions using automata models in widely different applications. Also, as mentioned earlier, these models operate in the reinforcement learning framework and thus can naturally handle randomness in the evaluatory signal, $\mathcal{R}$. In fact, these methods are particularly effective mainly when there is much noise in the examples from which the system is required to learn.

Another interesting aspect of these models is that we can build complex learning machines using simple components. The simplest building block is a learning automaton with finitely many choices of action and it learns the optimal action by interacting with a random environment. Starting with such simple components we can build structures containing many automata to tackle a variety of learning situations. In this chapter we concentrate on a single automaton and in the next two chapters discuss some of the ways in which we can construct multiautomata systems.

1.2 Learning Automata

The learning automaton is a simple model for adaptive decision making in unknown random environments. Intuitively, its operation could be considered similar to the process of learning by an organism placed in such environments. Common examples of such situations are those of a child learning to make proper movements and a person finding the best route from home to office. The organism tries different actions and chooses new actions on the basis of the response of the environment to previous actions. The mechanism for such adaptive selection of actions (or equivalently, decisions) is represented by the learning automaton. The problem of learning the proper action is complicated by the fact that the environmental responses are not completely reliable as they are random and the relevant probability distributions are unknown.

The model is well represented by the following analogy with a probabilistic teacher. Consider a teacher and a student. The student is posed a question and is given several alternative answers. The student can pick one of the alternatives, following which the teacher responds yes or no. This response is probabilistic – the teacher may say yes for a wrong alternative and vice versa. The student is expected to learn the correct alternative through such repeated interactions. While the problem is ill-posed with this generality, it becomes solvable with an added condition. It is assumed that the probability of the teacher saying 'yes' is maximum for the correct alternative.

In the above analogy, the student represents the automaton and the teacher represents the random environment. The alternatives are the actions of the automaton. The correct alternative corresponds to the best action which elicits the favorable response from the environment with maximum probability. The learning process results in the identification of the best action.

Such a learning model is useful in many applications involving adaptive decision making. For example, in a communication network, each router has to choose one of the finitely many alternatives available in order to optimize some measure of the performance of the network as a whole. However, it is unreasonable to assume that the router can be provided with the full status information of the relevant portion of the network. Hence, it would be attractive to have an algorithm that can learn to make good choices based on some 'noisy' feedback regarding the 'goodness' of the choice. This fits into the automata model. In a pattern recognition problem, the classifier is required to decide on the class label of any pattern input to it. Often, the only information available for design of a classifier is a set of training examples. The optimal decision rule is to be learnt using these examples. This can also be posed as a problem of learning to choose one of the available actions based on some stochastic feedback on the goodness of each choice made. A general class of problems where this type of learning model is useful is that of optimizing a function based only on noisy measurements of function values. Suppose we want to learn the optimal value of a parameter vector, θ, of some system so as to maximize a performance index $f(\theta)$. Further, suppose that we do not know the function $f(\cdot)$ or its derivatives. The only available information is: given any specific θ, we can obtain a noise-corrupted value of $f(\theta)$ (may be through an auxiliary simulation). This is a generic problem that can be solved using the automata models. Here, all possible values for θ constitute the actions and the noise-corrupted value of the performance index constitutes the response from the environment. The problem is that of learning the optimal action through repeated interactions with the environment. Some of these applications would be discussed in detail in the later chapters.

The learning process in the context of the learning automaton proceeds as follows. Each time it interacts with the environment, the automaton randomly chooses an *action* (from the action set) based on some probability distribution. (Initially, each action can be selected with equal probability). Following the response of the environment to a selected action, the automaton updates its *action probability distribution*. A new action is then selected according to the updated probability distribution. The response of the environment is elicited for this action and the procedure is repeated. The updating algorithm for the action probability distribution is called the *learning algorithm* or the *reinforcement scheme*. If this algorithm is properly designed, the selection probability of the best action converges to unity while the other action probabilities go to zero. Such convergence can be regarded as learning in an ideal sense. It will be seen later on that close approximations to the above could be achieved in practice.

A strong point of the automaton approach is the complete generality of the concept of the action set. For example, the actions could be parameter values in a system, queues in a queuing network, alternative routes in a communication network, or possible class labels in pattern recognition. In fact, the action set may be finite or continuous as the situation demands. This enables a common approach, employing the same mathematical framework, to diverse learning problems.

Another attractive feature of this approach is that the learning automaton could be regarded as a simple unit from which complex systems such as hierarchies, games and feedforward networks could be constructed. These could be designed to handle complicated learning problems. Moreover, convergence results exist for such collections of learning automata and they provide a basis for confident design. In this book we would be concerned mainly with learning algorithms that result in good performance in the context of systems of multiple automata.

Yet another facet of the learning automaton model is that it is a general approach to stochastic optimization. Given such an optimization problem, one could design a learning automata system which can solve it to the required accuracy. In fact, in this context, the approach could be regarded as an evolutionary method in as much as the action probability distribution can be regarded as an implicit representation of a population of possible solutions and it evolves as a result of environmental responses tending toward the optimum.

While the above considerations apply in a general way to the learning automaton approach, specific situations require specific types of learning automata. Depending on the nature of the environment and the learning problem posed by it, a variety of learning automata have been proposed and studied. We make a beginning with the basic concepts and properties of these Learning Automata (LA) models in the next few sections. In this chapter we would be discussing only single learning automata. Models consisting of many automata are considered in the later chapters.

1.3 The Finite Action Learning Automaton (FALA)

In all automata models, interaction with the environment and hence the evolution of action probabilities proceed in discrete time steps. At each instant, k, $k = 0, 1, 2, \ldots$, the automaton chooses an action, $\alpha(k)$, at random based on its current action probability distribution. This action is input to the environment which responds with a *reaction* or *reinforcement*, $\beta(k)$. This response or reinforcement from the environment is the input to the automaton. Now, the automaton updates its action probability distribution and the cycle of random choice of action, eliciting reinforcement and updating action probabilities, repeats. A general block diagram of an automaton operating in a random environment is shown in Fig. 1.1

In this section we discuss automata with finitely many actions. We call these as finite action-set learning automata or finite action learning automata (FALA). Automata with a finite set of actions were proposed earliest among the family of learning automata.

1.3.1 The Automaton

Formally, an FALA can be described by the quadruple $(A, B, \mathcal{T}, \mathbf{p}(k))$ where $A = \{\alpha_1, \alpha_2, \cdots, \alpha_r\}$, is the *finite* set of actions, B is the set of all possible inputs or reinforcements to the automaton, $\mathcal{T}$ is the learning algorithm for updating action probabilities and $\mathbf{p}(k)$ is the action probability vector[2] at instant k given by

$$\mathbf{p}(k) = [p_1(k), p_2(k) \ldots p_r(k)]^T. \quad (1.3.2)$$

---

[2] All vectors in this book are column vectors and a superscript 'T' denotes transpose

The Finite Action Learning Automaton (FALA)

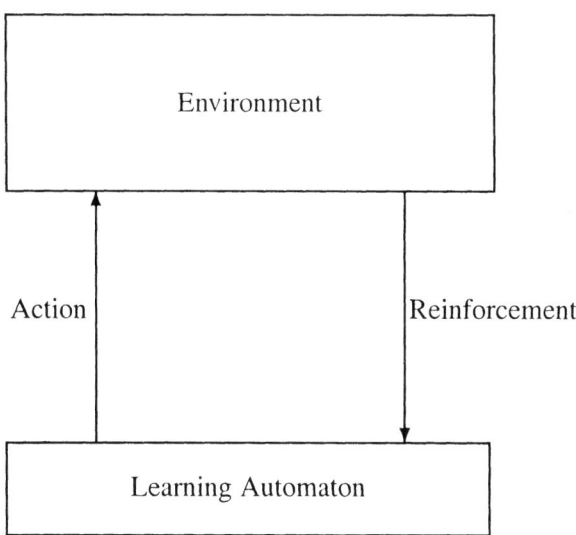

Figure 1.1. Block diagram of a Learning Automaton

We have $p_i(k) \geq 0$, $\forall i, k$ and $\sum_{i=1}^{r} p_i(k) = 1$ $\forall k, k = 0, 1, 2, \cdots$. Here $p_i(k)$ is the probability with which action α_i is chosen at instant k. That is,

$$p_i(k) = \text{Prob}[\alpha(k) = \alpha_i \mid \mathbf{p}(k)], \ i = 1. \ldots, r. \quad (1.3.3)$$

The learning algorithm for updating $\mathbf{p}(k)$ is generally of the form,

$$\mathbf{p}(k+1) = \mathcal{T}(\mathbf{p}(k), \alpha(k), \beta(k)) \quad (1.3.4)$$

$\mathcal{T}$ is some function of the current action probabilities, $\mathbf{p}(k)$, the action selected at k, $\alpha(k)$, and the response of the environment to the selected action, namely, $\beta(k)$. Since $\mathbf{p}(k+1)$ depends on $\alpha(k)$ and $\beta(k)$, both of which are random, $\mathbf{p}(k+1)$ is itself a random quantity. Thus $\{\mathbf{p}(k), k = 0, 1, 2, \ldots\}$ is a random process whose evolution is governed by the learning algorithm.

1.3.2 The Random Environment

The learning problem is embedded in the environment in which the learning automaton operates. The characteristics of the environment determine the nature of the learning problem.

The set of possible responses from the environment, B, may be discrete or continuous. It is always assumed that the reinforcement, $\beta(k) \in B$, is a real number. Traditionally, three types of environments are distinguished based on the nature of B. In the P–model, reinforcement is binary with $B = \{0, 1\}$ where 0 corresponds to unfavorable response (often called 'penalty') and 1 corresponds to favorable response (often called 'reward'). In the Q–model, we have $B = \{b_1, \ldots, b_q\}$, $b_i \in [0, 1], \forall i$ and $q < \infty$. Finally, in the S–model environment, the reinforcement is taken to be a real number with $B = [0, 1]$. In this book we assume the general case of $\beta(k)$ being a real number in $[0, 1]$. In some cases even this restriction can be removed and $\beta(k)$ could be any nonnegative real number. It is always assumed that higher values of reinforcement are more desirable. Thus, a general goal of any automata system is to maximize the reinforcement received.

In an FALA, as the name suggests, the action set is finite and we always consider the case $r \geq 2$. The action $\alpha(k) \in A$ forms the input to the environment and $\beta(k) \in B$ forms the output of the environment. For each action, the response or reinforcement, $\beta(k)$ is a random variable with an associated probability distribution. That is, associated with each action α_i is a probability distribution $\mathcal{F}_i$ over B, $i = 1, \ldots, r$, and this distribution is discrete or continuous based on the nature of B. When $\alpha(k) = \alpha_i$, $\beta(k)$ is a random variable with distribution $\mathcal{F}_i$. If $\mathcal{F}_i$ do not change with time k, then the environment is called stationary; otherwise it is nonstationary. In any learning problem, these distributions $\mathcal{F}_i$ are unknown (at least initially).

Let
$$d_i = E[\beta(k)|\alpha(k) = \alpha_i] \quad i = 1, \cdots r. \quad (1.3.5)$$

Although d_i is the mean value of the distribution $\mathcal{F}_i$, it is often referred to as *reward probability* of action α_i as it reduces to
$$d_i = \text{Prob}[\beta(k) = 1|\alpha(k) = \alpha_i] \quad (1.3.6)$$

in the binary case when $\beta(k) \in \{0, 1\}$. The environment is characterized by the set of reward probabilities
$$\{d_1, d_2, \ldots, d_r\}$$

and these remain independent of k in stationary random environments. We generally assume environments to be stationary in the development which follows.

The essence of the learning problem is that the reward probabilities are unknown to the automaton. The only input the automaton gets is the stochastic reinforcement in response to every action choice made.

1.3.3 Operation of FALA

The operation of FALA consists of repetitions of the steps: random choice of action, obtaining environmental reinforcement and updating of action probabilities. As the FALA operates in its environment, the learning algorithm generates a Markov process $\{\mathbf{p}(k)\}$. Let $S_r \subset \Re^r$ be the unit simplex defined by

$$S_r = \left\{ \mathbf{p} \mid \mathbf{p} = [p_1, p_2, \cdots p_r]^T; \ 0 \leq p_i \leq 1 \ \forall i, \sum_{i=1}^{r} p_i = 1 \right\}. \quad (1.3.7)$$

Then S_r is the state space of the Markov process $\{\mathbf{p}(k)\}$. Let $\mathbf{e}_i$ be the r-dimensional vector whose i^{th} component is unity and all others are zero. Then the set of all $\mathbf{e}_i (i = 1, \cdots r)$ is the set of vertices, V_r, of the simplex S_r. The interior of S_r is the set of all $\mathbf{p}$ in S_r with $0 < p_i < 1 \ \forall i$, and is denoted by S_r^o.

The asymptotic behavior of an FALA (which is determined by the asymptotic properties of the process $\mathbf{p}(k)$) is of much interest from the viewpoint of learning. We discuss below some ways of characterizing what is desired from the learning algorithm.

The goal of learning is to maximize reinforcement in some sense. Hence the main problem associated with an FALA is to identify the best action which is the one that yields maximum expected reinforcement. Let

$$d_\ell = \max_i \{d_i\} \quad (1.3.8)$$

Then the action α_ℓ is called the optimal action. Since we want the automaton to choose the optimal action with high probability, we define a desirable behavior for the automaton as follows. If

$$\lim_{k \to \infty} \mathbf{p}(k) = \mathbf{e}_\ell \quad w.p.1 \quad (1.3.9)$$

the FALA is called *optimal*. It may be noted that $\mathbf{p}(k)$ is a random process and hence whenever we talk about its convergence, we have to specify the specific type of convergence desired. In the above, the convergence is *with probability one* (denoted as *w.p.1*).

Identifying the optimal action is equivalent to solving the following optimization problem

$$\max_i \ E[\beta(k) \mid \alpha(k) = \alpha_i].$$

As mentioned earlier, what characterizes the automata approach is that the search for the optimal action is conducted in the space of all probability distributions over the action space rather than on the action space. Thus, through its learning algorithm, the automaton is actually solving the following optimization problem

$$\max_{\mathbf{p}} \ E[\beta(k) \mid \mathbf{p}(k) = \mathbf{p}]$$

where the maximization is over all $\mathbf{p}$ in the space S_r defined by (1.3.7). The maximum in the above optimization problem is attained at $\mathbf{p} = \mathbf{e}_\ell$ where $\mathbf{e}_\ell$ is the unit vector with ℓ^{th} component unity and all others zero. It is in this sense that the automaton identifies the optimal action.

It is difficult to achieve optimality (as defined by (1.3.9)) except under very special circumstances. One has to be satisfied with ϵ-*optimality* defined as follows.

DEFINITION 1.3.1 *An FALA is said to be ϵ-*optimal, *if*

$$\liminf_{k \to \infty} p_\ell(k) > 1 - \epsilon \quad w.p.1 \qquad (1.3.10)$$

can be achieved for any given $\epsilon > 0$ by a proper choice of the parameters in the learning algorithm.

As mentioned earlier, the general goal of an automaton is to maximize reinforcement by varying the action probabilities. Hence, a quantity of interest is

$$W(k) = E[\beta(k)|\mathbf{p}(k)] \qquad (1.3.11)$$

which is called the *Average Reward* at k. Note that the average reward, $W(k)$ is a random quantity because it depends on $\mathbf{p}(k)$.

We can characterize optimality and ϵ-optimality using $W(k)$ also. An equivalent characterization of optimal FALA is

$$\lim_{k \to \infty} W(k) = d_\ell, \quad w.p.1. \qquad (1.3.12)$$

Similarly, ϵ-optimality can be defined by replacing (1.3.10) with

$$\liminf_{k \to \infty} E[W(k)] > d_\ell - \epsilon. \qquad (1.3.13)$$

1.4 Some Classical Learning Algorithms

As explained above, the objective of a learning automaton is to identify the optimal action even though it has no knowledge of the distributions of the random reinforcement received in response to different action choices. What the automaton does is to update the action probability distribution based on the reinforcement received through a learning algorithm. A variety of learning algorithms can be used by FALA to achieve their objective. In this section and the next, we briefly describe a few typical learning algorithms for FALA.

1.4.1 Linear Reward–Inaction (L_{R-I}) Algorithm

This is one of the earliest FALA algorithms and is known to be very effective in many applications. The L_{R-I} algorithm updates the action probabilities as described below.

Some Classical Learning Algorithms

Let $\alpha(k) = \alpha_i$. Then the action probability vector $\mathbf{p}(k)$ is updated as

$$p_i(k+1) = p_i(k) + \lambda \beta(k)(1 - p_i(k))$$
$$p_j(k+1) = p_j(k) - \lambda \beta(k)p_j(k), \text{ for all } j \neq i \quad (1.4.14)$$

The above algorithm can be rewritten in vector notation as

$$\mathbf{p}(k+1) = \mathbf{p}(k) + \lambda \beta(k)(\mathbf{e}_i - \mathbf{p}(k)). \quad (1.4.15)$$

Here $\mathbf{e}_i$ is the unit vector with i^{th} component unity where the index i corresponds to the action selected at k, (that is, $\alpha(k) = \alpha_i$), and λ is the learning (or step-size) parameter satisfying $0 < \lambda < 1$. This algorithm is called linear because the RHS has only linear terms in components of $\mathbf{p}(k)$. As is easy to see from (1.4.15), we move $\mathbf{p}(k)$ a little towards $\mathbf{e}_i$ to obtain $\mathbf{p}(k+1)$. The name 'reward–inaction' comes from the nature of this update in the special case of $\beta(k) \in \{0,1\}$ (that is, in a P-model environment). When $\beta(k) = 1$ (that is, when we receive a favorable response from the environment) we increase the probability of the selected action and decrease others (that is, reward the selected action). When $\beta(k) = 0$ (that is, when we get an unfavorable response) we leave the probabilities unchanged (which is termed inaction). The increments or decrements in components of $\mathbf{p}(k)$ are done so as to be consistent with the requirement that $\mathbf{p}(k+1)$ should be a probability vector whenever $\mathbf{p}(k)$ is. (Note that for $\mathbf{p}(k)$ to be a probability vector, we should have $0 \leq p_i(k) \leq 1, \sum_{i=1}^{r} p_i(k) = 1$).

The L_{R-I} algorithm has been extensively used as it is simple and has several nice properties, some of which are listed below [NT89].

1. It is ϵ-optimal in all stationary random environments.

2. $E[p_\ell(k)]$ is monotonically increasing with k for any choice of the initial action probability distribution, $\mathbf{p}(0)$, as long as it is in S_r^o, the interior of the simplex S_r. Thus, at least, $p_\ell(k)$, the probability of the optimal action (which is associated with the largest reward probability d_ℓ) moves in the right direction in the expected sense. Similarly if d_s is the smallest reward probability, $E[p_s(k)]$ is monotonically decreasing with k for all $\mathbf{p}(0) \in S_r^o$.

3. Actually, property 2 listed above is a consequence of the fact that $\{p_\ell(k)\}$ is a submartingale and $\{p_s(k)\}$ is a supermartingale in all stationary random environments.[3] It is moreover true that $\{W(k)\}$ is a submartingale in all stationary random environments. It can be shown that

$$E[W(k+1)|\mathbf{p}(k)] > W(k) \quad \forall \mathbf{p}(k) \in S_r^o. \quad (1.4.16)$$

---
[3] See Appendix A for a definition of submartingales and supermartingales.

This property is referred to as *Absolute Expediency*. Thus the L_{R-I} algorithm is absolutely expedient. It is important as this property ensures that the performance of FALA is incrementally improving (in an expected sense) at every instant k irrespective of the environment and the values of action probabilities. It follows from the above that $E[W(k)]$ is monotonically increasing in all stationary random environments.

4 The set of unit vectors $\{\mathbf{e}_i, i = 1, \ldots, r\}$ forms the set of absorbing states for the Markov process $\{\mathbf{p}(k)\}$. Furthermore, $\mathbf{p}(k)$ converges to this set with probability 1.

5 Although $\mathbf{p}(k)$ can converge to any $\mathbf{e}_i$ in a sample path, it converges to $\mathbf{e}_\ell$ with a probability higher than the probability with which it converges to other $\mathbf{e}_i$. Bounds on such convergence probabilities can be computed in terms of the reward probabilities $\{d_i\}$, the learning parameter λ and the initial action probability $\mathbf{p}(0)$. The ϵ-optimality property of the algorithm implies that the probability of $\mathbf{p}(k)$ converging to $\mathbf{e}_\ell$ can be made arbitrarily close to 1 by choosing λ sufficiently small.

The Equivalent ODE for L_{R-I}

While the above properties have been obtained using the martingale and Markov properties of the process $\mathbf{p}(k)$, there is another general method of analysis for such learning algorithms. This method, to be called the ODE method, is elaborated in Appendix A. This is based on the fact that for small values of the learning parameter, λ, the behavior of the algorithm can be approximated (in a precisely defined sense) by an associated ordinary differential equation (ODE). Appendix A explains the procedure for deriving such approximating ODE for a general class of learning algorithms and shows how the method can be used for automata algorithms including L_{R-I}. Theorem A.1 in Appendix A states the precise nature of this approximation.

As shown in Appendix A (see Section A.3.1), the ODE associated with the L_{R-I} algorithm has the form

$$\frac{dx_i}{dt} = \sum_{j=1}^{r} x_i x_j (d_i - d_j), \ i = 1, \ldots, r, \tag{1.4.17}$$

with the constraints $\sum_{j=1}^{r} x_j = 1$ and $x_i \geq 0, \forall i$. In the solution of the above ODE, x_i approximates p_i. A simple analysis of the RHS of the above ODE shows that when $\{d_i\}$ are distinct, the unit vectors $\mathbf{e}_i$ are the only equilibrium points. Among these, only $\mathbf{e}_\ell$ is asymptotically stable whereas the other $\mathbf{e}_i$ are unstable. This indicates the tendency of $\mathbf{p}(k)$ to move towards $\mathbf{e}_\ell$ which gives rise to the ϵ-optimality property of the algorithm.

Some Classical Learning Algorithms 17

1.4.2 Other Linear Algorithms

There are several other algorithms having the same general philosophy as L_{R-I}, which have been well analyzed and used. (See [NT89] for detailed discussion of such algorithms). Here, we briefly explain one other algorithm called the linear reward–penalty (L_{R-P}) algorithm. Consider an r-action automaton. A general linear algorithm for updating the action probabilities is as follows.

If $\alpha(k) = \alpha_i$ then $\mathbf{p}(k)$ is updated as

$$\begin{aligned} p_i(k+1) &= p_i(k) + \lambda_1 \beta(k)(1 - p_i(k)) - \lambda_2(1 - \beta(k))p_i(k) \\ p_j(k+1) &= p_j(k) - \lambda_1 \beta(k) p_j(k) \\ &\quad + \lambda_2(1 - \beta(k))\left(\frac{1}{r-1} - p_j(k)\right), \ j \neq i. \end{aligned} \quad (1.4.18)$$

Compared to L_{R-I}, the main difference here is that, even when the reinforcement is unfavorable (that is, $\beta(k) = 0$) the action probabilities are updated. (Recall that under L_{R-I}, in such a situation the probabilities are left unchanged). If we take $\lambda_2 = 0$ in the above, we get back the L_{R-I} algorithm. For the L_{R-P} algorithm, we take $\lambda_1 = \lambda_2$.

Under the L_{R-I} algorithm, we have seen that the action probability vector converges to a unit vector. As can be seen from (1.4.18), in the case of the L_{R-P} algorithm, the unit vector cannot be a stable point of the process $\mathbf{p}(k)$. This is because, there is always a chance that the probability of the selected action is decreased. (Note that, except in trivial cases, the reward probability of any action would be strictly less than unity). It can be shown that under this L_{R-P} algorithm, the action probabilities converge in distribution and, if the learning step-size is sufficiently small, then the limiting value of p_i, probability of ith action, would essentially be proportional to $\frac{1}{1-d_i}$ where d_i is the reward probability of ith action [NT89].

Another similar algorithm is the linear reward ϵ-penalty ($L_{R-\epsilon P}$) algorithm which is obtained by choosing $\lambda_2 = \epsilon \lambda_1$ in (1.4.18) with $\epsilon < 1$ being a small number. The properties of $L_{R-\epsilon P}$ algorithm are in between those of L_{R-I} and L_{R-P}. (See [NT89] for details).

All these algorithms can be said to be based on a simple 'reward-penalty' idea. That is, whenever the selected action results in favorable reinforcement we increase the action probability; otherwise we decrease it. The increase and decrease can be by different step-sizes (with the extreme case being a step-size of zero). The naming of the algorithms as reward-penalty or reward-inaction is based on this. In a similar vein, we can also formulate a linear inaction-penalty algorithm.

All these schemes of updating the action probabilities are linear in the components of the action probability vector. Nonlinear algorithms have also been advocated to enhance the performance of FALA and for ensuring properties such as absolute expediency which in turn lead to ϵ-optimality. Rigorous results for choosing nonlinear functions are available [NT89].

1.4.3 Estimator Algorithms

In algorithms such as L_{R-I}, the updating of action probabilities at an instant depends only on the action selected and reinforcement obtained; it does not explicitly depend upon the history of past actions and reinforcements. Estimator algorithms are a class of learning algorithms where the updating of action probabilities depends on the past history of actions and reinforcements. These algorithms make use of estimates of the characteristics of the random environment in which the FALA is operating. Such estimates are computed online during the operation of FALA. The idea here is to make use of these estimates to improve the performance in terms of speed and accuracy of learning. Prominent among estimator algorithms is the Pursuit algorithm described below.

Pursuit Algorithm

Pursuit algorithm uses the history of actions selected and reinforcements obtained to compute estimates of the reward probabilities of actions. We denote the estimate of d_i at k by $\hat{d}_i(k)$. The basic approach of the pursuit algorithm is the so-called 'certainty–equivalence' principle. Let $\hat{d}_{M(k)}$ be the highest estimated reward probability at instant k. If we take the estimates to be the true values, we should set $p_{M(k)} = 1$ and the rest of action probabilities to zero. In other words, set $\mathbf{p}(k+1) = \mathbf{e}_{M(k)}$. However, since the estimates could be erroneous, the algorithm moves $\mathbf{p}(k)$ towards $\mathbf{e}_{M(k)}$ by a small amount determined by a learning parameter. Thus, the action probability vector 'pursues' the estimated optimal vector and hence the name of the algorithm.

At each k, the algorithm computes the current estimates of the reward probabilities, $\hat{d}_i(k)$ and these are in turn used to update $\mathbf{p}(k)$. For obtaining the estimates, the algorithm maintains two more vectors $(Z_1(k) \ldots Z_r(k))^T$ and $(\eta_1(k) \ldots \eta_r(k))^T$. The number of times action α_i is chosen till k is given by $\eta_i(k)$ while $Z_i(k)$ gives the total reinforcement obtained in response to action α_i till k. The algorithm uses $\alpha(k)$ and $\beta(k)$ to update $\eta_i(k)$ and $Z_i(k)$ and these are used to obtain the estimates $\hat{d}_i(k)$. The details are given below.

Let $\alpha(k) = \alpha_i$. Then we update $Z_i(k)$, $\eta_i(k)$ and obtain the estimates, $\hat{d}_i(k)$ as follows.

$$\begin{aligned} Z_i(k) &= Z_i(k-1) + \beta(k) \\ Z_j(k) &= Z_j(k-1), \; \forall j \neq i \\ \eta_i(k) &= \eta_i(k-1) + 1 \end{aligned}$$

Some Classical Learning Algorithms

$$\eta_j(k) = \eta_j(k-1), \forall j \neq i$$

$$\hat{d}_i(k) = \frac{Z_i(k)}{\eta_i(k)}, i = 1, \ldots r.$$

As can be seen from above, $\hat{d}_i(k)$ gives the *average* reinforcement obtained with action α_i till k. Note that we need $\eta_i(k) > 0$ for the $\hat{d}_i(k)$ as above to be well defined. For this, we may choose each action a few times at the beginning to get initial values of the estimates. Alternately, we can keep $\hat{d}_i(k)$ as zero till a time instant such that $\eta_i(k) > 0$. That is, until an action is chosen (in the normal operation of FALA), we do not start updating the corresponding reward probability estimate.

Now the action probabilities are updated as

$$\mathbf{p}(k+1) = \mathbf{p}(k) + \lambda(\mathbf{e}_{M(k)} - \mathbf{p}(k)) \quad (1.4.19)$$

where $0 < \lambda \leq 1$ is the learning parameter and the index $M(k)$ is determined by

$$\hat{d}_{M(k)} = \max_i \hat{d}_i(k) \quad (1.4.20)$$

In the above, ties in finding the index of maximum estimated reward probability are arbitrarily resolved.

It may be noted that in the pursuit algorithm, updating of action probabilities does not depend on the identity of the selected action. This is in contrast with algorithms such as L_{R-I}. The action selected and reinforcement obtained are used for updating the estimates of the reward probabilities. Then the action probabilities are updated so as to increase the probability of the current estimated best action.

Another interesting aspect of the pursuit algorithm is that the updating of $\mathbf{p}(k)$ does not directly involve $\beta(k)$, the environment response. One advantage of this feature is that $\beta(k)$ need no longer be constrained to remain in $[0, 1]$, a factor of considerable utility in applications.

The particular advantage of the pursuit algorithm is that it is an order of magnitude faster than algorithms like L_{R-I}. It however retains the properties of L_{R-I} such as ϵ-optimality in all stationary random environments and convergence of $\mathbf{p}(k)$ to the set of unit vectors w.p.1.

The convergence property of the algorithm can be stated as follows.

THEOREM 1.1 *A learning automaton using the pursuit algorithm (1.4.19) has the following property. Given any $\epsilon, \delta \in (0,1)$, there exist $K^*, \lambda^*, K^* > 0$, & $0 < \lambda^* \leq 1$ such that,*

$$Pr[p_\ell(k) > 1 - \epsilon] > 1 - \delta \quad (1.4.21)$$

$\forall k > K^*$ and $\forall \lambda$ such that $0 < \lambda < \lambda^*$.

The proof of the theorem is given in Appendix B. The proof rests on the following ideas.

1. If, after some finite time, the estimates of reward probabilities remain locked in a sufficiently small interval around the true values, $p_\ell(k)$ will approach unity as $k \to \infty$.

2. If λ is sufficiently small, each action will be chosen enough number of times and estimates will be as close as desired to actual values of reward probabilities.

The details are given in Appendix B.

1.4.4 Simulation Results

To illustrate the behavior of the L_{R-I} and the pursuit algorithms, we present some empirical experiments conducted on the following environments.

E2: A 2-action environment with reward probabilities: $\{0.6, 0.2\}$.

E5: A 5-action environment with reward probabilities: $\{0.35, 0.8, 0.5, 0.6, 0.15\}$.

E10: A 10-action environment with reward probabilities: $\{0.1, 0.45, 0.84, 0.76, 0.2, 0.4, 0.6, 0.7, 0.5, 0.3\}$.

It may be noted that environment $E10$ represents the most difficult learning problem here because the highest and the next highest reward probabilities are close and also because it has the highest number of actions.

Fig. 1.2 shows the evolution of the optimal action probability with time, in each of the three environments for the two algorithms. The results shown here are averages over 500 experiments. For the pursuit algorithm, ten trials were initially made to get initial estimates of the reward probabilities. The learning parameter chosen for each algorithm is the largest value so that in all 500 experiments the algorithm converged to the optimal action. The value of λ for L_{R-I} in the environments E2, E5 and E10 are 0.15, 0.025 and 0.007 respectively, and the corresponding values of λ for pursuit algorithm are 0.55, 0.015 and 0.001. From the figure, it is easily seen that the (averaged) optimal action probability increases monotonically. The pursuit algorithm converges faster than L_{R-I}. It is observed that there is a reduction in speed of convergence with increase in the number of actions.

1.5 The Discretized Probability FALA

There is always a need to reduce the memory overheads and improve the speed of operation of a learning automaton. Any such improvement will en-

The Discretized Probability FALA

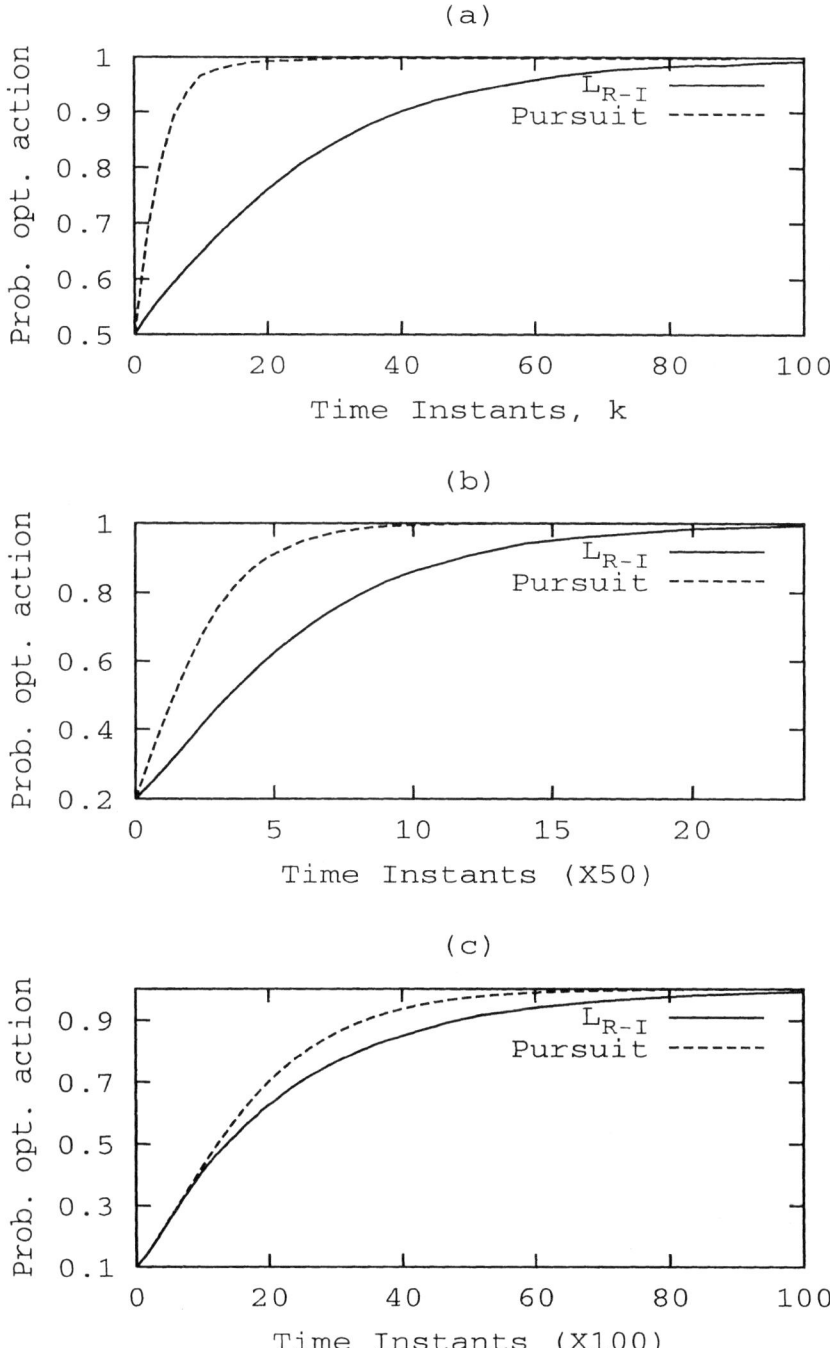

Figure 1.2. Performance of L_{R-I} and pursuit algorithms in the three environments: (a). Environment E2, (b). Environment E5, (c). Environment E10

large the areas of application. One of the ways in which such an improvement can be effected is by discretizing the action probability space of the LA [TO79].

In discretized automata models we restrict the action probabilities to a finite number of values in the interval $[0, 1]$. The number of such values denotes the level of discretization and is a design parameter. The values are generally spaced equally in [0,1] and in such a case the corresponding learning algorithm is said to be linear. If the values are spaced unequally, the learning algorithm is said to be nonlinear.

1.5.1 DL_{R-I} Algorithm

Every linear learning algorithm considered earlier for FALA, can be discretized in this manner. For instance, the discretized L_{R-I} algorithm (designated DL_{R-I}) can be stated as follows. Here we assume $\beta(k) \in \{0, 1\}$.

Let N be the resolution parameter indicating the level of discretization. The smallest change in any action probability is then chosen as $\Delta = 1/rN$ where r is the number of actions. The updating is given below.

If $\alpha(k) = \alpha_i$, and $\beta(k) = 1$,

$$\begin{align} p_j(k+1) &= \max\{p_j(k) - \Delta, 0\} \quad \text{for all } j \neq i \\ p_i(k+1) &= 1 - \sum_{j \neq i} p_j(k+1) \end{align} \quad (1.5.22)$$

If $\beta(k) = 0$, for any $\alpha(k)$,

$$p_j(k+1) = p_j(k) \quad \forall j.$$

The DL_{R-I} algorithm has been shown to be ϵ-optimal in all stationary random environments, in the two-action case. That is, as $N \to \infty$ the probability of the algorithm converging to a nonoptimal action goes to zero. The additional benefit of discretization is that DL_{R-I} is generally faster than L_{R-I}. The speedup occurs mainly when action probabilities are nearing 1 or 0. In the L_{R-I} algorithm, whenever an action probability, p_i is increased, the increase is proportional to $(1-p_i)$. Hence, as the probability approaches one, the rate of increase becomes small. In contrast, in the discretized algorithm, the increase is by a integral multiples of a fixed step and this results in faster convergence.

Discretized L_{R-P} algorithm is also available [Oom86]. Nonlinear discretized LA algorithms are not much used as they pose problems when the number of actions is higher than 2.

1.5.2 Discretized Pursuit Algorithm

The estimator algorithms for FALA can also be discretized. For instance, the discrete version of the pursuit algorithm is as follows.

In the discretized pursuit algorithm, the estimates $\hat{d}_i$ are obtained exactly in the same manner as in the continuous version. However, the action probabilities are changed by integral multiples of a fixed value. As in other discretized algorithms, Δ is the smallest step by which probability of any action can be changed. The normal choice is $\Delta = 1/rN$ where N is the resolution parameter. In the discretized pursuit algorithm, at each instant k, we first update the estimates $\hat{d}_i$ as explained in Section 1.4.3. Then, the action probabilities are updated as follows.

$$p_j(k+1) = \max\{p_j(k) - \Delta, 0\} \quad \text{for all } j \neq M(k)$$
$$p_i(k+1) = 1 - \sum_{j \neq i} p_j(k+1) \quad \text{for } i = M(k). \quad (1.5.23)$$

It may be recalled that $M(k)$ corresponds to the index such that

$$\hat{d}_{M(k)} = \max_i \hat{d}_i(k). \quad (1.5.24)$$

The discretized pursuit algorithm is ϵ-optimal with respect to the resolution parameter N. That is, by taking N sufficiently large, we can have the probability of optimal action go to unity with sufficiently large probability. Simulation studies have shown that the discretized algorithm is faster than its continuous counterpart [OA01].

Attempts have been made to superpose the Reward-Inaction concept of L_{R-I} on the pursuit algorithm. That is, action probabilities are updated only when the environment response $\beta(k)$ is 1. This feature has been seen to give a further speedup on the basis of simulation studies [LO92]. However, the improvement takes place under an additional requirement. Suppose the environment response $\beta(k)$ is continuous i.e. it takes values over an interval. In the original pursuit algorithm, these values can be used as they are in forming estimates $\hat{d}_i(k)$. However, when the reward - inaction feature is imposed, there is need to characterize reward and penalty by recognizing a threshold for $\beta(k)$. This introduces an additional design factor.

Stochastic Estimator Algorithms

A further development in improving the estimator algorithms in general and pursuit algorithm in particular, is the concept of stochastic estimator algorithms. The idea is to impose a perturbation of decreasing magnitude on the estimates $\hat{d}_i(k)$ before making the selection of the apparently best action. As earlier, let $Z_i(k)$ denote the total reinforcement obtained with action α_i till k and let $\eta_i(k)$ denote the number of times α_i is chosen till k. Let $\hat{d}_i(k) = \frac{Z_i(k)}{\eta_i(k)}$

be the usual estimate. The stochastic estimator of the ith reward probability at k is defined by $u_i(k) = \hat{d}_i(k) + \xi_i(k)$ where $\xi_i(k)$ is a random number uniformly distributed in the interval $(-\frac{\gamma}{\eta_i(k)}, \frac{\gamma}{\eta_i(k)})$ and γ is a parameter of the algorithm. Now the index $M(k)$ is defined by

$$u_{M(k)}(k) = \max_i u_i(k).$$

After determining the index $M(k)$ as above, the probability updating is done in the same way as earlier. As is easy to see, u_i are obtained by adding a random perturbation to the old estimates $\hat{d}_i$. The variance of the perturbation decreases as η_i, the number of times α_i is chosen, increases. The motivation for this is that when η_i is small, $\hat{d}_i$ may not be very reliable; but as η_i becomes large $\hat{d}_i$ comes close to the unknown d_i. Since we have such perturbations of the estimates $\hat{d}_i$, it may not be necessary to select all actions a few times to initialize the estimates as in the original pursuit algorithm. It can be shown that this algorithm is ϵ-optimal in all stationary environments. It is also seen through simulations that use of these stochastic estimators results in good improvement in the speed of convergence of the algorithm [PSP03].

1.6 The Continuous Action Learning Automaton (CALA)

The learning task could frequently involve learning a real-valued parameter such as a gain in a control system or a weight in a neural network. Since an automaton learns the 'optimal' action, the actions of the automaton could be the possible values of the parameter. To use an FALA for such a learning task, we have to discretize the range of possible values of the parameter to obtain a *finite* set of actions. Such discretization may not be feasible in all situations. Either the level of discretization may be too coarse for the problem or with a finer discretization, the number of actions becomes too large and the system becomes too slow. A natural option would be to work with a continuous space of actions. In this section we describe one such model called the continuous action learning automaton (CALA).

The set of actions of a CALA is the real line. The CALA chooses a real number x at each instant k based on the current action probability distribution which is a normal distribution with mean $\mu(k)$ and standard deviation $\sigma(k)$. We denote this normal distribution by $N(\mu(k), \sigma(k))$. The CALA updates $\mu(k)$ and $\sigma(k)$ at each instant based on the reinforcement received from the environment.

The objective of CALA is to learn the value of x for which $E[\beta_x]$ is maximum, where β_x is the response of the environment when the action chosen is x. Thus the learning problem handled by CALA is as follows. Consider a function $f : \Re \to \Re$, such that only noisy values of $f(x)$ are available for measurement for any given x. Noisy values correspond to β_x and we have

$f(x) = E[\beta_x | x]$. The function $f(x)$ here is analogous to the reward probabilities d_i of an FALA. Hence, the optimal action now is an x that maximizes $f(\cdot)$. To learn this, the CALA starts with an initial action probability distribution, $N(\mu(0), \sigma(0))$ and, through a learning algorithm, updates its action probability distribution after each interaction with the environment. The natural objective of such a learning algorithm is to make $N(\mu(k), \sigma(k))$ converge to $N(x_0, 0)$ where x_0 maximizes $f(x) = E[\beta_x | x]$. To make the asymptotic behavior of the algorithm analytically tractable, we cannot let $\sigma(k)$ converge to zero. Hence, we use another parameter $\sigma_L > 0$ (with σ_L sufficiently small) and keep the objective of learning as $\sigma(k)$ converging to σ_L and $\mu(k)$ converging to a local maximum of $f(x)$.

To sum up, CALA learns the value of x which maximizes the function $f(x)$ in a scenario where the function $f(x)$ is unknown, only noisy values of the function (for any given x) are available for measurement and the probability distribution that characterizes the relation between the noisy measurement β_x and $f(x)$ is unknown. (For example, the value of β_x could be $f(x)$ plus some zero-mean noise or β_x could be a 0 – 1 random variable with its probability of taking value 1 being $f(x)$. In both cases we have $f(x) = E[\beta_x | x]$).

The CALA operates as follows. At every k, it chooses a real number $x(k)$ at random based on its current action probability distribution $N(\mu(k), \phi(\sigma(k)))$ where ϕ is defined by (1.6.27) below. Then the CALA obtains responses from the environment for two actions, namely, $\mu(k)$ and $x(k)$. Let these reinforcements be $\beta_{\mu(k)}$ and $\beta_{x(k)}$ respectively. Then $\mu(k)$ and $\sigma(k)$ are updated according to the following learning algorithm.

$$\mu(k+1) = \mu(k) + \lambda \frac{(\beta_{x(k)} - \beta_{\mu(k)})}{\phi(\sigma(k))} \frac{(x(k) - \mu(k))}{\phi(\sigma(k))} \quad (1.6.25)$$

$$\sigma(k+1) = \sigma(k) + \lambda \frac{(\beta_{x(k)} - \beta_{\mu(k)})}{\phi(\sigma(k))} \left[\left(\frac{x(k) - \mu(k)}{\phi(\sigma(k))} \right)^2 - 1 \right]$$
$$- \lambda K (\sigma(k) - \sigma_L) \quad (1.6.26)$$

where,

$$\phi(\sigma) = \sigma_L \quad \text{for} \quad \sigma \leq \sigma_L$$
$$= \sigma \quad \text{for} \quad \sigma > \sigma_L > 0 \quad (1.6.27)$$

and λ is the learning parameter controlling the step size ($0 < \lambda < 1$), K is a large positive constant and σ_L is the lower bound on σ. The iterations continue till $\mu(k)$ does not change appreciably and $\sigma(k)$ is close to σ_L.

The idea behind the updating scheme given by (1.6.25) – (1.6.26) is as follows. If $x(k)$ gets 'better' response from the environment than $\mu(k)$, then,

$\mu(k)$ is moved towards $x(k)$; otherwise it is moved away from $x(k)$. Division by $\phi(\sigma(k))$ is to standardize the corresponding quantities. For updating $\sigma(k)$, whenever an action choice $x(k)$ that is away from the mean $\mu(k)$ by more than one standard deviation results in better response from the environment or when an action choice within one standard deviation from the mean is worse, we increase the variance; otherwise we decrease it. The last term on the RHS in (1.6.26) is a penalizing term for $\sigma(k)$ so as to push it towards σ_L. However, it may be noted that the updating given by (1.6.26) does not automatically ensure that $\sigma(k) > \sigma_L$, $\forall k$ or even that $\sigma(k) > 0$, $\forall k$. That is why we use the function ϕ and always use the projected version of $\sigma(k)$, namely, $\phi(\sigma(k))$, as the standard deviation in the action probability distribution. This implies that the variance in the distribution $N(\mu(k), \phi(\sigma(k)))$ remains above a small nonzero value σ_L^2 and helps in avoiding premature convergence of the learning process. The price paid for this feature is that $\mu(k)$ never settles down to a constant but will always be perturbed slightly, the extent of perturbation depending on the value of σ_L.

For implementing the learning algorithm as given by (1.6.25) and (1.6.26), the CALA needs to obtain two reinforcements from the environment at each instant, namely, $\beta_{x(k)}$ and $\beta_{\mu(k)}$. (In the context of a function optimization application, this implies two function evaluations per iteration). This is in contrast to FALA which elicits only one reinforcement per iteration. From the discussion in the previous paragraph, it is clear that the algorithm is using $\beta_{\mu(k)}$ as a 'baseline' to get a better idea of the significance of $\beta_{x(k)}$. However, we note here that the algorithm can work with only one reinforcement per iteration also. The analytical results regarding the behavior of the algorithm presented in the next section continue to be valid even if we take $\beta_{\mu(k)} \equiv 0$ in (1.6.25) and (1.6.26). However, the algorithm as given here converges faster and that is why we utilize two reinforcements per iteration.

1.6.1 Analysis of the Algorithm

In this section we explain how we can analyze the asymptotic behavior of the CALA algorithm. As we did earlier for the L_{R-I} algorithm, we follow an approach based on deriving an associated Ordinary Differential Equation (ODE). We can write the CALA algorithm in the form

$$Z(k+1) = Z(k) + \lambda G(Z(k), \xi(k)) \qquad (1.6.28)$$

where $Z(k) = [\mu(k)\ \sigma(k)]^T$ is the 'state' at k and G symbolically represents the updating of the two components of $Z(k)$ as given by (1.6.25) and (1.6.26). Here we take $\xi(k) = [x(k)\ \beta_{x(k)}\ \beta_{\mu(k)}]^T$ and the updating depends on both $Z(k)$ and $\xi(k)$. It may be noted that $\xi(k)$ is random and the probabilities of its taking different values depends on $Z(k)$. Thus the general learning algorithm as represented by (1.6.28) is a stochastic difference equation and we are inter-

The Continuous Action Learning Automaton (CALA)

ested in the asymptotic properties of $Z(k)$. The general approach to analyzing such algorithms is explained in Appendix A. What we do is to derive an ODE whose solution provides a good approximation to the behavior of the algorithm when λ is sufficiently small. The approximating ODE for this CALA algorithm is derived in section A.3.2 and this ODE is given by equation (A.3.37) in Appendix A. This ODE approximates the behavior of the CALA algorithm well under the assumptions $B1$–$B3$ (stated in section A.3.2) on the function $f(x)$. (Recall that $f(x) = E[\beta_x \mid x]$). These assumptions essentially impose some smoothness constraints on this function and these are not particularly restrictive.

The main features of the algorithm that result in the specific form of this ODE are as follows. The updating in (1.6.25)–(1.6.26) is essentially a stochastic gradient following procedure. Given any $\mu \in \Re, \sigma > 0$, define the average reinforcement as

$$J(\mu, \sigma) = \int_{\Re} f(x) dN(\mu, \sigma)$$
$$= \int_{\Re} f(x) \frac{1}{\sigma\sqrt{2\pi}} \exp\left(-\frac{(x-\mu)^2}{2\sigma^2}\right) dx \quad (1.6.29)$$

where $N(\mu, \sigma)$ denotes the Gaussian density function with mean μ and standard deviation σ. Let

$$F_1(\mu, \sigma, x, \beta, \beta') = \frac{(\beta - \beta')}{\phi(\sigma)} \cdot \frac{(x - \mu)}{\phi(\sigma)} \quad (1.6.30)$$

and

$$F_2(\mu, \sigma, x, \beta, \beta') = \frac{(\beta - \beta')}{\phi(\sigma)} \left[\left(\frac{(x - \mu)}{\phi(\sigma)}\right)^2 - 1\right]. \quad (1.6.31)$$

Note that these functions capture the prominent terms in the learning algorithm given by (1.6.25)–(1.6.26). Specifically, in terms of the above functions, the learning algorithm can be rewritten as

$$\mu(k+1) - \mu(k) = \lambda F_1(\mu(k), \sigma(k), x(k), \beta_{x(k)}, \beta_{\mu(k)})$$
$$\sigma(k+1) - \sigma(k) = \lambda F_2(\mu(k), \sigma(k), x(k), \beta_{x(k)}, \beta_{\mu(k)}) - \lambda K[\sigma(k) - \sigma_L]$$

Then, as proved in section A.3.2 (see equations (A.3.35) and (A.3.36) in Appendix A), we have

$$E[F_1(\mu(k), \sigma(k), x(k), \beta_{x(k)}, \beta_{\mu(k)}) \mid \mu(k) = \mu, \sigma(k) = \sigma] = \frac{\partial J}{\partial \mu}(\mu, \phi(\sigma))$$
$$(1.6.32)$$

$$E[F_2(\mu(k),\sigma(k),x(k),\beta_{x(k)},\beta_{\mu(k)})|\mu(k)=\mu,\sigma(k)=\sigma] = \frac{\partial J}{\partial \sigma}(\mu,\phi(\sigma))$$
(1.6.33)

Equations (1.6.32) and (1.6.33) show that the expectation of the changes in $\mu(k), \sigma(k)$ conditioned on their current values, is proportional to the gradient of the average reinforcement, J. This is called the stochastic gradient following property of the algorithm.

The ODE associated with the CALA algorithm (which is derived in section A.3.2) is given by

$$\frac{d\mu}{dt} = \frac{\partial J}{\partial \mu}(\mu,\phi(\sigma)) \tag{1.6.34}$$

$$\frac{d\sigma}{dt} = \frac{\partial J}{\partial \sigma}(\mu,\phi(\sigma)) - K(\sigma - \sigma_L) \tag{1.6.35}$$

The precise sense in which this ODE approximates the algorithm is explained in Appendix A. (See Theorem A.1). Intuitively, from the solution of the ODE we can infer the behavior of $\mu(k), \sigma(k)$ for large k if λ is sufficiently small.

Given the stochastic gradient following property of the algorithm as explained earlier, the ODE associated with the algorithm is as expected. In (1.6.34), (1.6.35), the RHS of the ODE is the gradient of J evaluated at the point $(\mu, \phi(\sigma))$. (The additional term in (1.6.35) is due to the penalty term used in the updating of $\sigma(k)$). Thus, we can expect μ and σ to converge to some maxima of J. However, our objective is to find a maximum of f and not of J. Observe that J is the expected value of the reinforcement, where the expectation is taken with respect to both a) the normal distribution with parameters μ and σ using which actions are chosen and b) the unknown distribution from which the reinforcement β_x is obtained. Apart from this, we have to take into account the penalizing term for σ in (1.6.35). From (1.6.29), it is intuitively clear that $J(\mu,\sigma)$ would be close to $f(\mu)$ if σ is sufficiently small. Hence, we can expect that for sufficiently small σ, $\frac{\partial J}{\partial \mu}(\mu,\sigma)$ would be close to $f'(\mu) \triangleq \frac{\partial f}{\partial x}|_{x=\mu}$ and hence maxima of J would be close to maxima of f. Relation between maxima of f and constrained maxima of J (to which the solutions of the approximating ODE and hence the algorithm converge) can be established. It can be proved that the algorithm converges to a close approximation of an isolated local maximum of f. We do not present any rigorous proofs here and they are available in [SST94, San94]. The main part of proof consists of establishing the following.

- Given any $\Delta > 0$ and any compact set $\widetilde{K} \in \Re$, there exists a $\sigma^* > 0$ such that

$$\sup_{\mu \in \widetilde{K}} \left| \frac{\partial J}{\partial \mu}(\mu,\sigma) - f'(\mu) \right| < \Delta$$

for any $0 < \sigma < \sigma^*$.

This result shows that the convergence of $\frac{\partial J}{\partial \mu}$ to $f'(\mu) = \frac{df(\mu)}{d\mu}$ as $\sigma \to 0$ is uniform over all $\mu \in \widetilde{K}$ for any compact set $\widetilde{K}$.

- If σ_L is sufficiently small, there exists at least one zero of $\frac{\partial J}{\partial \mu}$ in the neighborhood of any isolated zero of $f'(\mu)$.

- Given any value of $K > 0$ in the learning algorithm, there exists at least one equilibrium point for the ODE given by (1.6.34) – (1.6.35).

From (1.6.34), any equilibrium point of the ODE should satisfy $\frac{\partial J}{\partial \mu}(\mu, \phi(\sigma)) = 0$. Suppose we can establish that in any equilibrium point of the ODE (μ, σ), we would have $\phi(\sigma)$ close to σ_L. Then, the above results imply that if we choose σ_L sufficiently small, then, the ODE and hence the algorithm would converge to a maximum of J which is close to a maximum of f. Suppose (μ, σ) is an equilibrium point of the ODE. If $\sigma < \sigma_L$ then obviously $\phi(\sigma)$ is close to σ_L as required. Suppose, $\sigma > \sigma_L$. Since (μ, σ) is an equilibrium point, we must have

$$\frac{\partial J}{\partial \sigma}(\mu, \phi(\sigma)) - K(\sigma - \sigma_L) = 0.$$

If we can choose $K \gg \frac{\partial J}{\partial \sigma}(\mu, \phi(\sigma))$ uniformly for all μ, σ, then the above holds only for some σ close to σ_L. By our assumptions on the function $f(x)$ as explained in section A.3.2, such a choice of K exists. Thus, we can conclude that in any equilibrium point of the ODE (μ, σ), we would have $\phi(\sigma)$ as close to σ_L as desired by taking K sufficiently large.

Putting all this together, we can conclude that by choosing a small value of $\sigma_L > 0$, a small step size $\lambda > 0$ and sufficiently high $K > 0$, we can ensure that the iterates $\mu(k)$ of the CALA algorithm will be close to a maximum of the function f with a high probability after a long enough time.

At the beginning of this section, we motivated the CALA model using the example of learning optimal value of a parameter. If we use an FALA, we have to discretize the range of values of the parameter. By using a CALA, we can avoid discretizing the parameter. However, as the analysis above shows, with this learning algorithm, we can learn only a local maximum. On the other hand, by using an FALA we would learn the global maximum (within the finite set of discretized values).

In the next subsection we illustrate the behavior of the CALA algorithm through some simulations.

1.6.2 Simulation Results

The behavior of the CALA algorithm is studied in connection with a specific stochastic optimization task here. The task is to find a minimum[4] of the penalized Shubert function [APPZ85] which is notorious for having many minima in a relatively small interval.
The penalized Shubert function is

$$f(x) = \sum_{i=1}^{5} i \cos((i+1)x + 1) + u(x, 10, 100, 2)$$

where $u(\cdot, \cdot, \cdot, \cdot)$ is the penalizing function given by,

$$u(x, a, k, m) = \begin{cases} k(x-a)^m & \text{if } x > a \\ 0 & \text{if } |x| \leq a \\ k(-x-a)^m & \text{if } x < -a \end{cases}$$

This function has 19 minima within the region $[-10, 10]$ and three of these are global minima. The global minimum value of the function is approximately equal to -12.87 and it is attained at the points (close to) $-5.9, 0.4$ and 6.8. The next best minima are (close to) $-7, -0.6$ and 5.6 with function value equal to -8.51. This function is frequently used as one of the bench mark problems to test global optimization algorithms [APPZ85, GQ90].

We use the CALA algorithm for this optimization task. Different initial values of the mean and variance, $\mu(0)$ and $\sigma(0)$, are tried. In the algorithm we take $\sigma_L = 0.01$ and $K = 5$.

In the first set of simulations *no noise* was added to the function evaluations. Thus for any x, $\beta_x = f(x)$. The simulation was run up to a maximum of 3000 iterations for each initial condition, with step size $\lambda = 0.0003$. The mean value of the Gaussian distribution used in LA always converged to a minimum of the Shubert function within the interval $[-10, 10]$. *For the same initial value of mean but with an increased value of initial variance, the algorithm could converge to a global minimum in some simulations.* See Table 1.6.1 for the results of some typical runs of the simulation. (Note that the updating of $\sigma(k)$ as given by (1.6.25) can result in negative values for σ, but only the projected value (namely, $\phi(\sigma)$) is used as the standard deviation in the action probability distribution).

In another set of simulations the function evaluations were corrupted by a zero mean noise randomly generated in the range $[-0.5, 0.5]$ with uniform distribution. The simulations were carried up to 8000 iterations with step size $\lambda = 0.0002$. The results of the simulations for various initial conditions are

---
[4]could be converted to an equivalent maximization problem by a change of sign

The Continuous Action Learning Automaton (CALA)

Initial Values			After 3000 Iterations		Final function Value
$\mu(0)$	$\sigma(0)$	Step(λ)	$\mu(3000)$	$\phi(\sigma(3000))$	
3	5	3×10^{-4}	1.5	0.01	-3.578
3	6	3×10^{-4}	0.423	0.01	-12.87
-10	5	3×10^{-4}	-7.09	0.01	-8.5
-10	7	3×10^{-4}	-5.86	0.01	-12.87
10	5	3×10^{-4}	7.784	0.01	-3.58
10	7	3×10^{-4}	5.473	0.01	-8.513
7	3	3×10^{-4}	6.707	0.01	-12.87

Table 1.6.1. Simulation results for Shubert Function with *no noise* added. Global minimum has a value -12.87. Step size $\lambda = 0.0003$, $\sigma_L = 0.01$ and $K = 5.0$.

Initial Values			After 8000 Iterations		Final function Value
$\mu(0)$	$\sigma(0)$	Step(λ)	$\mu(8000)$	$\phi(\sigma(8000))$	
4	6	2×10^{-4}	2.534	0.01	-3.578
4	10	2×10^{-4}	0.4038	0.01	-12.87
8	3	3×10^{-4}	5.36	0.85	-8.5
8	5	2×10^{-4}	6.72	0.01	-12.87
-10	5	2×10^{-4}	-7.1	0.01	-8.5
-10	6	2×10^{-4}	-5.87	0.01	-12.87

Table 1.6.2. Simulation results *with noise* added to the Shubert function evaluations.

shown in Table 1.6.2. From the table it is seen that even with noisy measurements the algorithm converges to a minimum.

1.6.3 Another Continuous Action Automaton

As mentioned earlier, in the early developments of the field, learning automaton was defined to have only finitely many actions. (These are what are called FALA here). In this section we have considered CALA which are automata whose actions sets are continuous.

When the set of actions is finite, the action probability distribution is a probability mass function and is represented by a vector of numbers. Every possible probability distribution on the action set can be so represented. However, when the action set is continuous, we need to deal with probability density functions and hence the issue of how to represent the action probability distribution becomes important. In the CALA model described here, we have chosen a specific parameterized density (namely, the Gaussian density) and then updating of the action probability distribution is done by updating the parameter values. Using such a parametric representation, we have presented a learning algorithm which ensures convergence to an optimal action.

A natural question that can be asked now is whether it is possible to have a learning automaton with a continuous action set whose action probability distribution is represented in a non-parametric manner. There is such a model and it is called continuous action reinforcement learning automaton (CARLA) [HFGW97]. The action set of CARLA is an interval on real line. Let $A = [x_{min}, x_{max}]$ denote the action set. At each instant, the automaton chooses an action (which is a real number) based on its current action probability distribution, gets a reinforcement feedback from the environment and, using this, updates its action probability distribution. Let $p_k(x)$ denote the density function for choosing actions at instant k. To start with, $p_0(x)$ is a uniform density over A. Let $\alpha(k) \in A$ and $\beta(k) \in [0, 1]$ denote the action selected and reinforcement obtained at k. Then the action probability distribution is updated as

$$p_{k+1}(x) = \begin{cases} \gamma[p_k(x) + \beta(k)H(x, \alpha(k))] & \text{if } x \in A \\ 0 & \text{otherwise} \end{cases} \quad (1.6.36)$$

in which $H(x, z)$ is a symmetric Gaussian type function given by

$$H(x, z) = \eta \exp\left(-\frac{(x - z)^2}{2\sigma^2}\right), \quad (1.6.37)$$

where η and σ are parameters of the algorithm. In equation (1.6.36), the factor γ is a normalizing constant and its value is determined by the requirement that $\int p_{k+1}(x)\, dx = 1$. We can think of $H(x, z)$ as a neighborhood function that essentially spreads the effect of the current reinforcement in the neighborhood of the current action. The action probability distribution at $k + 1$ is a combination of previous action probability distribution and the neighborhood function. The weightage given to the neighborhood function is determined by the current reinforcement. As the iterations proceed, we are essentially putting small Gaussian functions at each of the selected actions which resulted in good reinforcement. As can be seen from (1.6.36), it is not possible to represent the action probability distribution through some parametric representation. In the implementation, it is stored, through numerical integration, as values of cumulative distribution function at regular intervals. However, since the action choice has to be continuous, the distribution function is linearly interpolated from these values for choosing an action.

It is seen through simulation experiments that this updating essentially results in the action probability distribution eventually peaking at an x where the expected value of reinforcement is a maximum. At present, there are no analytical results to characterize the asymptotic behavior of this algorithm. The algorithm is found useful in control applications [HFGW97].

1.7 The Generalized Learning Automaton (GLA)

In the automata models considered so far, the only input to the automaton is the reinforcement from the environment. At each instant, the automaton chooses an action, gets environmental response and updates the action probability distribution. The objective is to learn *the optimal* action that gets maximum expected value of reinforcement. In an optimization problem, for example, the actions are the possible values of the parameter to be learnt and optimal action is the one where the performance index attains a maximum.

However, there are many tasks where it is natural to formulate the learning problem in such a way that an action is optimal or not only in the context of the *state* of the environment made available to the learning system as an additional input. This input would be called *context vector*. Such context vector input could be important in many situations.

For example, consider a control problem. In a discrete-time control system, at each instant the best value of the control signal has to be selected and this naturally depends on the state of the system at that instant. If an LA is used to select the value of the control signal, the optimal action (or the desired value of the control signal) will depend on the state vector which forms the context vector for the problem.

Similar requirements can be there in a pattern recognition problem also. If we think of decisions regarding classification of a pattern as possible actions here, then whether or not a given action is optimal depends on the feature vector input which forms the context vector in this example.

In both these examples, the goal is not that of learning the identity of one optimal action. The goal is to learn an optimal mapping to associate different context vectors (state of the system or feature vector) with actions (desired control or classification decision). We want to learn this mapping through interactions with the environment in the form of selecting actions (in response to context vector input from environment) and receiving the reinforcement feedback from the environment. This type of learning has been called associative reinforcement learning.

For tackling such problems, there must be a provision for the LA to take into account the additional input in the form of context vector while selecting an action. The Generalized Learning Automaton (GLA) model considered in this section represents one of the methods for handling context vector inputs. The GLA is an extended learning automaton where there is provision for an extra input.

The GLA model, where the automaton can take an additional input, is also useful in many other situations. Often the process of learning proceeds in stages; i.e., learning is a multistage decision process. For instance, in pattern classification using decision trees, the feature space is subdivided into sequentially smaller regions at successive nodes of the tree. At each node in the tree a decision is to be made regarding which way the pattern is to be directed to complete the classification. The final classification decision is arrived at only after the feature vector traverses through all the nodes in a path in the tree. In many such learning problems it is advantageous to configure the individual learning elements into a network. In such cases, if we use LA as the individual learning elements, then, actions of some LA can form inputs to other LA. Thus once again, we need models to tackle additional inputs.

Before proceeding with the details of GLA models, we note here that it is possible to construct systems consisting of the FALA considered earlier so that the system as a whole can tackle context vector inputs. Such systems would be discussed in Chapter 3.

A GLA is described by the tuple $\langle \mathcal{X}, Y, B, \mathbf{u}, g, \mathcal{T} \rangle$. Here $\mathcal{X}$ is the set of all context vectors that can be input to the GLA, $Y = \{y_1, \ldots, y_r\}$ is the set of outputs or actions of GLA[5], B is the set of values that the reinforcement signal can take (which is usually taken to be the interval [0,1]), g is called the probability generating function, and $\mathbf{u}$ is the internal state (of the GLA) which is a vector of real numbers. $\mathcal{T}$ is the learning algorithm that updates $\mathbf{u}$. We denote individual context vectors (which are elements of $\mathcal{X}$) by $\mathbf{X}$ or $\mathbf{X}_i$.

The probability of the GLA taking action y_i when its internal state is $\mathbf{u}$ and the context vector input from the environment is $\mathbf{X}$ is given by

$$\text{Prob}[\alpha(k) = y_i | \mathbf{u}, \mathbf{X}] = g(\mathbf{X}, y_i, \mathbf{u}) \tag{1.7.38}$$

where the function g satisfies the conditions: $g(\mathbf{X}, y_i, \mathbf{u}) \geq 0$, $\forall\ y_i, \mathbf{u}, \mathbf{X}$, and $\sum_{i=1}^{r} g(\mathbf{X}, y_i, \mathbf{u}) = 1$, $\forall\ \mathbf{u}, \mathbf{X}$. The GLA operates as follows. At each instant k, the environment provides the GLA with a context vector $\mathbf{X}(k)$. The GLA takes an action $\alpha(k) \in Y$ based on its current internal state $\mathbf{u}(k)$ where the action choice is random and is governed by the probabilities as given by (1.7.38). Then the environment supplies the GLA with a reinforcement $\beta(k)$ and the GLA updates its state using the learning algorithm. In a GLA the internal state, $\mathbf{u}$, is analogous to the action probability vector in an FALA. Thus, at each instant k, the learning algorithm updates $\mathbf{u}(k)$ based on the current values of $\mathbf{X}(k), \mathbf{u}(k)$, the reinforcement signal $\beta(k)$ and the action chosen by the GLA, $\alpha(k)$. Dependence of the updating on the context vector, $\mathbf{X}(k)$, is the main characteristic of GLA.

The motivation for defining GLA is to be able to tackle associative reinforcement learning problems directly. Hence, with the same state vector $\mathbf{u}$, the probabilities with which a GLA chooses different actions can (and most often, would) be dependent on the context vector. That is why the probability generating function of the GLA is dependent on both the state $\mathbf{u}$ and the context

---

[5]The actions are denoted by y_i rather than α_i as the GLA learns an input-output mapping where the inputs are context vectors $\mathbf{X}$.

vector $\mathbf{X}$. The state $\mathbf{u}$ (along with the function g), is a *representation* for a mapping from the set of context vectors to a set of action probability distributions. Thus the goal of learning is to reach a value of $\mathbf{u}$ so that the GLA can map context vectors to appropriate actions with a high probability.

A simple example of a probability generating function for a GLA is as follows. Suppose the context vector $\mathbf{X}$ belongs to $\Re^n$. We choose the internal state $\mathbf{u}$ to be also n-dimensional. Suppose there are only two actions so that $Y = \{y_1, y_2\}$. Then a probability generating function for the GLA could be

$$g(\mathbf{X}, y_1, \mathbf{u}) = 1 - g(\mathbf{X}, y_2, \mathbf{u}) = \frac{1}{1 + \exp(-\mathbf{X}^T \mathbf{u})}. \qquad (1.7.39)$$

It is easy to see that learning an 'optimal' $\mathbf{u}$ with such a GLA is like learning an 'optimal' linear discriminant function in a 2-class pattern recognition problem.

In general, if there are r actions, then we can take the internal state to be a set of r vectors, $\mathbf{u}_i \in \Re^n$. That is, now $\mathbf{u} = (\mathbf{u}_1^T \ldots \mathbf{u}_r^T)^T$. The probability generating function can be

$$g(\mathbf{X}, y_i, \mathbf{u}) = \frac{\exp(-\mathbf{X}^T \mathbf{u}_i)}{\sum_j \exp(-\mathbf{X}^T \mathbf{u}_j)}. \qquad (1.7.40)$$

The goal for a GLA is to learn the desired mapping from context vectors to actions. As in the case of FALA, we can define reward probabilities for actions as expectations of the reinforcement. However, here the reinforcement depends on both the the action chosen as well as the context vector. Hence, the reward probabilities would also be functions of the context vector. Define,

$$d(\mathbf{X}, y) = E[\beta(k) \mid \alpha(k) = y, \mathbf{X}(k) = \mathbf{X}]. \qquad (1.7.41)$$

Let $O : \mathcal{X} \to Y$ represent the optimal mapping of context vectors to actions. Then this mapping is characterized by the property

$$d(\mathbf{X}, O(\mathbf{X})) = \max_y \{d(\mathbf{X}, y)\}$$

Since we want to learn such a mapping, we want $\mathbf{u}(k)$ to converge to a $\mathbf{u}$ so that for each $\mathbf{X} \in \mathcal{X}$, $g(\mathbf{X}, y, \mathbf{u})$ should be high when $y = O(\mathbf{X})$ and low for all other y. Then the GLA would be choosing the 'optimal' action based on the context vector with a high probability. However, ensuring that the GLA learns in this sense is very difficult or even impossible. Since the probability generating function is fixed, chosen by us, and since the environment is unknown, it is not possible to know whether there exists a vector $\mathbf{u}$ so that the desired mapping can be represented by a GLA. (For example, with a 2-action GLA using the probability generating function given by (1.7.39), any internal state vector **u** represents a hyperplane that separates the set of context vectors, for which action y_1 has higher probability than that of action y_2, from the rest. This GLA can learn only such **u** vectors and such a separation of context vectors into two parts by a hyperplane may or may not be consistent with the desired mapping from context vectors to actions). That is, a priori, there is no way to know whether there exists a **u** which can result in GLA choosing the optimal action in response to every context vector. Hence to be able to prove that a GLA learns the optimal mapping, we have to simply assume that an appropriate **u** exists which would often mean making very stringent assumptions on the problem. (For example, assuming that the set of context vectors is linearly independent). A better choice for the objective of learning is to maximize $f(\mathbf{u}) = E[\beta(k)|\mathbf{u}(k) = \mathbf{u}]$. This would ensure learning of the best mapping that is allowed after having chosen a probability generating function. In case, the desired mapping is representable by the GLA, then an algorithm that maximizes $f(\mathbf{u})$ would also result in learning of this optimal mapping. We illustrate this in an example later on.

1.7.1 Learning Algorithm

The learning algorithm updates the state **u** based on the current values of **u**, **X**, the reinforcement signal, $\beta(k)$, and the action selected, $\alpha(k)$. The earliest algorithm suggested is the REINFORCE algorithm [Wil88] which has the following form.

$$u_i(k+1) = u_i(k) + \lambda \beta(k) \frac{\partial \ln g(\mathbf{X}(k), \alpha(k), \mathbf{u}(k))}{\partial u_i} \quad (i = 1, \ldots) \tag{1.7.42}$$

where u_i is the ith component of **u**. The second term is the gradient term which uses natural logarithm for normalizing the value of the gradient of g with respect to the magnitude of g.

A basic property of the REINFORCE algorithm is the stochastic gradient following property:

$$E[u_i(k+1) - u_i(k)|\mathbf{u}(k)] = \lambda \frac{\partial E[\beta(k)|\mathbf{u}(k)]}{\partial u_i} \quad (i = 1, \ldots) \tag{1.7.43}$$

This result follows from the algorithm and properties of conditional expectation as shown below. Let $\Delta u_i(k) = u_i(k+1) - u_i(k)$. By the chain rule for conditional expectations, we have

$$E[\Delta u_i(k) \mid \mathbf{u}(k)] = \int \sum_{y_j \in Y} E[\Delta u_i(k) \mid \mathbf{X}(k) = \mathbf{X}, \alpha(k) = y_j, \mathbf{u}(k)]$$
$$\text{Prob}[\alpha(k) = y_j \mid \mathbf{X}(k) = \mathbf{X}, \mathbf{u}(k)] P_e(d\mathbf{X})$$

where $P_e(d\mathbf{X})$ is the probability distribution of context vectors and the integral is over the space of context vectors. Using algorithm (1.7.42), we can write the

The Generalized Learning Automaton (GLA)

RHS of the above equation as

$$\int \sum_{y_j} E\left[\lambda\beta(k)\frac{\partial \ln g(\mathbf{X}, y_j, \mathbf{u}(k))}{\partial u_i}\bigg|\mathbf{X}, y_j, \mathbf{u}(k)\right] g(\mathbf{X}, y_j, \mathbf{u}(k)) P_e(d\mathbf{X})$$

Since $\beta(k)$ is the only random variable within the expectation, the above can be rewritten as

$$\lambda \int \sum_{y_j} E[\beta(k)|\mathbf{X}, y_j, \mathbf{u}(k)]\frac{\partial g(\mathbf{X}, y_j, \mathbf{u}(k))}{\partial u_i} P_e(d\mathbf{X})$$

as $g(\mathbf{X}, y_j, \mathbf{u}(k))$ cancels with the corresponding factor in $\frac{\partial \ln g}{\partial u_i}$. In the above, $E[\beta(k)|\mathbf{X}, y_j, \mathbf{u}(k)]$ is independent of $\mathbf{u}(k)$ since $\mathbf{X}, y_j$ are already specified. Hence it can be taken inside the partial derivative. Thus

$$E[\Delta u_i(k)|\mathbf{u}(k)] = \lambda \int \sum_{y_j} \frac{\partial}{\partial u_i}\{E[\beta(k)|\mathbf{X}, y_j, \mathbf{u}(k)]g(\mathbf{X}, y_j, \mathbf{u}(k))\}P_e(d\mathbf{X}).$$

Applying the chain rule for conditional expectations in reverse,

$$E[u_i(k+1) - u_i(k)|\mathbf{u}(k)] = \lambda\frac{\partial}{\partial u_i}E[\beta(k)|\mathbf{u}(k)]$$

thus establishing the stochastic gradient property.

REMARK 1.7.1 *This property shows that the algorithm moves the state towards values which increase $E[\beta(k)|\mathbf{u}(k)]$. However, one major drawback of the algorithm is that it may give rise to unbounded behavior when the maximization of $E[\beta(k)]$ is attained as $\mathbf{u} \to \infty$. A modification of the algorithm which ensures boundedness is presented below.* ∎

Modified REINFORCE Algorithm

$$\begin{aligned} u_i(k+1) &= u_i(k) + \lambda\beta(k)\frac{\partial \ln g(\mathbf{X}(k), \alpha(k), \mathbf{h}(\mathbf{u}(k)))}{\partial u_i} \\ &+ \lambda K_i[h_i(u_i(k)) - u_i(k)] \quad i = 1, 2, \ldots r \quad (1.7.44) \end{aligned}$$

where $\mathbf{h}(\mathbf{u}) = [h_1(u_1), h_2(u_2) \cdots h_r(u_r)]^T$ with the functions h_i given by

$$h_i(\eta) = \begin{cases} L_i & \text{for } \eta \geq L_i \\ \eta & \text{for } |\eta| \leq L_i \\ -L_i & \text{for } \eta \leq -L_i \end{cases}$$

and $L_i, K_i > 0$ are constants.

It may be observed that the 3rd term on the RHS of (1.7.44) becomes positive when u_i has a large negative value and becomes negative when u_i has a large positive value. Thus the term pulls back u_i from becoming unbounded.

Analysis

The analysis of the algorithm can be done using the ODE method explained in Appendix A. All that we need to do is to verify the assumptions $A1$ to $A4$ stated in Section A.2.1. It can be seen that all conditions except $A3$ are readily satisfied. $A3$ is not satisfied because the function here that corresponds to the function g in Section A.2.1 is not globally Lipschitz but only Lipschitz on compact sets. Hence, we follow the method presented in Section A.4 (in Appendix A) so that being Lipschitz on compact sets is enough. A straightforward application of the procedure leads to the approximating ODE,

$$\frac{dz_i}{dt} = \frac{\partial f}{\partial z_i}(\mathbf{h}(z)) + K_i(h_i(z_i) - z_i) \quad i = 1, 2, \ldots r, \quad \mathbf{z}(0) = \mathbf{u}(0) \quad (1.7.45)$$

where $f(\mathbf{z}) = E[\beta|\mathbf{z}]$.

A look at the RHS of the ODE indicates that isolated local maxima of $f(\cdot)$ within the bounded region defined by $\mathbf{h}(\cdot)$ would be asymptotically stable equilibrium points and other equilibrium points in the interior would be unstable. Under certain assumptions on the local maxima, these results can be rigorously established and there is a one-to-one correspondence between local maxima of $f(\mathbf{u})$ and asymptotically stable equilibrium points of the ODE.

Another interesting property of the ODE is that the function $f(\cdot)$ is nondecreasing along any solution of the ODE. That is, $f(\mathbf{h}(\mathbf{u}(t))$ is nondecreasing for all t. Thus the ODE has a gradient ascent property. This property helps to rule out stable limit cycles unless these are also local maxima. These properties establish that barring pathological cases, the learning algorithm converges to local maxima of $f(\mathbf{u}) = E[\beta|\mathbf{u}]$ (See [Pha91, PT95] for details).

1.7.2 An Example

In this subsection we consider a simple example of a 3-action GLA to illustrate the algorithm. We first give the complete learning algorithm and then show results obtained on this example.

Since we are using a 3-action GLA, we cannot use the probability generating function given by (1.7.39). We use the function given by (1.7.40) which can be written as

$$g(\mathbf{X}, y_i, \mathbf{u}) = \frac{\exp(-\mathbf{X}^T \mathbf{u}_i)}{Z}. \quad (1.7.46)$$

where
$$Z = \sum_j \exp(-\mathbf{X}^T \mathbf{u}_j). \quad (1.7.47)$$

Since we have three actions, the internal state contains the three vectors $\mathbf{u}_i$, $i = 1, 2, 3$. That is, $\mathbf{u} = (\mathbf{u}_1^T\ \mathbf{u}_2^T\ \mathbf{u}_3^T)^T$. Writing the learning algorithm given by (1.7.44) in vector notation we have

$$\begin{aligned}\mathbf{u}(k+1) &= \mathbf{u}(k) + \lambda \beta(k) \frac{\partial \ln g}{\partial \mathbf{u}}(\mathbf{u}(k), \alpha(k), \mathbf{h}(\mathbf{u}(k))) \\ &\quad + \lambda \mathbf{K}[\mathbf{h}(\mathbf{u}(k)) - \mathbf{u}(k)] \end{aligned} \quad (1.7.48)$$

where $\mathbf{h}(\mathbf{u}(k)) = [h_1(u_1(k)) \cdots h_n(u_n(k))]^T$ with the functions h_i being the same as those in the algorithm (1.7.44), $\mathbf{K}$ is a diagonal matrix with diagonal entries being K_i as in (1.7.44) and the derivative with respect to $\mathbf{u}$ denotes the gradient. Let $\mathbf{u}_i = (u_{i1} \ldots u_{in})^T$ and let $\mathbf{X} = (x_1 \ldots x_n)^T$. Now using (1.7.46) and (1.7.47) we have

$$\begin{aligned}\frac{\partial}{\partial u_{ij}} \ln g(\mathbf{X}, y_i, \mathbf{u}) &= \frac{1}{g(\mathbf{X}, y_i, \mathbf{u})} \frac{\partial}{\partial u_{ij}} g(\mathbf{X}, y_i, \mathbf{u}) \\ &= \frac{1}{g(\mathbf{X}, y_i, \mathbf{u})} \frac{\partial}{\partial u_{ij}} \frac{\exp(-\mathbf{u}_i^T \mathbf{X})}{Z} \\ &= \frac{-x_j}{g(\mathbf{X}, y_i, \mathbf{u})} \exp(-\mathbf{u}_i^T \mathbf{X})[\frac{1}{Z} + \frac{-1}{Z^2}\exp(-\mathbf{u}_i^T \mathbf{X})] \\ &= (-x_j)[1 - \frac{1}{Z}\exp(-\mathbf{u}_i^T \mathbf{X})] \\ &= (-x_j)[1 - g(\mathbf{X}, y_i, \mathbf{u})] \end{aligned} \quad (1.7.49)$$

Thus, we can write in vector notation,

$$\frac{\partial}{\partial \mathbf{u}_i} \ln g(\mathbf{X}, y_i, \mathbf{u}) = -\mathbf{X}[1 - g(\mathbf{X}, y_i, \mathbf{u})]. \quad (1.7.50)$$

Proceeding along similar lines, we can show that, for $j \neq i$,

$$\frac{\partial}{\partial \mathbf{u}_j} \ln g(\mathbf{X}, y_i, \mathbf{u}) = -\mathbf{X}[0 - g(\mathbf{X}, y_i, \mathbf{u})]. \quad (1.7.51)$$

Now, using (1.7.50) and (1.7.51), we can write the full learning algorithm as follows.
Suppose $\alpha(k) = y_i$. Then $\mathbf{u}$ is updated as

$$\begin{aligned}\mathbf{u}_i(k+1) &= \mathbf{u}_i(k) + \lambda \beta(k)(-\mathbf{X}(k))[1 - g(\mathbf{X}(k), y_i, \mathbf{h}(\mathbf{u}(k)))] \\ &\quad + \lambda \mathbf{K}[\mathbf{h}(\mathbf{u}_i(k)) - \mathbf{u}_i(k)]\end{aligned}$$

$$\begin{aligned}\mathbf{u}_j(k+1) &= \mathbf{u}_j(k) + \lambda \beta(k)(-\mathbf{X}(k))[0 - g(\mathbf{X}(k), y_i, \mathbf{h}(\mathbf{u}(k)))] \\ &\quad + \lambda \mathbf{K}[\mathbf{h}(\mathbf{u}_j(k)) - \mathbf{u}_j(k)], \forall j \neq i.\end{aligned} \quad (1.7.52)$$

In the example we consider, the context vectors come from the subset $[-1, 1] \times [-1, 1]$ of $\Re^2$. This square is divided into three triangular regions by the lines $x_2 - 2x_1 = 1$ and $x_2 + 2x_1 = 1$. (See Fig. 1.3(d)). In the three regions the three different actions are optimal. From the probability generating function given by (1.7.46), it is easily seen that the set of context vectors where probability of y_1 is the highest is given by $\{\mathbf{X} \mid \mathbf{u}_1^T \mathbf{X} > \mathbf{u}_2^T \mathbf{X} \text{ and } \mathbf{u}_1^T \mathbf{X} > \mathbf{u}_3^T \mathbf{X}\}$. Hence we essentially want $\mathbf{u}_i^T \mathbf{X} = 0$ to represent a general hyperplane in $\Re^2$. As is usually done, e.g., in pattern recognition, we take the context vectors $\mathbf{X}$ to be 3-dimensional with the third component set to 1 and we take each of the three $\mathbf{u}_i$ also as 3-dimensional. Thus, we take $\mathbf{X} = (x_1 \ x_2 \ 1)^T$ and $\mathbf{u}_i = (u_{i1} \ u_{i2} \ u_{i3})^T$, $i = 1, 2, 3$.

We generated a thousand points in $[-1, 1] \times [-1, 1]$ using uniform distribution and used them for making our context vectors. At every iteration we choose a context vector (from this set) with uniform distribution. The reinforcement β is taken to be binary valued. In each triangular region, the probability that β would be 1 is 0.9 for the optimal action and 0.15 for the other two actions. We used the probability generating function given by (1.7.46). The initial values of $\mathbf{u}_i$ are arbitrary but it is ensured that the line $\mathbf{u}_i^T(0)\mathbf{X} = 0$ cuts the square $[-1, 1] \times [-1, 1]$.

The typical results obtained with the GLA algorithm are shown in Fig. 1.3. The (d) part of the figure shows the three regions of the problem on which the examples are superposed. The (a) – (c) parts of the figure show the regions as learnt by the GLA. In these figures, the solid lines represent the actual regions while the dashed lines denote the regions as learnt by GLA. For example, for region R1, the dashed lines represent $(\mathbf{u}_1 - \mathbf{u}_2)^T \mathbf{X} = 0$ and $(\mathbf{u}_1 - \mathbf{u}_3)^T \mathbf{X} = 0$. In the (a) part of the figure, the actual region R1 consists of those points which are to the left of both the solid lines while the region as learnt by GLA consists of those points which are to the left of both the dashed lines. One can see from this figure that the region R1 is learnt well by the GLA though the GLA represented it using different lines. Similarly, in the (b) part of the figure the region is between the two lines while in (c) part the region is to the right of both lines. As can be seen, in this problem the probability generating function chosen is such that the GLA can represent the needed mapping between context vectors and actions and the GLA learnt the regions well.

Figure 1.4 illustrate how the learning proceeds. For this we generated another hundred random points in the space of context vectors in each region. That is, there are three sets of hundred points each. At each iteration k, using the $\mathbf{u}_i(k)$, we calculated the average probability of each of the actions for these sets of points. These average probabilities are plotted against the iteration number in the figure. The three subplots in the figure show evolution of the probability of the optimal action in each region. After some initial noisy

The Parameterized Learning Automaton (PLA) 41

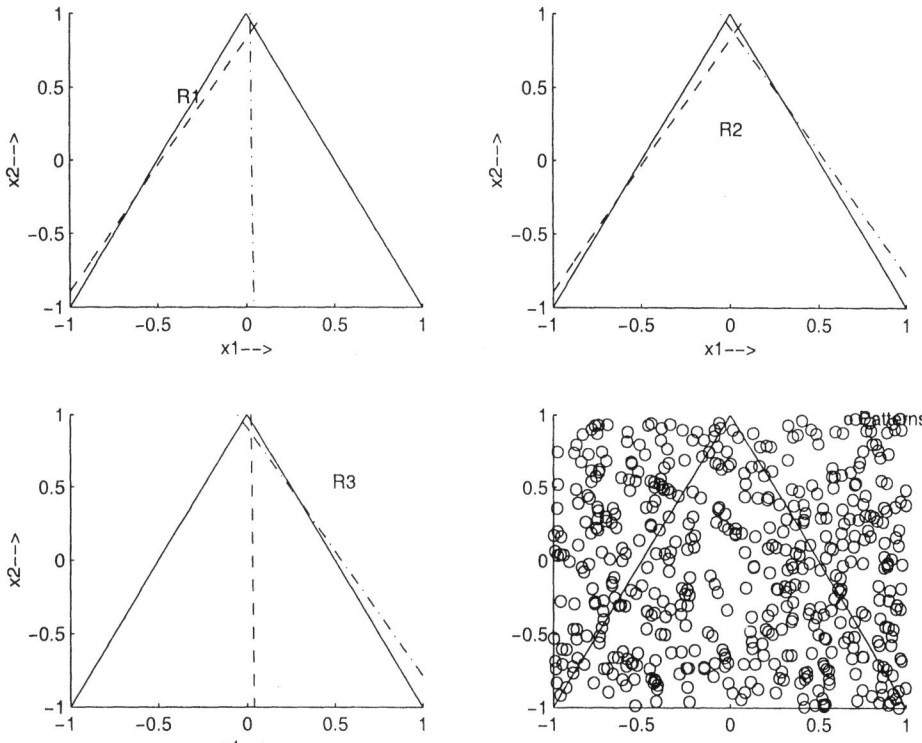

Figure 1.3. The example problem for GLA. In region R1 action y_1 is optimal, in R2 y_2 is optimal and in R3 y_3 is optimal. In (a), (b) and (c) the dashed lines together represent the regions learnt by GLA and solid lines represent the actual underlying regions. In (a) the relevant region is the set of points to the left of both lines; in (b) the relevant region is the one between the two lines and in (c) the relevant region is the one to the right of both lines. The (d) part shows the geometry of the problem along with the training examples

dynamics, the probability of optimal action increases to a high value in each region.

It may be noted that this problem can be thought of as a 2-feature 3-class pattern recognition problem where the classes are separable by a pair of lines. Since the reinforcement can take value 1 for the non-optimal decision also, this corresponds to learning a classifier under noisy samples. Thus this example illustrates how a GLA can be used as a pattern classifier.

1.8 The Parameterized Learning Automaton (PLA)

One feature of the GLA described in the previous section is that the internal state of the automaton is a vector of real numbers which is updated at each instant. Given a context vector, the probabilities of selecting different actions are

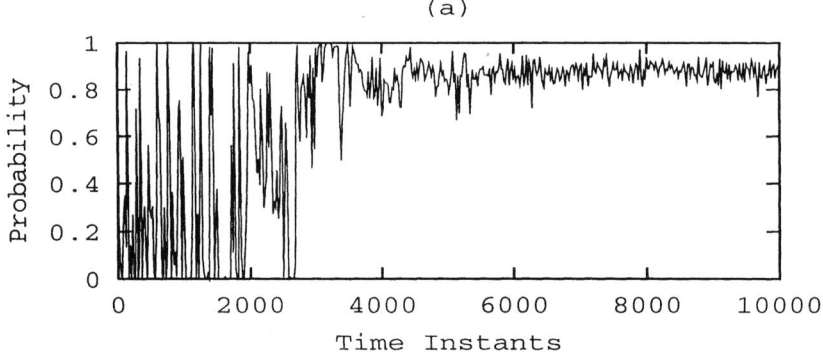

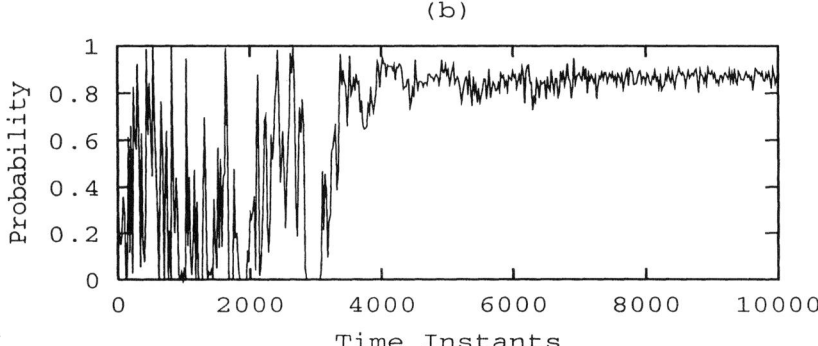

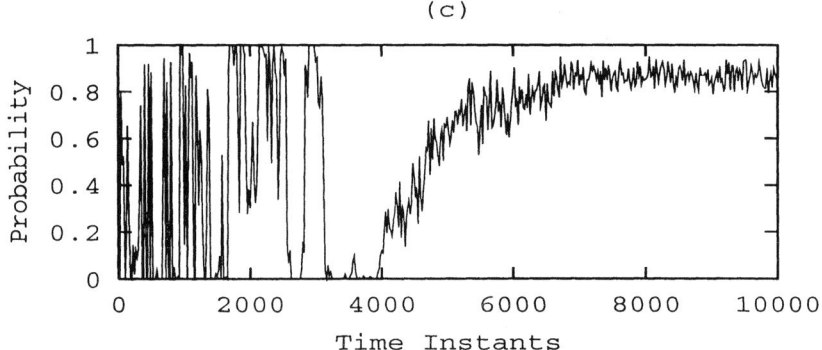

Figure 1.4. The example problem for GLA. Evolution of average probability of (a) action y_1 in region R1, (b) action y_2 in region R2 and (c) action y_3 in region R3, on the test set of 100 points.

The Parameterized Learning Automaton (PLA)

computed from this state using a probability generating function. This feature can be introduced in FALA also, even in the absence of context vectors. The action probabilities can be regarded as parameterized in terms of real numbers which are updated by the learning algorithm. Such an FALA is referred to as a Parameterized Learning Automaton (PLA).

A PLA will have an internal state vector $\mathbf{u}$, of real numbers which is not necessarily a probability vector. The probabilities of various actions are calculated, based on the value of $\mathbf{u}$, using a probability generating function g. For example, the ith action probability of a PLA could be computed as

$$p_i = g(\mathbf{u}, i) = \frac{\exp(u_i)}{\sum_{j=1}^{r} \exp(u_j)} \qquad (1.8.53)$$

where u_i is the ith component of $\mathbf{u}$. In the above, we have taken the state vector $\mathbf{u}$ to be r-dimensional where r is the number of actions of the automata. The main point about a PLA is that the action probabilities will be within the probability simplex irrespective of the values taken by $u_i(i = 1, \ldots, r)$.

It may be noted that we are using, with a little abuse of notation, the same symbol, namely, g, to denote the probability generating function of a GLA as well as a PLA. The probability generating function of a GLA has three arguments while that of a PLA has only two. Since the distinction is always available from context, this should not lead to any confusion.

In the case of a GLA, the internal state and probability generating function are together determining a representation for the mapping of context vectors to actions. That is the reason why the probability generating function of a GLA is a function of the context vector also. However, in the case of PLA the internal state is only a parameterization of the action probabilities. The action probability vector $\mathbf{p}$ of an FALA on the one hand and the internal state $\mathbf{u}$ plus the probability generating function g of a PLA together on the other hand, are both equivalent ways to represent a probability distribution over the finite set of actions. Thus, except for the parameterization of action probabilities, a PLA is essentially the same as an FALA. The need for such a parameterization can be seen as follows.

In the type of learning problems considered in this book, we are generally interested in maximizing $E[\beta|P]$ where P consists of the action probabilities of all the automata in the learning system. In multiautomata systems that we shall consider later, there will be several maxima of $E[\beta|P]$, many of them being local maxima only. Such a distinction between local and global maxima does not exist in the case of a single FALA as it always converges to the global maximum (that is, the action with highest reward probability) when designed properly. In multiautomata systems however, the use of algorithms such as L_{R-I} will only ensure convergence to a local maximum. In such cases one could change the learning algorithm (using ideas similar to Simulated Annealing) so that it converges to global maximum. This would entail imposing a random perturbation while updating the action probability vector $\mathbf{p}(k)$ so that the learning process moves out of local maxima. Introducing a random term directly in the updating of $\mathbf{p}(k)$ is difficult for two reasons. Firstly it is cumbersome to ensure that the resulting vector after updating remains a probability vector. Secondly, the resulting diffusion would be on a manifold rather than the entire space thus making the analysis hard. Parameterization in terms of the real vector $\mathbf{u}$, avoids the above difficulties. We consider such algorithms in Chapter 3.

1.8.1 Learning Algorithm

The goal of the learning algorithm for a PLA is to maximize the expected value of reinforcement which is the same as that for an FALA. However, as in the case of a GLA, there is a possibility of the state $\mathbf{u}$ becoming unbounded. To avoid such a situation, the goal of learning is set as solving the following optimization problem.

$$\text{Maximize } f(\mathbf{u}) = E[\beta|\mathbf{u}] \quad (1.8.54)$$

subject to $|u_i| \leq L_i$, $i = 1, \ldots, r$ where $L_i > 0$ are constants. These could be chosen sufficiently large so that there is insignificant difference between the constrained and unconstrained maxima. Following the development for the GLA, a learning algorithm for a PLA which can be shown to converge to a local maximum is as follows.

$$u_i(k+1) = u_i(k) + \lambda \beta(k) \frac{\partial \ln g(u_i(k), \alpha(k))}{\partial u_i(k)} + \lambda h'(u_i(k)) \quad (1.8.55)$$

where $h'(\cdot)$ is the derivative of $h(\cdot)$ defined as

$$\begin{aligned} h(x_i) &= -K(x_i - L_i)^{2J} \text{ for } x_i \geq L_i \\ &= 0 \quad \text{ for } |x_i| \leq L_i \\ &= -K(x_i + L_i)^{2J} \text{ for } x_i \leq -L_i \end{aligned}$$

K is a positive constant and J is a positive integer. If we use the probability generating function given by (1.8.53), the above updating of $\mathbf{u}$ would essentially be the same as updating the action probabilities through the L_{R-I} algorithm.

This algorithm has the same general structure as that of a GLA. The second term on the RHS of (1.8.55) is a gradient term and it is similar to the term employed in the GLA algorithm. The third term is meant to keep $|u_i|$ bounded and it has a different form from the term used in the case of GLA. It is similar to a penalty term introduced in many constrained optimization problems.

Analysis of the learning algorithm

Following the procedure outlined in Appendix A, it can be shown that under the above learning algorithm, the asymptotic behavior of $\mathbf{u}(k)$ is well approximated by the solution of the following ODE.

$$\frac{dz_i}{dt} = s_i(\mathbf{z}); \quad i = 1,\ldots,r, \quad \mathbf{z}(0) = \mathbf{u}(0) \qquad (1.8.56)$$

where

$$s_i(\mathbf{z}) = p_i \sum_j (d_i - d_j) p_j + h'(z_i) \qquad (1.8.57)$$

In (1.8.57) d_i is the reward probability of the ith action and p_i, $i = 1,\ldots,r$, are given by

$$p_i = \frac{\exp(z_i)}{\sum_{k=1}^{r} \exp(z_k)}.$$

It may be noted that the above ODE is very similar to the approximating ODE for the L_{R-I} algorithm given by (1.4.17).

Let $d_\ell = \max_i d_i$ as before and let $d_l - d_i \geq w > 0$ for all $i \neq \ell$. Then, a basic property of the above ODE is the following.

Let $\mathbf{z}(t)$ be the solution of the ODE (1.8.56). There exists a $T > 0$ such that for all $t \geq T$ and large enough $L_i, i = 1,\ldots,r$,

$$L_\ell \leq z_\ell(t) \leq L_\ell + V$$
$$-L_i - V \leq z_i(t) \leq -L_i \qquad (i \neq \ell)$$

where

$$V = (2JK)^{(-1/(2J-1))}$$

The proof of the above result follows from an upper bound on $h'(z_i)$ and consequent restrictions on $\dot{z}_i$ [Pha91]. The points to be noted are that the solutions are bounded and that when sufficient time has elapsed, $z_\ell(t)$ will have a large positive value and $z_i(t)$ $(i \neq \ell)$ will have large negative values. In turn, this means that, for large enough k, $p_\ell(k)$ is close to 1 and other $p_i(k)$ are close to zero, thus resulting in ϵ-optimality.

Thus, introduction of parameterization does not disturb the properly of ϵ-optimality. However, the computational overheads increase as action probabilities p_i have to be computed at each instant from the parameters u_i using the probability generating function.

The relevance of PLA is in games and networks of LA where the distinction between local and global maxima is important. Hence PLA algorithms suitable for achieving global maxima are taken up in Chapter 3.

1.9 Multiautomata Systems

We have described single learning automata of various types in the previous sections. In an FALA the set of actions available is finite while in a CALA the action set is the real line. The GLA allows for context vector input also so that the automaton can learn to choose actions as appropriate in different contexts. The PLA allows for parametcrization of the action probabilities.

In the rest of this book we will be describing systems consisting of many automata which are useful for tackling different learning problems. In this section we briefly explain some of the reasons why such multiautomata systems are needed.

An FALA is suitable for learning the optimal action from a finite set of alternatives available. However, when the number of actions is large (even though finite), learning becomes slow. There are essentially two reasons for this. The initial action probabilities are generally set as $\frac{1}{r}$ where r is the number of actions and these probabilities become small when r is large. A large number of iterations are thus needed for the best action probability to increase from $\frac{1}{r}$ to unity. The second reason is the computational overhead of large number of probability updatings needed at each instant. Also, with large number of actions, it is often the case that the d_i are closer and hence it becomes difficult to distinguish the optimal action from others. This would mean that we have to employ a smaller value of λ thus resulting in larger number of iterations for convergence.

While no absolute upper limit can be placed on the number of actions per automaton, as this depends on the speed of convergence needed and the characteristics of the environment, a useful rule of thumb in many realistic situations would be to limit the number to about 10. The question then is to find a way to handle situations where the number of actions needed exceeds this number.

For instance, let us say the number of actions is 100 and each automaton can have a maximum of 10 actions. Then the actions can be numbered from 00 to 99 and two LA can be used with LA1 choosing the first digit from 0 to 9 and LA2 choosing the second digit. Thus individually each automaton selects one of 10 actions while the combination selects one of 100. There would be only 20 action probabilities to be updated at each instant rather than 100. Also, each action probability would have an initial value of $\frac{1}{10}$ rather than $\frac{1}{100}$. Thus one way to handle large number of actions is to employ more than one automaton. The two LA in this example could be regarded as forming a team and this is the subject matter of the next chapter.

A natural circumstance which leads to the above situation of a large number of actions is that of learning the optimal value of a continuous parameter. The actions here could correspond to possible values of the parameter and when we want to use an FALA we need to discretize the range of possible values to form a finite action set. Higher accuracy demands a finer discretization and hence larger number of actions.

In the above, we could have used a single CALA to learn the optimal value of a single continuous parameter. However, we would be naturally led to considering multiple automata when the problem involves learning optimal values of many parameters. Even when it is feasible to discretize all the parameters, it is natural to use at least one LA per parameter as otherwise the number of combinations of parameter values increases geometrically making it practically impossible for a single FALA to handle the problem. When we want to use CALA also we need many automata because one CALA is needed for learning one parameter.

Thus another common reason for employing many automata together is that we want to learn multiple quantities. These two situations described above where multiautomata systems are useful, represent simple ways to organize a number of automata to solve a problem together. There are other reasons for considering multiautomata systems which are briefly indicated below. Some of the complex multiautomata systems that we consider in this book are motivated by such considerations.

A very attractive feature of learning automata is the complete generality regarding what constitutes an action. Hence, in a multiautomata system, actions of different automata may correspond to different types of quantities in the problem domain. Thus, another reason for employing a multiautomata system is the flexibility it offers in designing a solution best suited for the application on hand. We will consider a feedforward network of automata for pattern recognition in Chapter 4 that illustrates this point.

In multistage decision problems, one action is to be selected at each stage and a single LA or a group of LA is needed at each stage for this selection. The goodness or otherwise of the selections made by all the automata involved would be known only at the end of the decision process. Such problems would again need a collection of LA. Several LA are needed also for problems involving distributed information processing such as adaptive decision making in distributed communication networks.

We begin our study of multiautomata systems by considering games of automata in Chapter 2. We then go on to consider very general structures of multiautomata systems in the form of networks of groups of automata in Chapter 3.

1.10 Supplementary Remarks

A very readable account of the different approaches to machine intelligence can be found in [Gra88, BA91]. A good discussion of learning through reinforcement feedback is available in [Wal75]. Details of different techniques for machine learning can be found in, for example, [For89, Mit97, HTF01]. The unifying view of learning from examples as a special kind of optimization problem is discussed in [Tsy71, Tsy73]. A good reference on statistical learning theory and the problem of the generalization abilities of a learning machine is [Vap97].

Investigation of Learning Automata models began with the work of Tsetlin [Tse62, Tse73]. These early models were referred to as deterministic and stochastic automata operating in random environments. The early models consisted of finite state automata where the state transitions are probabilistic and dependent on random reinforcement from the environment. The state transition graph is fixed and hence these are also referred to as Fixed Structure Stochastic Automata (FSSA). The FSSA have many interesting properties. A good review of these early models can be found in [NT89]. Such FSSA are useful in applications (see, e.g., [OM88]) and an example of a recent FSSA model is the Star automaton [EK02]. In its current form the FALA models presented here roughly correspond to Variable Structure Stochastic Automata introduced in [VV63]. Some of the early investigations of the automata models include [FM66, CS68, Fu70]. The term *Learning Automata* became popular because of a survey paper by Narendra and Thathachar [NT74] though earlier uses of this term are known [Vis72, TP72]. A number of books and survey papers on this topic have appeared since then [Lak81, NT89, NP94, PN97, NL77, TS02].

Automata with finite action sets (the FALA models) represent the first and most extensively studied LA models. The book [NT89] contains extensive study of many linear and nonlinear learning algorithms for a single FALA. Many of these LA models are initially influenced by the work in psychology [BM55, AB65, Nor68].

The L_{R-I} algorithm was introduced by Shapiro and Narendra [SN69]. Good analysis of many such early linear models can be found in [Nor72]. The concept of absolute expediency is due to Lakshmivarahan and Thathachar [LT73] who have also derived bounds on the probability of a FALA converging to the optimal action [LT76].

The idea of estimator algorithms in general and the pursuit algorithm in particular is due to Thathachar and Sastry [TS85, TS86a, TS87]. As these estimator algorithms converge much faster than other families of FALA algorithms, estimator algorithms became popular and are extended in many directions [OL90, Pap94b, Pap94a, OA01, AO02]. The pursuit algorithm also has some interesting finite time results[RS96]. Another related algorithm is [SK89].

The idea of discretized probability automata is due to Thathachar and Oommen [TO79]. Many FALA algorithms have been discretized in this way [Oom86, OC88, OL90]. The discretized pursuit algorithm and its variants are found to be attractive from the point of view of rate of convergence [AO02].

The concept of CALA was introduced in [San94, SST94]. As explained here, the CALA updating scheme essentially results in following the gradi-

Supplementary Remarks

ent of the criterion function in an expected sense. In this sense, it is similar to the stochastic gradient following algorithms used in connectionist systems [Wil92]. The CARLA, which is another model of a learning automaton with a continuous action set, is discussed in [HFGW97].

The concept of GLA was introduced in [Pha91]. The REINFORCE algorithm was proposed by Williams [Wil88, Wil92]. Barto and Anandan were among the first to suggest that LA should be extended to take an additional input in the form of a context vector [BA85]. However, in the method proposed by them, they had to assume that the set of all context vectors is linearly independent. This is a severe restriction (e.g., there can be only n distinct context vectors if each is an n-dimensional vector) in almost all applications. The GLA approach allows for a better way to handle context vector input to automata. Analysis of GLA algorithm is available in [Pha91, PT95].

The PLA were also introduced in [Pha91] with the main motivation being design of algorithms that can converge to global optima for systems containing many automata such as teams and networks of LA. Analysis of PLA algorithm is available in [TP95b].

Certain multiautomata systems such as hierarchies and games of FALA have been considered in [NT89]. Some recent results on hierarchies of FALA are available in [BM02].

Chapter 2

GAMES OF LEARNING AUTOMATA

2.1 Introduction

In Section 1.2 we briefly indicated that learning automata are useful in applications that involve optimization of a function which is not completely known in the sense that only noise corrupted values of the function for any specific values of arguments are observable. Suppose we want to find the maximum of a function $f : \Re \rightarrow \Re$ given only noise corrupted observations. That is, given any specific $x \in \Re$, we can observe or calculate only $\tilde{f}(x)$, with $E[\tilde{f}(x) \mid x] = f(x)$ where E denotes expectation. One way of using a FALA for the problem is the following. We first discretize the domain of f into finitely many intervals. We choose the midpoints of each of the intervals as the possible actions of the automaton. The reaction or reinforcement for any choice of action would be (a suitably normalized version of) the noise corrupted value of the function at the value of argument corresponding to that action. Now the optimal action would correspond to the point in the domain where the function $f(x)$ attains the maximum value and we can use any of the FALA algorithms discussed in Chapter 1 to learn the maximum. It may be noted that this method finds the global maximum. However, the method would find the maximum only to some level of approximation dependent on the level of quantization employed. Alternatively, we can use a CALA (whose action set is the real line) and learn a maximum of the function. In such a case we do not need to discretize the domain; however we can learn only a local maximum. (See the discussion in Section 1.6). If the function is defined over N variables (i.e., if the domain of f is $\Re^N$) then the point where a maximum is attained would be a vector of N components. In the case where we discretize the domain of f we could, in principle, use a single FALA by making the actions correspond to points in $\Re^N$. However, the resulting number of actions would be generally very large. Hence, in practice, it would be better to use one automaton for learning one component of the maximum point. In the case where we cannot discretize the domain, we need to use as many CALA as there are values to learn. Thus, in most such applications we need a system consisting of many automata. The main focus of this book is on such multiautomata systems and the associated learning algorithms. In this chapter we discuss one configuration for a multiautomata system where the automata would be viewed as players involved in a game.

Before starting our discussion of games of automata, we briefly outline the general strategy underlying any design of multiautomata systems for applications involving optimization of a function as discussed above.

Let $\mathcal{D}$ be the space over which the function of interest is defined. We then design an automata system so that the set of all possible action tuples (of all automata in the system) can be put in one to one correspondence with $\mathcal{D}$. Thus, when all the automata choose their actions, effectively the system chooses a point in $\mathcal{D}$. (For example, when the function to be optimized is defined on $\Re^n$, we can have a system consisting of n number of CALA. Now any choice of a tuple of actions by this set of automata corresponds to a point in $\Re^n$). The state of the automata system, given by the action probability distributions of all automata in the system, would now correspond to a probability distribution over $\mathcal{D}$. The learning algorithm searches in the space of all such probability distributions to identify the points in $\mathcal{D}$ where the objective function is optimized. The main advantage of this method is that the algorithm need not presuppose any specific algebraic structure on the set $\mathcal{D}$. This is because the algorithm searches in a space of probability distributions over $\mathcal{D}$ and this space has the needed algebraic structure.

The model we consider in this chapter is a stochastic game of automata. Consider a game situation involving N players. Each of the players has a (finite or infinite) set of actions to choose from. Each play of the game consists of each of the players choosing a specific action from their respective action sets. Then each of the players gets a payoff from the environment and these payoffs (which may be different for different players) are random. The game we consider is of incomplete information and thus the players have no knowledge of the probability distributions that determine their payoff for different choices of actions. We are interested in identifying the optimal actions of the players (under some precisely defined notion of the solution to the game). In our automata game, each player is represented by an automaton and his actions constitute the action set of the automaton. Hence the action probability distribution represents the so called *mixed strategy* being used by the player at any given instant. We are interested in designing some efficient learning algorithms so that, by repeatedly playing the game and using the payoffs obtained to update the action probability distributions, the automata can, asymptotically, converge to a solution of the game. Since some of the players can have finitely many actions while others can have a continuum of actions, our automata players will include both Finite Action Learning Automata(FALA) as well as Continuous Action Learning Automata(CALA). In this chapter we consider such an automata game where each of the FALA use the L_{R-I} (cf. Section 1.4.1) algorithm while each of the CALA use the algorithm specified in Section 1.6.

It is easy to see how the above model can be used for the stochastic optimization problem illustrated earlier. Suppose we want to determine a maximum of the function $f : \Re^N \rightarrow [0, 1]$ based only on observations $\tilde{f}(\mathbf{x})$ where $E[\tilde{f}(\mathbf{x})|\mathbf{x}] = f(\mathbf{x})$. Let us call the arguments of f as parameters. We can use a team of N automata for this problem. If we can discretize the domain of f we use FALA; otherwise we use CALA. In either case, actions of automata represent possible values for parameters. Thus any tuple of actions corresponds to a point in the domain of f and we supply the value of $\tilde{f}$ at that point as the *common* reinforcement to all automata. As we have seen in Chapter 1, the generic goal of any learning automaton is to maximize the expected value of the reinforcement received. Now, if we have a learning algorithm so that all the automata together converge to an action tuple that maximizes the expected value of the reinforcement, then, we have a method to solve this stochastic optimization problem. Such optimization techniques are useful in many applications such as system identification, pattern classification etc.

The automata game model that we discuss in this chapter is more general. We consider an automata game where some of the players are FALA while others are CALA. Now a tuple of actions chosen by the automata can be denoted as $(\boldsymbol{\alpha}, \mathbf{x})$ where $\boldsymbol{\alpha}$ denotes the tuple of actions chosen by the FALA members while $\mathbf{x}$ denotes the tuple of actions chosen by the CALA members. It may be noted that components of $\boldsymbol{\alpha}$ can take only finitely many values (since the action set of FALA is finite) while components of $\mathbf{x}$ are real-valued. The expected value of reinforcement is a function of the action tuple chosen and we can represent it by $F(\boldsymbol{\alpha}, \mathbf{x})$. Note that this function F has some arguments that are discrete while the others are continuous variables. It is to emphasize this distinction that we have used the symbol F rather than f as earlier. Now once again, this automata team would represent a method to optimize a function F (which is defined over some discrete and some continuous variables) given only noisy observations of the function values (which are used to generate the reinforcement in response to choice of actions by the team). Such optimization problems come up naturally in, e.g., Concept Learning. In addition to having both FALA and CALA in the team, the automata game considered in this chapter allows for different automata to have different payoffs. Then, instead of one function F we have as many functions as there are automata. We can denote the expected value of reinforcement received by the lth automaton as $F^l(\boldsymbol{\alpha}, \mathbf{x})$. It may be noted here that the reinforcement received by each automaton depends on the actions chosen by *all* automata. The stochastic optimization problem considered earlier corresponds to a special case of this game where all automata get the same payoff or reinforcement. The multiple payoff game can be useful, e.g., in multi-criteria optimization. In the next section we describe this automata game.

2.2 A Multiple Payoff Stochastic Game of Automata

Consider a collection of $N + M$ learning automata consisting of N Finite Action Learning Automata (FALA) and M Continuous Action Learning Automata (CALA) involved in a stochastic game with incomplete information. Let the action set of ith FALA be denoted by A_i with $|A_i| = r_i, 1 \leq i \leq N$. Let the jth CALA choose actions from the real line $\Re$, $1 \leq j \leq M$.

Let $\alpha_i(k) \in A_i$, denote the action chosen by the ith FALA, $i = 1, \ldots, N$, and let $x_j(k) \in \Re$, be the action chosen by the jth CALA, $j = 1, \ldots, M$, at the kth instant. We use $\boldsymbol{\alpha}(k) = (\alpha_1(k), \alpha_2(k), \ldots, \alpha_N(k))$ to denote the actions chosen by the FALA members and $\mathbf{x}(k) = (x_1(k), x_2(k), \ldots, x_M(k))$, to denote the actions chosen by the CALA members, at instant k. Let $Y(k) = (\boldsymbol{\alpha}(k), \mathbf{x}(k))$ denote the tuple of actions chosen by all automata involved in the game, at instant k. It may be noted that $Y(k) \in \mathcal{D}, \forall k$, where, the set $\mathcal{D}$ is defined by $\mathcal{D} \stackrel{\Delta}{=} (\prod_{i=1}^{N} A_i) \times \Re^M$. Each element of $\mathcal{D}$ is a tuple of action choices by the players, and as per our notation, we denote an arbitrary element of $\mathcal{D}$ by $(\boldsymbol{\alpha}, \mathbf{x})$. In response to any tuple of actions chosen, the environment provides a stochastic reinforcement (also called payoff) to each of the automata. Let β_l be the payoff to the lth automata player[1], $1 \leq l \leq N + M$. For $1 \leq l \leq N$, the lth player is the lth FALA and for $N + 1 \leq l \leq N + M$, the lth player is the $(l - N)^{th}$ CALA. This convention will be used throughout this section. It is assumed that β_l takes values in $[0, 1]$, for all l.

Define functions $F^l : \mathcal{D} \to [0, 1], 1 \leq l \leq N + M$, by

$$F^l(\boldsymbol{\alpha}, \mathbf{x}) = \mathrm{E}[\ \beta_l \mid i^{th} \text{ FALA chose } \alpha_i \in A_i \text{ and } j^{th} \text{ CALA chose } x_j \in \Re]. \tag{2.2.1}$$

F^l is called the payoff function for player l. The players only receive the payoff (or reinforcement) signal (namely, β_l) and they have no knowledge of the payoff functions.

DEFINITION 2.2.1 *We say* $(\boldsymbol{\alpha}^*, \mathbf{x}^*)$, $\boldsymbol{\alpha}^* = (\alpha_1^*, \ldots, \alpha_N^*)$; $\mathbf{x}^* = (x_1^*, \ldots, x_M^*)$, *is an* optimal point *of the game if*

---

[1]The subscript l is used in a generic sense in this chapter unlike the case in Chapter 1 where it denoted the index of optimal action

(1) For each i, $1 \leq i \leq N$,
$$F^i(\boldsymbol{\alpha}^*, \mathbf{x}^*) \geq F^i(\boldsymbol{\alpha}, \mathbf{x}^*),$$
for all $\boldsymbol{\alpha} = (\alpha_1^*, \ldots, \alpha_{i-1}^*, \alpha_i, \alpha_{i+1}^*, \ldots, \alpha_N^*)$ such that $\alpha_i \neq \alpha_i^*, \alpha_i \in A_i$.

(2) For each j, $1 \leq j \leq M$, $\exists \epsilon > 0$ such that
$$F^{N+j}(\boldsymbol{\alpha}^*, \mathbf{x}^*) \geq F^{N+j}(\boldsymbol{\alpha}^*, \mathbf{x}),$$
for all $\mathbf{x}$ such that $\mathbf{x} \in \mathcal{B}^M(\mathbf{x}^*, \epsilon)$, where $\mathcal{B}^M(\mathbf{x}^*, \epsilon)$ is an ϵ-ball in $\Re^M$ centered at $\mathbf{x}^*$.
■

REMARK 2.2.1 *In the above definition, Condition (1) implies that $\boldsymbol{\alpha}^*$ is a Nash equilibrium of the game matrix $F^l(\cdot, \mathbf{x}^*)$ indexed by $\alpha_i, 1 \leq i \leq N$. Condition (2) means that $\mathbf{x}^*$ is a local maximum of $F^l(\boldsymbol{\alpha}^*, \cdot)$.* ■

REMARK 2.2.2 *Consider a special case where $\beta_l = \beta$, and hence $F^l(\cdot, \cdot) = F(\cdot, \cdot)$, $\forall l$. Here all players get the same payoff and we call such a game as game with common payoff. Now $(\boldsymbol{\alpha}^*, \mathbf{x}^*)$ will be an optimal point if $\exists \epsilon > 0$ such that*
$$F(\boldsymbol{\alpha}^*, \mathbf{x}^*) \geq F(\boldsymbol{\alpha}, \mathbf{x}), \ \forall (\boldsymbol{\alpha}, \mathbf{x}) \in \mathcal{N}_\epsilon(\boldsymbol{\alpha}^*, \mathbf{x}^*)$$
where $\mathcal{N}_\epsilon(\boldsymbol{\alpha}^*, \mathbf{x}^*)$ is the ϵ-neighborhood of $(\boldsymbol{\alpha}^*, \mathbf{x}^*)$ in $\mathcal{D}$ defined by
$$\mathcal{N}_\epsilon(\boldsymbol{\alpha}^*, \mathbf{x}^*) = \{(\boldsymbol{\alpha}, \mathbf{x}) \in \mathcal{D} \ : \ \boldsymbol{\alpha} \text{ and } \boldsymbol{\alpha}^* \text{ differ in only one component}$$
$$\text{and } \mathbf{x} \in \mathcal{B}^M(\mathbf{x}^*, \epsilon)\}$$

Thus the optimal points correspond to local maxima of F where local maxima are defined with respect to the neighborhood as above. This model then would be useful for maximizing a regression function defined over both continuous and discrete variables. If we consider a team with only CALA then this corresponds to the classical problem of optimizing a regression function. The case of the team with only FALA corresponds to the problem of maximizing the function defined over a finite set based on noisy measurements. We will be discussing these special cases in detail in Sections 2.4, 2.5 and 2.6. ■

Now the learning problem is one of identifying optimal points of the game through repeated plays, that is, by repeatedly choosing actions and receiving respective payoffs. For this, (as usual!) each automaton maintains a probability distribution over its action set which is modified after each play using a learning algorithm.

2.2.1 The Learning Algorithm

Let[2] $\mathbf{p}_i(k) = [p_{i1}(k), \ldots, p_{ir_i}(k)]^T, 1 \leq i \leq N$, denote the action probability distribution of the ith FALA, where $p_{ij}(k) = \text{Prob}[\alpha_i(k) = \alpha_{ij}]$ and α_{ij} is the jth action in A_i. The action probability distribution of jth CALA at k^{th} instant is $N(\mu_j(k), \phi(\sigma_j(k)))$ which is a normal distribution with mean $\mu_j(k)$ and standard deviation $\phi(\sigma_j(k))$ (the function $\phi(\cdot)$ is given by (2.2.5) below). Let $\mathbf{c}_j(k) = (\mu_j(k), \sigma_j(k)) \in \Re^2, 1 \leq j \leq M$. Then the *state* of ith FALA is given by $\mathbf{p}_i(k)$ and the *state* of jth CALA is given by $\mathbf{c}_j(k)$.

Let $P(k) = [\mathbf{p}_1(k), \ldots, \mathbf{p}_N(k)]$ and $C(k) = [\mathbf{c}_1(k), \ldots, \mathbf{c}_M(k)]$. Then the state of the collection of automata, at instant k, is given by $S(k) = (P(k), C(k))$. Define the set $\mathcal{K}$ by: $\mathcal{K} \triangleq \prod_{i=1}^{N} S_{r_i} \times \prod_{j=1}^{M} \Re^2$ where S_{r_i} is the r_i-dimensional unit simplex (that is, the set of all r_i-dimensional probability vectors). It may be noted that $\mathcal{K} \subset [0,1]^{r_1 + \cdots + r_N} \times \Re^{2M}$ and we have $S(k) \in \mathcal{K}, \forall k$. It may be noted that any point in $\mathcal{K}$ represents a probability distribution over $\mathcal{D}$, the set of all possible action tuples.

At instant k, the ith FALA chooses $\alpha_i(k) \in A_i$ at random according to $\mathbf{p}_i(k), 1 \leq i \leq N$, and j^{th} CALA chooses $x_j(k) \in \Re$ at random according to $N(\mu_j(k), \phi(\sigma_j(k))), 1 \leq j \leq M$. Then the lth player gets two reinforcements from the environment: $\beta_l(k)$ and $\beta'_l(k), 1 \leq l \leq M + N$. $\beta_l(k)$ is the response to the action tuple $(\boldsymbol{\alpha}(k), \mathbf{x}(k))$ and $\beta'_l(k)$ is the response to the action tuple $(\boldsymbol{\alpha}(k), \boldsymbol{\mu}(k))$, where $\boldsymbol{\mu}(k) = (\mu_1(k) \ldots \mu_M(k))^T$. Then all the automata update their action probability distributions as described below.

The ith FALA, $1 \leq i \leq N$, updates its action probability distribution using the L_{R-I} learning algorithm as follows:

$$\mathbf{p}_i(k+1) = \mathbf{p}_i(k) + \lambda \beta'_i(k) [\mathbf{e}_{\alpha_i(k)} - \mathbf{p}_i(k)], \quad 1 \leq i \leq N, \quad (2.2.2)$$

where $\mathbf{e}_{\alpha_i(k)}$ is the unit vector of dimension r_i with $\alpha_i(k)$-th component unity and $\lambda \in (0,1)$ is the step size parameter of the algorithm.

The jth CALA, $1 \leq j \leq M$, updates its state as follows:

$$\begin{aligned}
\mu_j(k+1) &= \mu_j(k) + \lambda \mathcal{F}_1(\mu_j(k), \sigma_j(k), x_j(k), \beta_{N+j}(k), \beta'_{N+j}(k)) \\
\sigma_j(k+1) &= \sigma_j(k) + \lambda \mathcal{F}_2(\mu_j(k), \sigma_j(k), x_j(k), \beta_{N+j}(k), \beta'_{N+j}(k)) \\
&\quad - \lambda K[\sigma_j(k) - \sigma_L]
\end{aligned} \quad (2.2.3)$$

where $\mathcal{F}_1(\cdot), \mathcal{F}_2(\cdot)$ are defined as below:

$$\mathcal{F}_1(\mu, \sigma, x, \beta, \beta') = \left(\frac{\beta - \beta'}{\phi(\sigma)} \right) \left(\frac{x - \mu}{\phi(\sigma)} \right)$$

---

[2] Recall that all our vectors are column vectors and the superscript T denotes *transpose*.

A Multiple Payoff Stochastic Game of Automata 57

$$\mathcal{F}_2(\mu, \sigma, x, \beta, \beta') = \left(\frac{\beta - \beta'}{\phi(\sigma)}\right) \left[\left(\frac{x - \mu}{\phi(\sigma)}\right)^2 - 1\right] \quad (2.2.4)$$

with

$$\phi(\sigma) = (\sigma - \sigma_L)I\{\sigma > \sigma_L\} + \sigma_L \quad (2.2.5)$$

and $\sigma_L, K > 0, \lambda \in (0, 1)$ are parameters of the algorithm. (In (2.2.5), $I\{\cdot\}$ denotes the indicator function). Note that this learning algorithm is the same as the CALA algorithm considered in Section 1.6.

REMARK 2.2.3 *The parameter σ_L above is a lower bound on the standard deviation of the normal distribution from which CALA choose actions as explained in Section 1.6. This is needed to ensure proper convergence of the algorithm.* ∎

Goal of the Learning Algorithm

Before proceeding to analyze the asymptotic behavior of the learning algorithm presented above, we have to first decide what should be the goal of the learning algorithm. To appreciate the issues involved, first consider the case of a single FALA (cf. Section 1.3.3). We defined the reward probability as expectation of the reinforcement conditioned on the action chosen. The action for which this expectation is maximum, is the optimal action. However, the learning algorithm does not operate in the space of actions; it updates the action probability distribution. Hence the goal of the learning algorithm was to maximize the expected value of the reinforcement as a function of the action probability distribution. In the case of a single FALA, this maximum is attained when the action probability vector becomes a unit vector with unity in the position corresponding to the optimal action. It is in this sense that the FALA learns the optimal action. In the case of our automata game also, we would follow the same general philosophy.

The optimal points of the game (see definition 2.2.1) are analogous to the optimal action of a single automaton. In the case of a single automaton, there is only one reinforcement and hence only one function (expected value of the reinforcement) to be maximized. However, the game we consider is a multipayoff game and hence there are many functions to be simultaneously maximized. Thus, there is no longer an *obvious* way to define what the goal is. We have chosen one method of defining the objective (namely, optimal points) based on the notion of Nash equilibrium in finite matrix games. (As stated in Remark 2.2.2, in the special case of the game where all automata get the same reinforcement, this will become the usual objective of maximizing expected value of reinforcement). Optimal points correspond to action tuples such that the expected value of reinforcement (as a function of action tuple chosen) is 'locally' maximized in the sense explained earlier. There can be many optimal points and these can be viewed as the solutions of the game and we want the algorithm to identify such action tuples. However, the learning algorithm does not operate in the space of actions; it updates $S(k) = (P(k), C(k))$, which represents the action probability distributions of all the automata. Thus the goal of the learning algorithm has to be defined in terms of appropriately maximizing the expected value of reinforcement as a function of the action probability distributions, (P, C). This is what we do by defining functions g^l as given below and then defining what we mean by maximizing these functions.

Define functions $g^l : \mathcal{K} \to [0, 1]$, $1 \leq l \leq N + M$, by

$$g^l(P, C) = \text{E}[\,\beta_l \mid i^{th} \text{ FALA has action probability vector } \mathbf{p}_i, 1 \leq i \leq N \\ \text{and the state of } j^{th} \text{ CALA is } \mathbf{c}_j, 1 \leq j \leq M\,]. \quad (2.2.6)$$

where $P = (\mathbf{p}_1, \ldots, \mathbf{p}_N)$ and $C = (\mathbf{c}_1, \ldots, \mathbf{c}_M)$, $\mathbf{c}_i = (\mu_i, \sigma_i)$. Thus, $F^l(\cdot, \cdot)$ give expected payoffs as a function of the actions chosen by the automata while $g^l(\cdot, \cdot)$ give expected payoffs as a function of the action probability distributions.

Assuming $F^l(\boldsymbol{\alpha}, \mathbf{x})$ is integrable, using Fubini's theorem [Hal50], we get from (2.2.1), for $1 \leq l \leq N + M$,

$$g^l(P, C) = \int_{\Re^M} \sum_{j_1, \ldots, j_N} F^l(\boldsymbol{\alpha}, \mathbf{x}) \prod_k p_{k j_k} N(\boldsymbol{\mu}, \Sigma) d\mathbf{x}, \quad (2.2.7)$$

where $\boldsymbol{\alpha} = (\alpha_{j_1}, \ldots, \alpha_{j_N})$, $\boldsymbol{\mu} = (\mu_1, \ldots, \mu_M)$, and Σ is the $M \times M$ diagonal matrix with the ith diagonal entry being $(\phi(\sigma_i))^2$. (As a notation we represent such diagonal matrices by $diag(d_i)$, where d_i is the ith diagonal entry). $N(\boldsymbol{\mu}, \Sigma)$ is the multidimensional Gaussian density given by

$$N(\boldsymbol{\mu}, \Sigma) = \frac{1}{|\Sigma|^{\frac{1}{2}} (2\pi)^{\frac{M}{2}}} exp(-\frac{1}{2}(\mathbf{x} - \boldsymbol{\mu})^T \Sigma^{-1} (\mathbf{x} - \boldsymbol{\mu})).$$

For convenience, we use C and $(\boldsymbol{\mu}, \Sigma)$ interchangeably.

REMARK 2.2.4 *It may be noted that if all the diagonal entries of the diagonal matrix Σ are sufficiently small, then the RHS of equation (2.2.7) is well approximated by the value of the integrand at $\mathbf{x} = \boldsymbol{\mu}$, that is, $\sum F^l(\boldsymbol{\alpha}, \boldsymbol{\mu}) \Pi p_{k j_k}$. This is essentially due to the fact that, in the case of a CALA, when the variance of the Gaussian distribution (representing the action probability distribution of CALA) is small, expectation of reinforcement conditioned on the action probability distribution is close to that conditioned on the action selected being the mean of the distribution.* ■

A Multiple Payoff Stochastic Game of Automata

The optimal points given by Definition 2.2.1 give desirable values for the actions of automata. Since a learning algorithm updates only the action probability distributions, we now define desirable values for the action probability distributions.

DEFINITION 2.2.2 *We say* $(P^*, C^*) \in \mathcal{K}, P^* = (\mathbf{p}_1^*, \ldots, \mathbf{p}_N^*), C^* = (\boldsymbol{\mu}^*, \Sigma^*)$, *is a* maximal point *of the game if*

(1) For each i, $1 \le i \le N$,

$$g^i(P^*, C^*) \ge g^i(P, C^*),$$

for all $P \in S_{r_1} \times \cdots \times S_{r_N}$ such that $P = [\mathbf{p}_1^*, \ldots, \mathbf{p}_{i-1}^*, \mathbf{p}_i, \mathbf{p}_{i+1}^*, \ldots, \mathbf{p}_N^*]$, $\mathbf{p}_i \in S_{r_i}$ is a probability vector and $\mathbf{p}_i \ne \mathbf{p}_i^*$.

(2) For each j, $1 \le j \le M$, $\exists \epsilon > 0$ such that

$$g^{N+j}(P^*, (\boldsymbol{\mu}^*, \Sigma^*)) \ge g^{N+j}(P^*, (\boldsymbol{\mu}, \Sigma^*)),$$

for all $\boldsymbol{\mu}$ such that $\boldsymbol{\mu} \in \mathcal{B}^M(\boldsymbol{\mu}^*, \epsilon)$ where $\mathcal{B}^M(\boldsymbol{\mu}^*, \epsilon)$ is an ϵ-ball in $\Re^M$ centered at $\boldsymbol{\mu}^*$.

■

In the above definition, condition (1) implies that $(\mathbf{p}_1^*, \ldots, \mathbf{p}_N^*)$ is a Nash equilibrium in mixed strategies for the FALA part of the game (assuming that the state of the CALA part is held fixed at $(\boldsymbol{\mu}^*, \Sigma^*)$). Similarly the second condition implies that, if the action probability vectors of all FALA are held fixed at P^* and all entries of the diagonal matrix Σ^* are held fixed, then $\boldsymbol{\mu}^*$ is a local maximum of the expected value of reinforcement of all CALA.

Let $(\boldsymbol{\alpha}^*, \mathbf{x}^*)$ be an optimal point of the game. Let $\boldsymbol{\alpha}^* = (\alpha_1^*, \ldots, \alpha_N^*)$. Note that, for all i, $\alpha_i^* \in A_i$. Let $\mathbf{e}_i^*$ be a unit vector of dimension r_i with unity for the α_i^*-th component and zero everywhere else. Then, Definition 2.2.2 implies that the point in $\mathcal{K}$ given by (P^*, C^*) with $P^* = (\mathbf{e}_1^*, \ldots, \mathbf{e}_N^*)$ and $C^* = (\mathbf{x}^*, \Sigma^*)$ would be a maximal point if all entries in the diagonal matrix Σ^* are sufficiently small. (See Remark 2.2.4).

The optimal points of the game defined by Definition 2.2.1 are those tuples of actions in $\mathcal{D}$ which (in a special sense) "locally maximize" all payoff functions. The automata learning algorithm searches over the space of probability distributions over $\mathcal{D}$, represented by $\mathcal{K}$. Hence, a reasonable goal for the learning algorithm is to converge to maximal points in $\mathcal{K}$ where *all $g^l(\cdot)$* functions are simultaneously maximized in the sense of the above definition. As explained above, a maximal point where all the action probability vectors of FALA are unit vectors and the variances of the action probability distributions of CALA are sufficiently small, would essentially be the same as an optimal point.

The functions F^l give the expected value of reinforcement conditioned on the action tuple while the functions g^l give this expectation conditioned on the action probability distributions. We define below two more functions both of which are expectations of reinforcement; however, in one case the expectation is conditioned on the action tuple of FALA and action probability distributions of CALA while in the other case the expectation is conditioned on the action tuple of CALA and action probability distributions of FALA. Both these functions are useful in the analysis of the learning algorithm presented later on.

For $P = (\mathbf{p}_1, \ldots, \mathbf{p}_N)$, $C = (\mu, \Sigma)$, $\mathbf{x} \in \Re^M$ and $\boldsymbol{\alpha} \in \prod_i A_i$, we define two sets of functions $H^l, J^l, l = 1, \ldots, N + M$, by

$$H^l(P, \mathbf{x}) \triangleq E[\beta_l \mid P, \text{CALA part chose } \mathbf{x}]$$
$$= \sum_{j_1, \ldots, j_N} F^l((\alpha_{j_1}, \ldots, \alpha_{j_N}), \mathbf{x}) \prod_k p_{k j_k} \quad (2.2.8)$$

$$J^l(\boldsymbol{\alpha}, C) \triangleq E[\beta_l \mid \text{FALA part chose } \boldsymbol{\alpha}, C]$$
$$= \int_{\Re^M} F^l(\boldsymbol{\alpha}, \mathbf{x}) N(\mu, \Sigma) d\mathbf{x} \quad (2.2.9)$$

Then, we can write $g^l(\cdot, \cdot)$ given by (2.2.7) as

$$g^l(P, C) = \sum_{j_1, \ldots, j_N} J^l((\alpha_{j_1}, \ldots, \alpha_{j_N}), C) \prod_k p_{k j_k} \quad (2.2.10)$$
$$= \int_{\Re^M} H^l(P, \mathbf{x}) N(\mu, \Sigma) d\mathbf{x} \quad (2.2.11)$$

In the case of a single FALA, the set of all action probability vectors forms a simplex and the corners of this simplex are action probability vectors which are unit vectors. An action probability vector which is a unit vector essentially corresponds to a single action and in this sense elements of the action set can be viewed as also belonging to the set of all probability distributions over the set of actions. This fact is important because it enables a learning algorithm, searching in the space of probability distributions over the action set, to converge, so to say, to an action. Such a neat representation is not possible in the case of the automata game model that we are considering. Here, $\mathcal{D}$ is the space of all possible action tuples and $\mathcal{K}$ is the space of probability distributions over $\mathcal{D}$. However, we do not have points in $\mathcal{K}$ which are equivalent to points in $\mathcal{D}$. Since we do not allow the variance of the Gaussian distribution to go to zero, there is no action probability distribution of a CALA which corresponds to an action choice by the CALA. All the same, it is useful to define what we call corners of $\mathcal{K}$ to get a rough correspondence between some points in $\mathcal{K}$ and points in $\mathcal{D}$.

A Multiple Payoff Stochastic Game of Automata 61

Consider an $S = (P, C) \in \mathcal{K}$ where $P = [\mathbf{e}_{\alpha_1}, \ldots, \mathbf{e}_{\alpha_N}]$ and $C = (\boldsymbol{\mu}, \Sigma)$. (Here, $\mathbf{e}_j$ is a unit vector of appropriate dimension with j^{th} component unity). We call such an S as a *corner* point of $\mathcal{K}$. We denote such a corner point also by $(\boldsymbol{\alpha}, C)$, where $\boldsymbol{\alpha} = (\alpha_1, \ldots, \alpha_N)$. It is easy to see that for each corner $S = (P, C)$ associated in this way with $(\boldsymbol{\alpha}, C)$, we have $g^l(P, C) = J^l(\boldsymbol{\alpha}, C)$.

A corner $(\boldsymbol{\alpha}, C)$ where $C = (\boldsymbol{\mu}, \Sigma)$ with all diagonal entries of the diagonal matrix Σ sufficiently small, would be a good approximation of the point $(\boldsymbol{\alpha}, \boldsymbol{\mu})$ in $\mathcal{D}$. Since corner points are, in this sense, related to points in $\mathcal{D}$, it is worthwhile to characterize those corners which could approximate optimal points in $\mathcal{D}$. This is what we do in the next definition.

DEFINITION 2.2.3 *We say* $(\boldsymbol{\alpha}^*, C^*), \boldsymbol{\alpha}^* = (\alpha_1^*, \ldots, \alpha_N^*); C^* = (\boldsymbol{\mu}^*, \Sigma^*)$, *is a* modal point *of the game if*

(1) For each i, $1 \leq i \leq N$,

$$J^i(\boldsymbol{\alpha}^*, C^*) \geq J^i(\boldsymbol{\alpha}, C^*),$$

for all $\boldsymbol{\alpha} = (\alpha_1^*, \ldots, \alpha_{i-1}^*, \alpha_i, \alpha_{i+1}^*, \ldots, \alpha_N^*)$ *such that* $\alpha_i \neq \alpha_i^*, \alpha_i \in A_i$.

(2) For each j, $1 \leq j \leq M$, $\exists \epsilon > 0$ *such that*

$$J^{N+j}(\boldsymbol{\alpha}^*, C^*) \geq J^{N+j}(\boldsymbol{\alpha}^*, (\boldsymbol{\mu}, \Sigma^*)),$$

for all $\boldsymbol{\mu}$ *such that* $\boldsymbol{\mu} \in \mathcal{B}^M(\boldsymbol{\mu}^*, \epsilon)$ *where* $\mathcal{B}^M(\boldsymbol{\mu}^*, \epsilon)$ *is an ϵ-ball in $\Re^M$ centered at* $\boldsymbol{\mu}^*$.
∎

REMARK 2.2.5 *Suppose* $(\boldsymbol{\alpha}^*, C^*)$, $C^* = (\boldsymbol{\mu}^*, \Sigma^*)$, $\Sigma^* = diag(\sigma_i^*)$, *is a modal point. Then, by definition of J^l given by (2.2.9), $(\boldsymbol{\alpha}^*, \boldsymbol{\mu}^*)$ is arbitrarily close to an optimal point, if σ_i^* are sufficiently small. This is intuitively clear because, as the variance decreases, the integral in (2.2.9) is essentially given by* $F^l(\boldsymbol{\alpha}, \boldsymbol{\mu})$.
∎

REMARK 2.2.6 *If $(\boldsymbol{\alpha}^*, C^*)$ is a modal point, then we call the corresponding corner point (P^*, C^*), a* pure *maximal point. It can be noted that the definition of pure maximal point makes sense only for corner points of $\mathcal{K}$.*
∎

REMARK 2.2.7 *Consider a special case of the game where all players get the same payoff, that is, $F^l(\cdot, \cdot) = F(\cdot, \cdot), \forall l$. This function, defined over $\mathcal{D}$, is a function of some continuous and some discrete variables. The optimal points of the game now correspond to 'local' maxima of F. The game formulation presented in the previous section allows for a more general setting of multiple payoffs and the optimal points are like Nash equilibria. The goal of the* automata algorithm is to identify these optimal points. For this, the automata approach is to search in the space of probability distributions over $\mathcal{D}$. It is easy to see that the state of the team at k, $S(k) = (P(k), C(k)) \in \mathcal{K}$, is a probability distribution over $\mathcal{D}$. Thus, the state, $S(k)$, can converge (if at all) only to a point in $\mathcal{K}$. It is for this reason that we defined maximal and modal points. In view of remarks 2.2.5 and 2.2.6, we want $S(k)$ to converge to some pure maximal point where the standard deviations of the normal distributions representing the action probability distributions of the CALA members of the team are sufficiently small. This is essentially what we prove in the next section. ∎

DEFINITION 2.2.4 (P^*, C^*)(resp. (α^*, C^*)) is a strict maximal point (resp. strict modal point) if (P^*, C^*) (resp. (α^*, C^*)) satisfies Definition 2.2.2 (resp. Definition 2.2.3) with strict inequalities for both conditions (1) and (2). ∎

In the next section, we analyze the learning algorithm used by the automata in this stochastic game and show that the algorithm essentially converges to pure maximal points. Consequently, by Remarks 2.2.5 and 2.2.6, the algorithm identifies the optimal points of the game to a good approximation. Before we present a formal analysis of our algorithm, we first give a simple overview of our main results.

The state of the collection of automata at instant k is given by $S(k) = (P(k), C(k)) \in \mathcal{K}$. The analysis of the long time behavior of $S(k)$ proceeds in two steps. The learning algorithm presented in the previous section specifies a stochastic difference equation which governs the evolution of $S(k)$. In the first step of the analysis we obtain an Ordinary Differential Equation(ODE) that approximates this difference equation. The general technique for obtaining such ODE approximations to automata algorithms is explained in Appendix A. This step of the analysis is the same as that used for the CALA, GLA and PLA algorithms in Chapter 1. We state the result about the approximating ODE for our algorithm in Theorem 2.1.

As explained in Appendix A (see Theorem A.1), the nature of our ODE approximation is such that the solution of the stochastic difference equation and that of the ODE would be arbitrarily close by taking the step-size, λ, in the algorithm sufficiently small. Thus the second and main part of our analysis here concentrates on characterizing the asymptotic solutions of the ODE. This is done in Theorem 2.2. We show that all strict pure maximal points are asymptotically stable (in the small) for the ODE and that if there is any equilibrium point of the ODE that is not a maximal point then it is unstable. We also show that in all equilibrium points of the ODE, the σ_i's of the action probability distributions of CALA will be arbitrarily close to σ_L. Thus, by taking σ_L in the learning algorithm sufficiently small, if the ODE solution (and hence the algorithm) converges to a pure maximal point in $\mathcal{K}$ then it should be a good approximation to an optimal point of the game. Then, in Theorem 2.3, we provide a sufficient condition to ensure that the ODE solution converges to an equilibrium point rather than, e.g., exhibit a limit cycle behavior. It will be seen that this sufficient condition is always satisfied for a game where all players get the same payoff.

2.3 Analysis of the Automata Game Algorithm

In this section we state and prove the main results regarding the asymptotic behavior of our algorithm. We make the following assumptions.

(A-1) For each player, conditioned on the action tuple chosen by the team, the reinforcement is independent of the past reinforcements. Further, conditioned on the action tuple, the reinforcements received by different players are independent. Thus, we assume that for any specific k, the random variables, $(\beta_j(k) - F^j(\alpha(k), \mathbf{x}(k)))$, $(\beta'_j(k) - F^j(\alpha(k), \mu(k)))$, $1 \leq j \leq N + M$, are zero mean independent random variables.

(A-2) $F^l(\alpha, \mathbf{x})$ is continuously differentiable with respect to $\mathbf{x}$ for every α, $l = 1, \ldots, N + M$.

(A-3) For every l and α, $\exists$ a constant $B < \infty$ such that

$$\sup_x \|\nabla_{\mathbf{x}} F^l(\alpha, \mathbf{x})\| < B.$$

(A-4) For every l and α, $F^l(\alpha, \cdot)$ has finite number of maxima, all in a compact set and has no maxima at infinity.

REMARK 2.3.1 *Assumptions (A-2), (A-3) and (A-4) specify smoothness and growth conditions on the payoff functions. Since (A-2) holds, if $F^l(\alpha, \cdot)$ has compact support then (A-3) is automatically satisfied. These restrictions are required mainly to satisfy some integrability conditions. The class of functions under these assumptions still includes a large family of functions of interest. It may be noted that our results hold after relaxing assumption (A-3) by demanding only that the magnitude of the gradient be bounded above by an exponential function in $\|\mathbf{x}\|$ as in [San94].* ∎

Before we begin the analysis we state some results which will be useful in the analysis to follow.

Define, for $P = (\mathbf{p}_1, \ldots, \mathbf{p}_N)$ and $C = (\mathbf{c}_1, \ldots, \mathbf{c}_M)$,

$$h_{lq}(P, C) = \text{E}[\, \beta_l \mid i^{th} \text{ FALA has action probability}$$
$$\text{distribution } \mathbf{p}_i, 1 \leq i \leq N, i \neq l,$$
$$l^{th} \text{ FALA chooses action } q; \text{ and}$$
$$\text{the state of } j^{th} \text{ CALA is } \mathbf{c}_j, 1 \leq j \leq M] \quad (2.3.12)$$

Using (2.2.6) and (2.2.10), we have

$$h_{lq}(P,C) = \sum_{j_s,\, s \neq l} J^l((\alpha_{j_1},\ldots,\alpha_{j_{l-1}},q,\alpha_{j_{l+1}},\ldots,\alpha_{j_N}),C) \prod_{k \neq l} p_{kj_k}. \tag{2.3.13}$$

Note that the summation above is over all indices: $j_1,\ldots,j_{l-1},j_{l+1},\ldots,j_N$. It can be noted from (2.2.10) and (2.3.13) that

$$g^l(P,C) = \sum_{q=1}^{r_l} h_{lq}(P,C) p_{lq} \tag{2.3.14}$$

From the definition of the functions h_{lq} given by (2.3.13), it is clear that for any l, the values of h_{lj}, $j = 1,\ldots,r_l$, are not dependent on the action probabilities of lth FALA, namely, p_{lj}. Hence, from (2.3.14), we have

$$h_{lq}(P,C) = \frac{\partial g^l(P,C)}{\partial p_{lq}}, \quad \forall l,\ 1 \leq q \leq r_l. \tag{2.3.15}$$

We list below two special properties of the h_{lq} functions which are used later on in the analysis.

LEMMA 2.1 (P^0,C^0), where $P^0 = (\mathbf{p}_1^0,\ldots,\mathbf{p}_N^0)$ and $C^0 = (\mathbf{c}_1^0,\ldots,\mathbf{c}_M^0)$, satisfies condition (1) of Definition 2.2.2 if and only if for each i, $1 \leq i \leq N$,

$$h_{is}(P^0,C^0) \leq g^i(P^0,C^0), \quad 1 \leq s \leq r_i.$$

COROLLARY 2.1 Let (P^0,C^0) (as in Lemma 2.1) satisfy condition (1) of Definition 2.2.2. Then, for each i, $1 \leq i \leq N$,

$$h_{is}(P^0,C^0) = g^i(P^0,C^0), \quad \forall s \text{ such that } p_{is}^0 > 0.$$

Both these results are just restatements of some of the properties of Nash equilibria [BO82, SPT94]. They can be proved directly from the definition and we skip the details.

Now consider the algorithm described in the previous section. We can write the state evolution of our algorithm as

$$S(k+1) = S(k) + \lambda G(S(k), \Psi(k)) \tag{2.3.16}$$

where $\Psi(k) = (\boldsymbol{\alpha}(k), \mathbf{x}(k), (\beta_1(k),\ldots,\beta_{N+M}(k)), (\beta_1'(k),\ldots,\beta_{N+M}'(k)))$ and $G(\cdot,\cdot)$ defines the updating given by equations (2.2.2) and (2.2.3).

Consider the following Ordinary Differential Equation (ODE)

$$\frac{dZ}{dt} = \mathcal{G}(Z), \quad Z(0) = S(0), \tag{2.3.17}$$

where
$$\mathcal{G}(Z) = E[G(S(k), \Psi(k)) \mid S(k) \equiv Z] \qquad (2.3.18)$$

In the following theorem we show that (2.3.17) is the approximating ODE for our algorithm.

THEOREM 2.1 *Under the learning algorithm (2.2.2)-(2.2.3) and under the assumptions (A-1) to (A-4) stated in section 2.3, $S(k)$ is well approximated by the unique solution of the ODE (2.3.17). That is, for any initial condition $S(0)$, given any finite T, ϵ, $\delta > 0$, there exists a $\lambda^* > 0$ such that for all $0 < \lambda \leq \lambda^*$,*

$$Prob[sup_{0 \leq k \leq \frac{T}{\lambda}} \|S(k) - Z(k\lambda)\| \geq \epsilon] \leq \delta. \qquad (2.3.19)$$

Proof: The general method for deriving the equivalent ODE is explained in Appendix A. Comparing (2.3.16) above with (A.2.10) in Appendix A, it is easy to see that, by taking $X_k = S(k)$, $\xi_k = \Psi(k)$ and $b = \lambda$ in (A.2.10), the automata team algorithm is in the same general form. Now, comparing (A.2.11) and (A.1.3) in Appendix A with (2.3.17) and (2.3.18) above, it is clear that this theorem follows directly from Theorem A.1 in Appendix A. Hence, all that we need to do is to verify all the conditions $A1$ to $A4$ of Section A.2.1 which are needed for applying that theorem in the appendix. As explained in Section A.3.1, all these conditions are easily satisfied for L_{R-I} algorithm for single FALA and the same thing is true for a team of FALA also as explained in Section A.3.3. For the CALA part, as explained in Sections A.3.2 and A.3.3, we need some additional assumptions. It is straightforward to see that the assumptions (A-1) to (A-4) stated earlier in this section are the ones needed so that we can apply Theorem A.1. ∎

2.3.1 Analysis of the Approximating ODE

As outlined at the end of Section 2.2.1, in the remaining part of the analysis we concentrate on characterizing the solutions of the ODE.

The equivalent ODE of our algorithm is given by (2.3.17). **To avoid unnecessary notational complexity we use S itself as the variable in the ODE instead of** Z. Thus the ODE would be $\frac{dS}{dt} = \mathcal{G}(S)$.

By our notation, $S = (P, C)$ where $P = (\mathbf{p}_1, \ldots, \mathbf{p}_N)$, $C = (\mathbf{c}_1, \ldots, \mathbf{c}_M)$, and $\mathbf{c}_j = (\mu_j, \sigma_j)$. Note that C represents the action probability distributions of all CALA. As already mentioned, we would also denote C by $C = (\boldsymbol{\mu}, \boldsymbol{\sigma})$ or $C = (\boldsymbol{\mu}, \Sigma)$ where $\boldsymbol{\mu} = (\mu_1, \ldots, \mu_M)^T$ and $\boldsymbol{\sigma} = (\sigma_1, \ldots \sigma_M)^T$. Whenever we use Σ it is a diagonal matrix with the ith diagonal entry being $\phi(\sigma_i)^2$. When the state of CALA is C, the joint density corresponding to action probability distributions of all CALA is $N(\boldsymbol{\mu}, \Sigma)$ which is an M-dimensional Gaussian density with a diagonal covariance matrix. In the following, whenever we use the expectation integral, $dN(\boldsymbol{\mu}, \Sigma)$ stands for $N(\boldsymbol{\mu}, \Sigma)d\mathbf{x}$.

The approximating ODE is a vector differential equation. It essentially consists of three sets of component equations corresponding to the $\mathbf{p}_i$'s, μ_j's and σ_j's respectively. We consider each set separately below.

1. $\frac{dP}{dt} = \mathcal{G}^p(S)$,, i.e.

$$\frac{dp_{iq}}{dt} = \mathcal{G}^p_{iq}(S), \ 1 \le q \le r_i, \ 1 \le i \le N \qquad (2.3.20)$$

where

$$\begin{aligned}
\mathcal{G}^p_{iq}(S) &= p_{iq}(1-p_{iq})E[\beta'_i|S, \alpha_i=q] + \sum_{s\ne q} p_{is}(-p_{iq})E[\beta'_i|S, \alpha_i=s], \\
&\quad \text{from (2.2.2) and (2.3.18)} \\
&= p_{iq}(1-p_{iq})h_{iq}(S) - \sum_{s\ne q} p_{is}p_{iq}h_{is}(S), \quad \text{from (2.3.12)} \\
&= p_{iq}\left(\sum_{s\ne q} p_{is}\right) h_{iq}(S) - \sum_{s\ne q} p_{is}p_{iq}h_{is}(S), \\
&= p_{iq}\sum_{s\ne q} p_{is}[h_{iq}(S) - h_{is}(S)] \\
&= p_{iq}\sum_{s} p_{is}[h_{iq}(S) - h_{is}(S)] \qquad (2.3.21)
\end{aligned}$$

Using (2.3.15), we can rewrite the above as

$$\mathcal{G}^p_{iq}(S) = p_{iq}\sum_{s} p_{is}\left[\frac{\partial g^l}{\partial p_{iq}}(S) - \frac{\partial g^l}{\partial p_{is}}(S)\right] \qquad (2.3.22)$$

2. $\frac{d\boldsymbol{\mu}}{dt} = \mathcal{G}^\mu(S)$,, i.e.

$$\frac{d\mu_j}{dt} = \mathcal{G}^\mu_j(S), \ 1 \le j \le M, \qquad (2.3.23)$$

where, using (2.2.3), (2.2.4) and (2.3.18), we have

$$\begin{aligned}
\mathcal{G}^\mu_j(S) &= E[\mathcal{F}_1(\mu_j(k), \sigma_j(k), x_j(k), \beta_{N+j}(k), \beta'_{N+j}(k)) \mid S], \\
&= \int_{\Re^M} E\left[\left(\frac{\beta_{N+j} - \beta'_{N+j}}{\phi(\sigma_j)}\right)\bigg| S, \mathbf{x}\right] \left(\frac{x_j - \mu_j}{\phi(\sigma_j)}\right) dN(\boldsymbol{\mu}, \Sigma) \\
&= \int_{\Re^M} \left[\left(\frac{H^{N+j}(P,\mathbf{x})}{\phi(\sigma_j)}\right)\right] \left(\frac{x_j - \mu_j}{\phi(\sigma_j)}\right) dN(\boldsymbol{\mu}, \Sigma) - 0,
\end{aligned}$$

by (2.2.8) and because

$$E\left[\beta'_{N+j}(k)|S,\mathbf{x}\right] = H^{N+j}(P,\boldsymbol{\mu}), \text{ is not dependent on } \mathbf{x}$$

and $\int \frac{x_j - \mu_j}{\phi(\sigma_j)} dN(\boldsymbol{\mu}, \Sigma) = 0$

$$= \int_{\Re^M} \frac{H^{N+j}(P,\mathbf{x})}{\phi(\sigma_j)} \left(\frac{x_j - \mu_j}{\phi(\sigma_j)}\right) dN(\boldsymbol{\mu}, \Sigma)$$

$$= \frac{\partial g^{N+j}}{\partial \mu_j}(S), \quad \text{using (2.2.11)}$$

Therefore,

$$\mathcal{G}_j^\mu(S) = \frac{\partial g^{N+j}}{\partial \mu_j}(S) \tag{2.3.24}$$

We mention in passing that the derivation given above shows that, even if we take $\beta'_{N+j} = 0$ in the learning algorithm, the ODE would be the same as above. Thus, as mentioned in Section 1.6, the asymptotic properties of the CALA algorithm remain unchanged even if we use only one reinforcement (or equivalently, only one function evaluation) per iteration.

3. $\frac{d\boldsymbol{\sigma}}{dt} = \mathcal{G}^\sigma(S)$,, i.e.

$$\frac{d\sigma_j}{dt} = \mathcal{G}_j^\sigma(S), \ 1 \leq j \leq M, \tag{2.3.25}$$

where, from (2.2.3) and (2.3.18), we have

$$\mathcal{G}_j^\sigma(S) = E\left[\mathcal{F}_2(\mu_j(k), \sigma_j(k), x_j(k), \beta_{N+j}(k), \beta'_{N+j}(k)) - K[\sigma_j - \sigma_L]|S\right]$$

We can write

$$\mathcal{G}_j^\sigma(S) = \psi_j^\sigma(S) - K[\sigma_j - \sigma_L]$$

where, using (2.2.4),

$$\psi_j^\sigma(S) = \int_{\Re^M} E\left[\frac{\beta_{N+j} - \beta'_{N+j}}{\phi(\sigma_j)}\bigg|S,\mathbf{x}\right]\left[\left(\frac{x_j - \mu_j}{\phi(\sigma_j)}\right)^2 - 1\right] dN(\boldsymbol{\mu}, \Sigma)$$

$$= \int_{\Re^M} \left(\frac{H^{N+j}(P,\mathbf{x})}{\phi(\sigma_j)}\right)\left[\left(\frac{x_j - \mu_j}{\phi(\sigma_j)}\right)^2 - 1\right] dN(\boldsymbol{\mu}, \Sigma) - 0,$$

by (2.2.8) and because

$$E\left[\beta'_{N+j}(k)|S,\mathbf{x}\right] = H^{N+j}(P,\boldsymbol{\mu}), \text{ is not dependent on } \mathbf{x}$$

$$= \int_{\Re^M} \frac{H^{N+j}(P,\mathbf{x})}{\phi(\sigma_j)}\left[\left(\frac{x_j - \mu_j}{\phi(\sigma_j)}\right)^2 - 1\right] dN(\boldsymbol{\mu}, \Sigma)$$

$$= \frac{\partial g^{N+j}}{\partial \sigma_j}(S), \text{ using } (2.2.11)$$

Therefore,
$$\mathcal{G}_j^\sigma(S) = \frac{\partial g^{N+j}}{\partial \sigma_j}(S) - K[\sigma_j - \sigma_L] \qquad (2.3.26)$$

Define Γ_1, Γ_2 and Γ_3, subsets of $\mathcal{K}$, as follows.

$$\Gamma_1 = \bigcap_{i,q} \{S \in \mathcal{K} \mid \mathcal{G}_{iq}^p(S) = 0\} \qquad (2.3.27)$$

$$\Gamma_2 = \bigcap_{j} \{S \in \mathcal{K} \mid \mathcal{G}_j^\mu(S) = 0\} \qquad (2.3.28)$$

$$\Gamma_3 = \bigcap_{j} \{S \in \mathcal{K} \mid \mathcal{G}_j^\sigma(S) = 0\} \qquad (2.3.29)$$

That is, Γ_1, Γ_2 and Γ_3 respectively denote the set of all equilibrium points of the three sets of (component) ODE's given by (2.3.20),(2.3.23) and (2.3.25). It is easy to see that the set of equilibrium points of ODE (2.3.17) is $\Gamma_1 \cap \Gamma_2 \cap \Gamma_3$. The following theorem characterizes the equilibrium points of the ODE's.

THEOREM 2.2 *Let Γ_1, Γ_2 and Γ_3 be as defined by (2.3.27),(2.3.28) and (2.3.29). Then,*

1 All corners of $\mathcal{K}$ belong to Γ_1.

2 All maximal points belong to $\Gamma_1 \cap \Gamma_2$.

3 An equilibrium point, $S^0 \in \Gamma_1 \cap \Gamma_2$, is unstable if it is not a maximal point.

4 All strict pure maximal points are asymptotically stable

5 If $(P,(\boldsymbol{\mu}, \boldsymbol{\sigma})) \in \Gamma_3, \boldsymbol{\mu} = (\mu_1, \ldots, \mu_M); \boldsymbol{\sigma} = (\sigma_1, \ldots, \sigma_M)$, then each $\sigma_j, 1 \leq j \leq M$, lies in a small neighborhood (determined by the parameter K of the learning algorithm (2.2.3)) around σ_L.

Proof:

1 Let $S^* = (P^*, C^*)$ be a corner of $\mathcal{K}$. We know that $P^* = [\mathbf{e}_{\alpha_1}, \ldots, \mathbf{e}_{\alpha_N}]$ for some $\alpha_i \in A_i, 1 \leq i \leq N$. Therefore,

$$\begin{aligned}
\mathcal{G}_{iq}^p(S^*) &= p_{iq}^* \sum_s p_{is}^*[h_{iq}(S) - h_{is}(S)], \ 1 \leq q \leq r_i, \ 1 \leq i \leq N, \\
&= 0,
\end{aligned}$$

since either $p_{iq}^* = 0$ or $p_{is}^* = 0, s \neq q$. Hence, $S^* \in \Gamma_1$.

Analysis of the Automata Game Algorithm

2 Let $S^* = (P^*, C^*)$ be a maximal point. We have to prove that $\mathcal{G}^p(S^*)$ and $\mathcal{G}^\mu(S^*)$ are zero. From (2.3.21), for any S^* we have

$$\begin{aligned}
\mathcal{G}^p_{iq}(S^*) &= p^*_{iq} \sum_s p^*_{is}[h_{iq}(S^*) - h_{is}(S^*)], \quad 1 \leq q \leq r_i, \; 1 \leq i \leq N \\
&= p^*_{iq} h_{iq}(S^*) \sum_s p^*_{is} - p^*_{iq} \sum_s p^*_{is} h_{is}(S^*), \\
&= p^*_{iq}[h_{iq}(S^*) - g^i(S^*)], \quad \text{by (2.3.14)} \quad (2.3.30)
\end{aligned}$$

Since S^* is a maximal point, it satisfies condition (1) of Definition 2.2.2. Hence, it follows by Corollary 2.1 and (2.3.30) that

$$\mathcal{G}^p_{iq}(S^*) = 0, \; 1 \leq q \leq r_i, \; 1 \leq i \leq N$$

implying $S^* \in \Gamma_1$.
Since S^* also satisfies condition (2) of Definition 2.2.2, it follows that $\frac{\partial g^{N+j}}{\partial \mu_j}(S^*) = 0$. Therefore, from (2.3.24),

$$\mathcal{G}^\mu_j(S^*) = 0, \quad 1 \leq j \leq M.$$

implying $S^* \in \Gamma_2$.
Hence, $S^* \in \Gamma_1 \cap \Gamma_2$.

3 Let $S^0 = (P^0, C^0)$ belonging to $\Gamma_1 \cap \Gamma_2$, not correspond to a maximal point. Therefore, at least one of the conditions (1) and (2) of Definition 2.2.2 is not satisfied by S^0.

(a) Suppose condition (1) is not satisfied. Now, by Lemma 2.1, $\exists i, s$ such that
$$h_{is}(P^0, C^0) > g^i(P^0, C^0) \quad (2.3.31)$$

By continuity of the functions involved, the inequality (2.3.31) will hold for all points S in a small neighborhood around S^0 in $\mathcal{K}$. This implies, by (2.3.30), that for all points S in this neighborhood, $\mathcal{G}^p_{is}(S) = \frac{dp_{is}}{dt}(S) > 0$ if $p_{is} \neq 0$. Hence in all sufficiently small neighborhoods of P^0, there will be infinitely many points starting from which the trajectories diverge from the neighborhood. Hence, S^0 is unstable.

(b) Suppose condition (2) is not satisfied. That is, μ^0 is not a local maximum of $g^{N+j}(P^0, \cdot)$ for some j. Since

$$\frac{d\mu_j}{dt} = \frac{\partial g^{N+j}}{\partial \mu_j},$$

it is obvious that S^0 will be unstable.

4 Let $S^0 = (P^0, C^0)$, $P^0 = [\mathbf{e}_{\alpha_1^0}, \ldots, \mathbf{e}_{\alpha_N^0}]$; $C^0 = (\boldsymbol{\mu}^0, \boldsymbol{\sigma}^0)$, be a strict pure maximal point. Then, we know that $Q^0 = (\boldsymbol{\alpha}^0, C^0)$ is a strict modal point (cf. Remark 2.2.6). Let Q^0 satisfy condition (2) of Definition 2.2.3 with $\epsilon = \epsilon^0$. Define a region S_ρ by

$$S_\rho = \{(P, (\boldsymbol{\mu}, \boldsymbol{\sigma}^0)) \in \mathcal{K} \mid P = [\mathbf{p}_1, \ldots, \mathbf{p}_N] \text{ and } \boldsymbol{\mu} \in \mathcal{B}^M(\boldsymbol{\mu}^0, \epsilon^0)\}$$

where each $\mathbf{p}_i$, $1 \leq i \leq N$, is a probability vector of dimension r_i close to $\mathbf{e}_{\alpha_i^0}$, and $\mathcal{B}^M(\mathbf{x}, \epsilon)$ denotes the open ball of radius ϵ centered at $\mathbf{x}$ in $\Re^M$.

Let $S = (P, C) \in S_\rho$. In $P = [\mathbf{p}_1, \ldots, \mathbf{p}_N]$, only $\sum_{i=1}^N (r_i - 1)$ components are independent since, for each i, $\sum_k p_{ik} = 1$. Choose $p_{iq}, q \neq \alpha_i$, as the independent components. For these, we get

$$\frac{dp_{iq}}{dt} = H_{iq}(S) + \text{ second and higher order terms in components of } P$$

where, by Taylor expansion of $h_{iq}(S)$ around S^0,

$$\begin{aligned} H_{iq}(S) &= p_{iq}[J^i((\alpha_1^0, \ldots, \alpha_{i-1}^0, q, \alpha_{i+1}^0, \ldots, \alpha_N^0), C) \\ &\quad - J^i((\alpha_1^0, \ldots, \alpha_N^0), C)], \\ &\quad \text{by (2.3.13) and (2.3.21)} \\ &< 0, \quad \forall q \neq \alpha_i^0, \end{aligned} \qquad (2.3.32)$$

since (P^0, C^0) is a strict pure maximal point. Define

$$\Phi(\sigma) = \frac{(\sigma - \sigma_L)^2}{2} I\{\sigma > \sigma_L\}, \quad \sigma \in \Re \qquad (2.3.33)$$

Consider the Lyapunov function defined over S_ρ, with $S = (P, (\boldsymbol{\mu}, \boldsymbol{\sigma}))$,

$$\begin{aligned} V(S) &= \sum_{j=1}^M [g^{N+j}(P^0, C^0) - g^{N+j}(P^0, \widehat{C_j})] \\ &\quad + \sum_{j=1}^M K\Phi(\sigma_j) + \sum_{i,q: q \neq \alpha_i} p_{iq}, \end{aligned}$$

where $\widehat{C_j} = ((\mu_1^0, \ldots, \mu_{j-1}^0, \mu_j, \mu_{j+1}^0, \ldots, \mu_M^0), \boldsymbol{\sigma}^0)$.
We have $V(S^0) = 0$ and $V(S) > 0$ since $S^0 = (P^0, C^0)$ is a strict pure maximal point. Now we have,

$$\dot{V} = -\sum_{j=1}^M \frac{\partial g^{N+j}}{\partial \mu_j} \dot{\mu}_j - \sum_{j=1}^M \frac{\partial g^{N+j}}{\partial \sigma_j} \dot{\sigma}_j$$

$$+ \sum_{j=1}^{M} K[\sigma_j - \sigma_L]\dot{\sigma}_j + \sum_{i,q:q\neq\alpha_i} \dot{p}_{iq}$$

$$= -\sum_{j=1}^{M}(\mathcal{G}_j^\mu)^2 - \sum_{j=1}^{M}\dot{\sigma}_j \left[\frac{\partial g^{N+j}}{\partial \sigma_j} - \sum_{j=1}^{M} K[\sigma_j - \sigma_L]\right] + \sum_{i,q:q\neq\alpha_i} \mathcal{G}_{iq}^p$$

$$= -\sum_{j=1}^{M}(\mathcal{G}_j^\mu)^2 - \sum_{j=1}^{M}(\mathcal{G}_j^\sigma)^2$$

$$+ \sum_{i,q:q\neq\alpha_i^0} p_{iq} \left[(J^i((\alpha_1^0,\ldots,\alpha_{i-1}^0,q,\alpha_{i+1}^0,\ldots,\alpha_N^0),C)\right.$$

$$\left. - J^i((\alpha_1^0,\ldots,\alpha_N^0),C)) + \text{higher order terms}\right]$$

$$< 0, \quad \text{from (2.3.32), (2.3.24) and (2.3.26)}$$

since $S^0 = (P^0, C^0)$ is a strict pure maximal point. Hence, S^0 is asymptotically stable.

5 Suppose $S = (P, (\boldsymbol{\mu}, \boldsymbol{\sigma})) \in \Gamma_3, \boldsymbol{\mu} = (\mu_1,\ldots,\mu_M); \boldsymbol{\sigma} = (\sigma_1,\ldots,\sigma_M)$. Then, $\mathcal{G}_j^\sigma(S) = 0, \forall j$.
Let $\delta' > 0$. For each j, $1 \leq j \leq M$, let K_j be a constant such that $K_j > \frac{1}{\delta'}\sup_S |\frac{\partial g^{N+j}}{\partial \sigma_j}|$. Note that we can always choose such a constant by assumption (A-3).[3] In the learning algorithm given by (2.2.3), K is a parameter. Suppose we choose K such that $K \geq \max_j K_j$. Now, we have the following.
If $\sigma_j < \sigma_L - \delta'$,

$$\mathcal{G}_j^\sigma(S) = \frac{\partial g^{N+j}}{\partial \sigma_j} - K_j[\sigma_j - \sigma_L]$$

$$> \frac{\partial g^{N+j}}{\partial \sigma_j} + K\delta'$$

$$> 0, \quad \text{by choice of } K$$

If $\sigma_j > \sigma_L + \delta'$,

$$\mathcal{G}_j^\sigma(S) = \frac{\partial g^{N+j}}{\partial \sigma_j} - K_j[\sigma_j - \sigma_L]$$

---

[3] We can show this as follows. The integration on the RHS of equation (2.2.7) is with respect to the normal density $N(\boldsymbol{\mu}, \Sigma)$. This can be changed to that with respect to the M-dimensional standard normal density by the usual coordinate transformation. Then, the only dependence of this integrand on σ_j would be in the argument of the function F^l. Now, by differentiating this expression under the integral sign, we see that partial derivatives of g^l can be bounded in terms of bounds on the partial derivatives of F^l. In the case of a single CALA, this is illustrated by equation (A.3.38) in Appendix A.

$$< \frac{\partial g^{N+j}}{\partial \sigma_j} - K\delta'$$
$$< 0, \quad \text{by choice of } K$$

Therefore, all zero crossings of $\mathcal{G}_j^\sigma(S)$ have to occur within $(\sigma_L - \delta', \sigma_L + \delta')$. Now, it follows that all zeros of $\mathcal{G}^\sigma$ have to lie in an open ball of radius δ' centered at σ_L. Hence, every point $(P, (\boldsymbol{\mu}, \boldsymbol{\sigma})) \in \Gamma_3$ is such that each component of $\boldsymbol{\sigma} = (\sigma_1, \ldots, \sigma_M)$ lies in a small neighborhood around σ_L. ∎

Since the set of equilibrium points of the ODE (2.3.17) is $\Gamma_1 \cap \Gamma_2 \cap \Gamma_3$, from Theorem 2.2, we can conclude that the solutions of the ODE will not converge to a point in $\mathcal{K}$ that is not a maximal point, and all strict pure maximal points are locally asymptotically stable. Moreover, by assertion 5 of Theorem 2.2, in any equilibrium point, $(P, \boldsymbol{\mu}, \boldsymbol{\sigma})$, of the ODE, each component of $\boldsymbol{\sigma}$ lies in a small neighborhood around σ_L and so asymptotically the action chosen by each CALA will be close to the mean of the action probability distribution of the corresponding CALA if σ_L is sufficiently small (see Remark 2.2.5). Thus, with small enough σ_L, the pure maximal point (P^0, C^0) to which the solution of the ODE (and hence the algorithm) converges will be arbitrarily close to an optimal point of the game (cf. Definition 2.2.1). Still, the theorem does not guarantee the convergence to a maximal point because the ODE may exhibit, e.g., limit cycle behavior. However, we give a sufficient condition under which convergence to a maximal point can be assured.

THEOREM 2.3 *Suppose there is a differentiable function,*

$$\Theta : \Re^{r_1 + \ldots + r_N + 2M} \to \Re,$$

such that for some constants $b_1, b_2, b_3 > 0$,

$$\frac{\partial \Theta}{\partial p_{iq}}(P, C) = -b_1 \, h_{iq}(P, C) \quad (2.3.34)$$

$$\frac{\partial \Theta}{\partial \mu_j}(P, C) = -b_2 \frac{\partial g^{N+j}}{\partial \mu_j}(P, C) \quad (2.3.35)$$

$$\frac{\partial \Theta}{\partial \sigma_j}(P, C) = -b_3 \left[\frac{\partial g^{N+j}}{\partial \sigma_j}(P, C) - K[\sigma_j - \sigma_L] \right] \quad (2.3.36)$$

for all $(P, C) \in \mathcal{K} \subset \Re^{r_1 + \ldots + r_N + 2M}$. *Let* Θ *be bounded below and suppose that there is a constant* $c > 0$ *such that*

$$\Lambda_0 = \{(P, C) \in \mathcal{K} \mid \Theta(P, C) \leq c\}$$

is a bounded set. Then, the automata team using the algorithm given by (2.2.2) and (2.2.3) converges to one of the maximal points, for any initial condition.

Analysis of the Automata Game Algorithm 73

Proof: We have

$$\frac{d\Theta}{dt} = \sum_{i,q} \frac{\partial \Theta}{\partial p_{iq}} \frac{dp_{iq}}{dt} + \sum_j \frac{\partial \Theta}{\partial \mu_j} \frac{d\mu_j}{dt} + \sum_j \frac{\partial \Theta}{\partial \sigma_j} \frac{d\sigma_j}{dt} \quad (2.3.37)$$

$$= -b_1 \sum_{i,q} h_{iq} p_{iq} \sum_s p_{is}[h_{iq} - h_{is}] - b_2 \sum_j \frac{\partial g^{N+j}}{\partial \mu_j} \frac{d\mu_j}{dt}$$

$$-b_3 \sum_j \frac{d\sigma_j}{dt}\left[\frac{\partial g}{\partial \sigma_j} - K[\sigma_j - \sigma_L]\right],$$

by (2.3.20) – (2.3.21) and (2.3.34) – (2.3.36)

$$= -b_1 \sum_i \sum_q \sum_s p_{iq} p_{is}[h_{iq}{}^2 - h_{iq} h_{is}] - b_2 \sum_j (\mathcal{G}_j^\mu)^2$$

$$-b_3 \sum_j (\mathcal{G}_j^\sigma)^2, \quad \text{by (2.3.24) – (2.3.26)}$$

$$= -b_1 \sum_i \sum_q \sum_{s>q} p_{iq} p_{is}[h_{iq} - h_{is}]^2 - b_2 \sum_j (\mathcal{G}_j^\mu)^2$$

$$-b_3 \sum_j (\mathcal{G}_j^\sigma)^2 \quad (2.3.38)$$

$$\leq 0$$

Thus Θ is nonincreasing along the trajectories of the ODE.

Since Θ is bounded below and Λ_0 is a bounded set, by [Vid78, Lemma 81, Chapter 4], asymptotically all the trajectories will be in the set $\mathcal{K}_1 = \{(P, C) \in \mathcal{K} | \frac{d\Theta}{dt}(P, C) = 0\}$.

It is easy to see from (2.3.38) that if $(P^0, C^0) \in \mathcal{K}_1$, then by (2.3.27),(2.3.28) and (2.3.29), $(P^0, C^0) \in \Gamma_1 \cap \Gamma_2 \cap \Gamma_3$. That is, (P^0, C^0) is an equilibrium point of the ODE (2.3.17). Thus the ODE has to converge to some equilibrium point. Now the theorem follows by noting that all equilibrium points that are not maximal points are unstable by Theorem 2.2. ∎

We can summarize the analysis presented in this section as follows. In view of Theorem 2.1, the behavior of the algorithm is well approximated by the solutions of the ODE. Thus, if starting from some initial point, the solution of the ODE converges to a point in $\mathcal{K}$ then we can say that asymptotically the algorithm would also be very close to that point. We would be implicitly using this connection between the algorithm and the ODE. *Thus, we talk about some point in $\mathcal{K}$ being asymptotically stable for the algorithm etc.*

Theorem 2.1 and Theorem 2.2 together provide an analytical result for the asymptotic behavior of our algorithm. Essentially what we have shown is that we can expect the algorithm to converge to a maximal point if the learning parameters are sufficiently small. If the sufficient conditions needed by Theorem 2.3 are met, the algorithm always converges to one of the maximal points of the game. Of these, the pure maximal points are stable and strict pure maximal points are asymptotically stable, by Theorem 2.2. We cannot, in general, conclude anything about the stability of other maximal points. Since the L_{R-I} algorithm we use for FALA's has unit vectors as absorbing states, in practice, it is found that the algorithm always converges to one of the pure maximal points.

As observed in Section 2.2, if (P^*, C^*) is a pure maximal point, it means that the corresponding (α^*, C^*) is a modal point. Hence our result also means that the algorithm converges to a modal point of the game. Now, by assertion 5 of Theorem 2.2, we know that all stationary points are such that each σ_i lies in a small neighborhood (determined by the parameter K) around σ_L. Hence, supposing that K is chosen sufficiently large and σ_L sufficiently small, by Remark 2.2.5, if (α^*, C^*), $C^* = (\mu^*, \sigma^*)$, is a modal point, then (α^*, μ^*) is arbitrarily close to an optimal point. Thus, the above results also imply that the algorithm converges (arbitrarily closely) to an optimal point.

2.4 Game with Common Payoff

An important special case of the general model discussed so far is a game with common payoff where all the players receive the same payoff after each play. For this special case, we have $\beta_i = \beta_j = \beta$, and hence $F^i = F^j = F, \forall i, j$, where F^i is as given by (2.2.1). Hence the payoff structure of the game can be defined by a single function $F : \mathcal{D} \to \Re$

$$F(\alpha, \mathbf{x}) = \mathrm{E}[\,\beta \mid i\text{-th FALA chose } \alpha_i \in A_i \text{ and } j\text{-th CALA chose } x_j \in \Re]. \tag{2.4.39}$$

Similarly we will have a single function $g(\cdot, \cdot)$ in place of $g^l(\cdot, \cdot)$ which would now be given by

$$\begin{aligned}g(P, C) &= \mathrm{E}[\,\beta \mid i^{th} \text{ FALA has action probability vector } \mathbf{p}_i, 1 \leq i \leq N \\ &\quad \text{ and the state of } j^{th} \text{ CALA is } \mathbf{c}_j, 1 \leq j \leq M]. \tag{2.4.40}\end{aligned}$$

There would be a single function $J(\cdot, \cdot)$ in place of $J^l(\cdot, \cdot)$ and similarly for H^l. The relations between g and the functions F, J and H would be the same as before. The functions h_{lq} would also be defined similarly. That is, h_{lq} would now be expectation of β (rather than of β_l) but the expectation being conditioned on the same things as earlier in equation (2.3.12). Equations (2.3.13), (2.3.14) and (2.3.15) continue to hold except for the fact that there would be no superscript on the functions g and J.

All automata in a common payoff game receive the same payoff and all of them are trying to maximize the expected value of the payoff. Thus, though

Game with Common Payoff 75

the game model is completely decentralized, effectively the automata are co-operating with each other. Hence we refer to them as a team of automata.

In this special case, we can strengthen our convergence result to say that the algorithm would always converge to a maximal point. To prove this result, we need to assume (A-5) given below in addition to the assumptions (A-1)–(A-4) on the payoff function $F(\cdot,\cdot)$.

(A-5) $F(\alpha,\cdot)$ vanishes outside a compact set in $\Re^M$.

THEOREM 2.4 *Consider a game with common payoff played by a team of FALA's and CALA's. Then the automata team using the algorithm given by (2.2.2) and (2.2.3) converges to one of the maximal points.*

Proof: Define the function Θ over $\mathcal{K}$ by

$$\Theta(P,C) = -g(P,C) + \sum_{j=1}^{M} K\Phi(\sigma_j)$$

where $\Phi(\cdot)$ is as given by (2.3.33).

We have from (2.2.10), for $1 \leq q \leq r_i, 1 \leq i \leq N$,

$$\frac{\partial \Theta}{\partial p_{iq}}(P,C) = -\sum_{j_s, s \neq i} J((\alpha_{j_1},\ldots,\alpha_{j_{i-1}},q,\alpha_{j_{i+1}},\ldots,\alpha_{j_N}),C) \prod_{k \neq i} p_{kj_k},$$
$$= -h_{iq}(P,C), \quad \text{from (2.3.13)} \tag{2.4.41}$$

By differentiating the Θ function given above, we have

$$\frac{\partial \Theta}{\partial \mu_j} = -\frac{\partial g}{\partial \mu_j} \tag{2.4.42}$$

$$\frac{\partial \Theta}{\partial \sigma_j} = -\frac{\partial g}{\partial \sigma_j} + K[\sigma_j - \sigma_L]$$

$$= -\left[\frac{\partial g}{\partial \sigma_j} - K[\sigma_j - \sigma_L]\right] \tag{2.4.43}$$

From assumption (A-5) g vanishes outside a compact set. Since the second term in the definition of Θ is strictly convex, we can choose c (which is the constant in the definition of Λ_0 used in Theorem 2.3) to be the value of Θ just outside the support of g. Then the set Λ_0 of Theorem 2.3 is bounded. Now the proof follows by applying Theorem 2.3. ∎

REMARK 2.4.1 *In the above proof, we have used Assumption (A-5) only to prove the boundedness of Λ_0. Since g is bounded and, by assumption (A-4), all* maxima of $F(\alpha,\cdot)$ are inside a compact set and there is no maxima at infinity, it seems plausible that we can establish this boundedness using only (A-1)–(A-4). However, assumption (A-5) is not very stringent in many applications because the compact set on which $F(\alpha,\cdot)$ is supported can be arbitrarily large and we do not need any information regarding this set for the algorithm. We can, for example, assign the payoff to be zero outside the domain space of the examples to ensure that (A-5) is satisfied.

As outlined in Section 2.1, this game with common payoff is useful in many applications that involve optimization of a regression functional. That is, the optimization problem is such that we can only get noise corrupted values of the objective function for any given values of the arguments. The attractive feature of the automata algorithm is that it does not need any gradient information. The algorithm performs a stochastic search in such a manner that it follows the gradient in an expected sense and converges to the optimal points. Such algorithms are useful in many applications such as system identification, adaptive control, pattern classification, concept learning etc. The utility of gradient-free techniques for optimization is that in many applications the objective function value at any given argument may have to be computed by an auxiliary simulation and thus gradient information may be difficult to obtain. We discuss the utility of such algorithms for pattern classification in more detail in Chapter 4.

Another special feature of this automata model is that the team consists of both FALA and CALA. This means that the functional to be optimized may be defined over some continuous and some discrete variables. As we shall see later, such problems occur naturally in concept learning.

2.5 Games of FALA

Traditionally, Learning Automata models always assumed finite action sets [NT89]. Thus there is extensive literature concerning games of FALA. (See Section 2.9 for references). In this section, we consider this important special case of a game involving only FALA.

In the model considered in Section 2.2, if we choose M=0, we get a game with only FALA players. Now the space $\mathcal{D}$ is finite and it consists of all possible tuples of action choices of automata. The payoff functions, F^l now would have only α as argument and $F^l(\alpha)$ is the expected value of the reinforcement to l^{th} FALA in response to the choice of action tuple α by the automata. Since $\mathcal{D}$ is finite, each of these payoff functions, F^l, can be represented by an hypermatrix of dimension $r_1 \times \ldots \times r_N$. (Recall that r_i is the number of actions available for the ith FALA). These N matrices, to be called payoff matrices, completely determine the payoff structure. However, these payoff matrices are *unknown* and each automaton only receives the stochastic reinforcement (β_l) after each play.

The functions g^l would now be functions of only P which consists of the action probability vectors of all the automata. The value of $g^l(P)$ is the expected value of reinforcement (to the lth FALA) conditioned on the state of the automata system being P. Now the space $\mathcal{K}$ would be the set of all possible tuples of action probability vectors (of the FALA) and $P(k) \in \mathcal{K}$, $\forall k$. As is easy to see from (2.2.9), now the function J^l would be same as F^l and hence is not relevant here.

For defining the optimal points of the game, only the first part of definition 2.2.1 is relevant now. As is easy to see from that definition, optimal points now correspond to Nash equilibria in pure strategies of this finite stochastic game. Similarly, the maximal points of the game would correspond to Nash equilibria in mixed strategies (cf. definition 2.2.2) and the modal points (cf. definition 2.2.3) are same as optimal points because J^l is same as F^l. Now the problem tackled by this model is one of learning Nash equilibria in a multiperson finite stochastic game, when the payoff is random and the game matrices corresponding to expected value of payoff are not known.

From the results presented earlier, we can conclude that the automata converge to a Nash equilibrium if each of them uses L_{R-I} algorithm. Specifically, we can specialize the results proved earlier to this case of FALA game and state it as follows.

THEOREM 2.5 *Consider a stochastic game as in Section 2.2 but with only N FALA. Suppose, as earlier, that each FALA updates its action probabilities with the L_{R-I} algorithm using its own reinforcement. Then the following are true regarding the asymptotic behavior of the algorithm.*

1. *All strict Nash equilibria in pure strategies (which would be corners of $\mathcal{K}$ corresponding to optimal points of the game) are asymptotically stable.*

2. *Any equilibrium point (of the approximating ODE) which is not a Nash equilibrium in pure or mixed strategies is unstable.* ∎

As is easy to see, this theorem is only a particularization of Theorem 2.2 to the special case of game of FALA. Thus this model would help in learning Nash equilibria by repeated plays of the game and would be useful in many multi-criteria optimization problems. An attractive feature of the method is that the algorithm is completely decentralized. The automata do not have to exchange any information. As a matter of fact, none of the automata 'knows' that it is part of a game and that its payoff is affected by what other players are doing!

An interesting application which can be formulated in this framework is the relaxation labeling problem as used in Image Processing (see [TS86b, SPT94]). Many problems in low level vision can be solved using this model. In this case, if the labeling problem satisfies a mild condition of symmetric compatibility functions, then we can show that the sufficient condition needed by Theorem 2.3 is satisfied and thus the algorithm always finds a pure maximal point [ST94] which corresponds to a *consistent labeling* in the jargon of such labeling problems.

2.5.1 Common Payoff Games of FALA

Now consider the special case of common payoff game of FALA. That is, $\beta_i = \beta$, $\forall i$. As explained in Section 2.4, now there is only one payoff function, F. Since we have only FALA here, this payoff function is defined on the finite set $\mathcal{D}$. (Each element of $\mathcal{D}$ is an N-tuple of actions). Since $\mathcal{D}$ is finite, the payoff function is represented by a hyper-matrix of dimension $r_1 \times \cdots \times r_N$, to be called the payoff matrix. Recall that the action set of jth automaton is given by $\{\alpha_{j1}, \ldots, \alpha_{jr_j}\}$, $j = 1, \ldots, N$. For any $\alpha = (\alpha_{1i_1}, \ldots \alpha_{Ni_N}) \in \mathcal{D}$, we can represent $F(\alpha)$ by the element $d_{i_1 i_2 \ldots i_N}$ of this payoff matrix. Thus, the entries of the payoff matrix are defined by

$$d_{i_1 i_2 \ldots i_N} = \text{E}[\, \beta \mid j\text{-th FALA chose } \alpha_{j i_j} \,]. \tag{2.5.44}$$

The objective of learning now is to find a 'local maximum' of this payoff function. From Remark 2.2.2, since we have only FALA here, a tuple of actions, α^*, is an optimal point of the game if the payoff at α^* is more than that at any action tuple α that differs from α^* in only one component. Thus, any optimal point corresponds to an element of the payoff matrix which is maximum over all hyper-rows that it belongs to. Such elements are called modes of the matrix. (In two dimensions, a mode corresponds to an element that is greater than or equal to every other element in its row and column).

Specifically, an element of the payoff matrix, $d_{i_1^* i_2^* \ldots i_N^*}$ is a *mode* of the payoff matrix if *all* the following inequalities hold simultaneously.

$$d_{i_1^* i_2^* \ldots i_N^*} \geq d_{i_1 i_2^* \ldots i_N^*}, \forall i_1$$
$$d_{i_1^* i_2^* \ldots i_N^*} \geq d_{i_1^* i_2 \ldots i_N^*}, \forall i_2$$
$$\cdot$$
$$\cdot$$
$$d_{i_1^* i_2^* \ldots i_N^*} \geq d_{i_1^* i_2^* \ldots i_{j-1}^* i_j i_{j+1}^* \ldots i_N^*} \; \forall i_j$$
$$\cdot$$
$$\cdot$$
$$d_{i_1^* i_2^* \ldots i_N^*} \geq d_{i_1^* i_2^* \ldots i_N}, \forall i_N$$

Now optimal points of the game correspond to modes of the payoff matrix. Also, since this is a common payoff game, by Theorem 2.4, we know that the team would converge to one of the maximal points. Further, under the

L_{R-I} algorithm, only unit probability vectors are the absorbing points of the algorithm. Thus, we can conclude that the team would converge to a mode of the payoff matrix. We can summarize this result by the following theorem.

THEOREM 2.6 *Consider a game with common payoff involving only FALA. The automata team using the L_{R-I} algorithm converges to a mode of the payoff matrix.* ∎

The payoff function F is defined over the finite set of all possible tuples of actions by the team and its value is the expectation of reinforcement. As explained earlier, a mode is like a local maximum of this function. Thus the L_{R-I} algorithm gives a decentralized way of finding such local maxima of a regression function defined over a finite set. Such a game of learning automata has been used extensively in many applications. Since we can use this model only when $\mathcal{D}$ is finite we need to quantize the variables (if any) that take continuous values. We illustrate this in the context of learning optimal pattern classifiers in Chapter 4.

However, there are applications where it is desirable that the team converges to the global maximum of F rather than a local maximum. The global maximum would correspond to the largest entry in the payoff matrix.

As briefly mentioned in Section 1.8, we can use a PLA in place of FALA and then we have available some learning algorithms that ensure convergence to global maximum. This is explained in Chapter 3 in a more general setting. Another way of ensuring that the automata team converges to the global maximum is to use estimator algorithms, such as pursuit algorithm, in place of L_{R-I}. We briefly explain below how pursuit algorithm can be used for an FALA team.

2.5.2 Pursuit Algorithm for a Team of FALA

As explained in Section 1.4.3, the main idea in estimator algorithms is to keep estimates of reward probabilities and use these in updating action probabilities. In a common payoff game, the entries of payoff matrix are what can be estimated using the reinforcement received at each instant. These estimates are denoted by $\hat{d}_{i_1 i_2 \ldots i_N}$. Recall from Section 1.4.3 that in the case of a single FALA, we had used two sets of variables, Z_i and η_i to estimate $\hat{d}_i$, the estimate of the reward probability of ith action. In that algorithm $Z_i(k)$ was the total reinforcement received till k in response to i^{th} action and $\eta_i(k)$ was the total number of times i^{th} action was selected till k. These quantities can be updated at each instant based on the action selected and reinforcement obtained. Then $\hat{d}_i$ is estimated as the ratio of Z_i and η_i. In the case of a team of FALA here, we do essentially the same thing. Here, we have one payoff matrix entry for every N-tuple of actions. Thus we now keep two hypermatrices, Z and η whose dimension would be the same as that of the payoff matrix. The element $Z_{i_1 i_2 ... i_N}(k)$ of the hypermatrix $Z(k)$ would give the total reinforcement obtained for the action tuple $(\alpha_{1i_1}, \alpha_{2i_2}, \ldots, \alpha_{Ni_N})$ till k while the element $\eta_{i_1 i_2 ... i_N}(k)$ of the hypermatrix $\eta(k)$ would give the number of times this tuple of actions is selected till k. These hypermatrices are updated at each instant and we use them to get estimates of the entries of payoff matrix. The details are as given below.

At instant k, each automaton chooses an action at random based on its current action probability vector. Suppose the jth automaton has chosen action α_{ji_j}. That is, $\alpha_j(k) = \alpha_{ji_j}$, $1 \leq j \leq N$. Let the reinforcement received at k be $\beta(k)$. Now the estimates of the elements of the payoff matrix are obtained as follows.

$$Z_{i_1 i_2 ... i_N}(k) = Z_{i_1 i_2 ... i_N}(k-1) + \beta(k)$$
$$Z_{j_1 j_2 ... j_N}(k) = Z_{j_1 j_2 ... j_N}(k-1),$$
$$\forall (j_1, j_2, \ldots, j_N) \neq (i_1, i_2, \ldots, i_N)$$

(2.5.45)

$$\eta_{i_1 i_2 ... i_N}(k) = \eta_{i_1 i_2 ... i_N}(k-1) + 1$$
$$\eta_{j_1 j_2 ... j_N}(k) = \eta_{j_1 j_2 ... j_N}(k-1),$$
$$\forall (j_1, j_2, \ldots, j_N) \neq (i_1, i_2, \ldots, i_N)$$

(2.5.46)

$$\hat{d}_{i_1 i_2 ... i_N}(k) = \frac{Z_{i_1 i_2 ... i_N}(k)}{\eta_{i_1 i_2 ... i_N}(k)}$$
$$\hat{d}_{j_1 j_2 ... j_N}(k) = \hat{d}_{j_1 j_2 ... j_N}(k-1),$$
$$\forall (j_1, j_2, \ldots, j_N) \neq (i_1, i_2, \ldots, i_N)$$

(2.5.47)

Next, for $1 \leq j \leq N$, the jth automaton calculates (based on the estimated payoff matrix) a vector, $\hat{\mathbf{E}}^j(k) = [\hat{E}_1^j, \ldots, \hat{E}_{r_j}^j]^T$, which will be used for updating the action probabilities. These vectors are obtained as follows. (Recall that α_{ji_j} is the action selected by j^{th} automaton at k).

$$\hat{E}_{i_j}^j(k) = \max_{l_i} \{\hat{d}_{l_1 l_2 ... l_{j-1} i_j l_{j+1} ... l_N}(k)\}$$
$$\hat{E}_l^j(k) = \hat{E}_l^j(k-1), \forall l \neq i_j$$

(2.5.48)

As can be seen from above, in general, for any l, $1 \leq l \leq r_j$, $\hat{E}_l^j$ is the maximum element from the estimated payoff matrix where the maximum is over all those elements whose j^{th} index is l. Thus, $\hat{E}_l^j$ can be thought of as

Games of FALA

the estimated reward probability for the lth action of the jth automaton (in the context of a team of automata). Next, for $1 \leq j \leq N$, the jth automaton updates its action probability distribution as follows.

$$\mathbf{p}_j(k+1) = \mathbf{p}_j(k) + \lambda(\mathbf{e}_{M_j(k)} - \mathbf{p}_j(k)) \qquad (2.5.49)$$

where $\mathbf{e}_{M_j(k)}$ is a unit vector with $M_j(k)$-th component unity and the index $M_j(k)$ is defined by

$$M_j(k) = \arg\max_l \hat{E}_l^j(k). \qquad (2.5.50)$$

Comparing (2.5.49) with (1.4.19) in Chapter 1, it is easy to see that $\hat{E}_l^j$ serve the same purpose here as the estimates of reward probabilities did in pursuit algorithm for a single FALA.

To sum up, the pursuit algorithm for the team is as follows. At each instant, all FALA choose actions based on their respective action probability vectors. Then the reinforcement from the environment (which is the common payoff to the team) is obtained. Knowing the actions selected and payoff, the hypermatrices Z and η are updated using (2.5.45) and (2.5.46). These are then used to update the estimates of payoff matrix entries using (2.5.47) and $\hat{E}_l^j$ are updated using (2.5.48). Finally, the action probability vectors are updated using (2.5.49).

The following theorem states the convergence result for a team of FALA using pursuit algorithm.

THEOREM 2.7 *In a common payoff game of FALA, if each automaton in the team uses pursuit algorithm as above with a sufficiently small value for the learning parameter, λ, then the team converges to the tuple of actions corresponding to the global maximum in the payoff matrix.* ∎

We have given the proof of convergence for the pursuit algorithm in the case of a single FALA in Appendix B. The proof of the above theorem is a straightforward extension of that proof and hence we omit the details.

As in the case of a single FALA, use of pursuit algorithm results in faster convergence of the team in a common payoff game also. In addition, here pursuit algorithm results in convergence to global maximum while L_{R-I} can ensure convergence only to a local maximum. However, there is a price to be paid for this improved performance. In the case of a team using L_{R-I}, the algorithm is completely decentralized. However, in the pursuit algorithm we need to obtain estimate of payoff matrix and hence, automata need to exchange information regrading actions selected by them. All estimates can be maintained centrally and this process needs to communicate with all automata. A second and more serious overhead is that of the memory needed to keep all the estimates. If each automaton has r actions then the payoff matrix has r^N

elements. Same is true of the matrices Z and η. This overhead can become prohibitive when r and N are large. Where such extra memory overhead is manageable, this algorithm is attractive because it results in faster convergence as well as convergence to global maximum. In some applications a reasonable approximation to the global maximum may be good enough. In such cases, it is possible to considerably reduce the memory overhead involved. We illustrate one such application in Chapter 6.

2.5.3 Other Types of Games

The model we considered in this chapter is a general multiperson game where different players get different payoffs. In the case of FALA game, as we have shown, the algorithm learns Nash equilibria. One of the special cases we have considered earlier is that of common payoff. In addition, there are other special types of games where automata can learn effective strategies. Some of these are indicated below.

A classical game model is the two person zero-sum game. Here the two players are in competition and one person can gain only at the expense of the other. In the context of automata games, this means that for any choice of a pair of actions by the two players if one player gets a favorable response (that is, $\beta_1 = 1$) then the other gets unfavorable response (that is, $\beta_2 = 0$). In such games, if both the automata players use L_{R-I} algorithm with sufficiently small learning parameter, then they converge to the optimal solution if the optimal solution is a saddle point in pure strategies. If the optimal solution of the game is a saddle point in mixed strategies, then automata players using the $L_{R-\epsilon P}$ algorithm can be made to converge to the optimal solution by careful choice of learning parameters [LN81, LN82].

An interesting two person non-zero sum game is the prisoner's dilemma game. If two FALA play an iterative prisoner's dilemma using the L_{R-I} scheme, then it can be shown that by interchanging the reward and penalty reinforcement, they can converge to the cooperative solution which is optimal. If the players can have knowledge of the actions selected after each iteration, use of estimator algorithms achieves the same goal at a faster rate of convergence [TS91]. These results also carry over to the N-person prisoners' dilemma [Rao84].

Another simple symmetric game played by several players is the Goore game [Tse73]. It is a cooperative game where all players have identical action sets of two actions each and the random payoff to players is based on the number of players choosing a particular action. The objective is to make the right number of players choose a particular action which results in maximum expected payoff. This game has been used to model control of group behavior and for discrete stochastic optimization of one variable. One can learn this optimal choice of actions in a Goore game using FALA. It has been shown

[TA97] that when the automata use the L_{R-I} algorithm, there is a one-to-one correspondence between the stable stationary points (of the equivalent ODE) of the algorithm and the Nash equilibria of the game.

2.6 Common Payoff Games of CALA

If we take $N = 0$ in our model, we get a game involving only CALA. Now we have $\mathcal{D} = \Re^M$ and thus all payoff functions are real valued functions defined over a finite dimensional Euclidean space. Thus, by Remark 2.2.2, the optimal points correspond to points in $\Re^M$ that locally maximize all the payoff functions. For a game of CALA, the interesting special case is that of a common payoff game.

Consider the special case of a common payoff game with only CALA. As explained earlier, the reinforcement, β, is the common payoff to all automata and there is only one payoff function, F. This function, F, is now a function of only the actions of CALA and thus is defined on $\Re^M$. Specifically, we have (see equation (2.2.1))

$$F(\mathbf{x}) = E[\beta_{\mathbf{x}} \mid \mathbf{x}] \qquad (2.6.51)$$

where $\beta_{\mathbf{x}}$ is the payoff received in response to the action tuple, $\mathbf{x}$ of the automata team.

From definition 2.2.1, the optimal points now correspond to local maxima of the real valued function, F, defined on $\Re^M$. The functions g and J would now be same and both would be functions of the action probability distributions of the CALA. Specifically,

$$g(\boldsymbol{\mu}, \Sigma) = E[\beta(k) \mid C(k) = (\boldsymbol{\mu}, \Sigma)] \qquad (2.6.52)$$

where $C(k)$ is the state of the CALA team (which consists of the means and variances of the action probability distributions of all CALA) as earlier. Now the maximal and modal points are the same and they correspond to local maxima of $g(\cdot, \Sigma)$. We can now specialize Theorem 2.4 as follows.

THEOREM 2.8 *Consider a common payoff game of CALA where all the automata use the algorithm given by (2.2.3). If the learning step-size, λ, is sufficiently small then the CALA team converges to a modal point.* ∎

As seen earlier in the analysis of the general automata game algorithm, any pure maximal point that the algorithm converges to is such that all the variances of the action probability distributions of CALA would be close to σ_L. Hence, by taking σ_L sufficiently small, in the common payoff game of CALA, the modal point to which the algorithm converges is a local maximum of g and it closely approximates an optimal point which is a local maximum of F.

Thus the class of problems tackled by a CALA team is as follows. Suppose we want to maximize a function $F : \Re^M \to \Re$ given only noisy observations

$\beta_\mathbf{x}$ such that $E[\beta_\mathbf{x} \mid \mathbf{x}] = F(\mathbf{x})$. Now, a team of CALA with M members can be used to obtain a good approximation to local maxima of F. Since only the noisy values, $\beta_\mathbf{x}$ are available and $F(\mathbf{x})$ is expectation of this (conditioned on $\mathbf{x}$), we would sometimes refer to F as the regression function.

Such optimization problems are relevant in many applications such as adaptive signal processing, pattern recognition and neural networks. As remarked earlier, the automata algorithm is a gradient-free technique for solving this stochastic optimization problem. In this CALA team algorithm we do not even estimate the gradient based on multiple observations of the value of the noise-corrupted objective function. As a matter of fact, in this algorithm, in each iteration we need only two (noisy) function values irrespective of M, the dimension of the space on which the objective function is defined. (It may be noted that for estimating the gradient in a naive manner, we would need $2M$ function values). What the analysis presented earlier says is that the algorithm essentially follows the gradient in an expected sense (even though the gradient is never estimated) thus converging to a local optimum. From the analysis given earlier, it can be shown that we can do with only one function evaluation per iteration. In the CALA algorithm, we used terms such as $(\beta_\mathbf{x} - \beta_\mu)$. Even if we take β_μ to be zero, all the results go through. Thus, one function evaluation per iteration would suffice. However, it is observed through simulations that such an algorithm takes much longer to converge.

The automata approach is useful even in deterministic problems (that is, when $F(\mathbf{x})$ is directly observable) especially if the gradient information is difficult to come by. For example, we can use this algorithm in place of back-propagation for learning the weights in feedforward neural networks. Then we do not need to calculate the gradient and thus the learning algorithm would be independent of the architecture of the network. (For example, the same algorithm would work whether we are using a feedforward or recurrent network). But the approach would be attractive mostly in situations where there is a lot of noise (e.g., the training samples are noisy).

There is one more aspect of the automata approach that is worth mentioning here. We shall see in Chapter 5 that these algorithms are easily parallelizable in a special sense. We can make the automata choose more than one action at any instant, elicit reactions for all these actions and then update the action probability distributions. This results in a good speed up of the algorithm.

The CALA team algorithm converges to the local maximum of the regression function. Hence an important question is how good such local maxima may be in different applications. As is illustrated in Section 1.6.2, by having a larger value for the initial variance, the CALA often converges to much better local maxima than the one closest to the initial μ. This is a distinct advantage for the method. In addition, it is possible to impose a random walk on the updating equations so that the algorithm can find the global maximum. Such an algorithm (similar to the global algorithms in Chapter 3) would have the same flavor as simulated annealing type algorithms and like them its rate of convergence would be slow. It is also true that in many applications the local maxima are good enough. (See section 2.7.2 and Chapter 4).

2.6.1 Stochastic Approximation Algorithms and CALA

A classical method for optimizing a real-valued function defined over $\Re^M$ using only noisy observations is that of Stochastic Approximation. A stochastic approximation scheme for solving this optimization problem without needing any gradient information is the so called Kiefer–Wolfowitz scheme. Another fundamental stochastic approximation scheme is the so called Robbins–Munro algorithm (which is essentially the same as the LMS algorithm which is popular in Neural Networks literature), where we assume that a noisy gradient value is also available. In this section we discuss the similarities (and differences) between the CALA algorithm and the stochastic approximation schemes.

Suppose we want to maximize a function[4] $f : \Re^M \to \Re$ based on noisy measurements $r(\mathbf{x})$ such that $E[r(\mathbf{x})|\mathbf{x}] = f(\mathbf{x})$. Then the Kiefer-Wolfowitz stochastic approximation algorithm is:

$$\mathbf{x}_{k+1} = \mathbf{x}_k + \eta_k \widehat{\nabla f}(\mathbf{x}_k) \qquad (2.6.53)$$

where the estimated gradient is given by

$$\widehat{\nabla f}(\mathbf{x}) = \frac{1}{2c_k} \sum_{i=1}^{M} (r(\mathbf{x} + c_k \mathbf{e}_i) - r(\mathbf{x} - c_k \mathbf{e}_i))\mathbf{e}_i \qquad (2.6.54)$$

where $\mathbf{e}_i \in \Re^M$ denotes the unit vector with i^{th} component unity and all others zero. If the step sizes η_k and c_k satisfy

$$\sum_k \eta_k = \infty; \quad \sum_k \eta_k c_k < \infty; \quad \sum_k \left(\frac{\eta_k}{c_k}\right)^2 < \infty; \quad \eta_k < \rho c_k^2 \qquad (2.6.55)$$

where $0 < \rho < \infty$, then it can be shown that the algorithm converges to a local maximum of f (both in mean square sense and with probability 1) under some conditions on the function f and the variance of noise in the observations. The usual choices for step sizes is $c_k = c_1/k^\gamma$, $\eta_k = \eta_1/k$ with $0 < \gamma < \frac{1}{2}$, $\eta_1 > 0$; $c_1 > 0$.

---

[4]In this section we use f to denote a general function which needs to be optimized based on noisy function evaluations. When we use a CALA team for the problem, this function corresponds to F. Similarly, we use $r(\mathbf{x})$ to denote the noisy function value which corresponds to $\beta_\mathbf{x}$ when we use the CALA team.

This stochastic approximation scheme needs $2M$ function evaluations per iteration and the algorithm is essentially implementing a search in $\Re^M$ for the local maximum. It is possible to considerably reduce the number of function evaluations by using some perturbation schemes using the so called Simultaneous Perturbation Stochastic Approximation (SPSA) [Spa92, Spa98]. We will not be considering the SPSA algorithms here.

Our algorithm (with an M-member CALA team) solves the same problem and can thus be considered an alternative to the classical stochastic approximation type algorithms. We need only two function evaluations per iteration irrespective of the value of M and our search is in the space of some probability distributions over $\Re^M$. In the stochastic approximation algorithm, due to the small step size, the successive points in $\Re^M$ visited by the algorithm would be close to each other and we can expect the algorithm to converge to the local optimum close to the starting point. In contrast, in our algorithm we can make long jumps because the points visited are chosen from some normal distributions. Even though the distributions used in successive iterations would be close to each other (which is due, once again, to the small step-size in the learning algorithm), the successive points in $\Re^M$ visited by the algorithm could be far away from each other. This can help the CALA team converge to a better local optimum than the one closest to the starting point; or even to the global optimum. This may be controlled through the initial variances of the action probability distributions in our algorithm. We illustrate all this through simulations in the next subsection.

Examples

We now present results on two simple function optimization problems to compare the performance of the CALA algorithm with Kiefer – Wolfowitz scheme. The first example is to find a minimum of penalized Shubert function which was considered in Section 1.6.2. This is a one dimensional problem and hence we need only one CALA. We consider it again here only to compare the CALA algorithm with stochastic approximation. We consider only the case where noise is added to the function evaluations.

For the Kiefer-Wolfowitz algorithm on this problem, we have chosen an initial step size (with $\eta_1 < 1$) in each simulation [5] and the constant $c_1 = 0.1$ in all the simulations. The step size was decreased as $\eta_k = \eta_1/k$ and we used $c_k = c_1/k^{0.25}$ for finite difference approximation in gradient computation. The results of the simulations are provided in Table 2.6.1. The table shows the typical value of the estimate of a minimum point at the end of 1000 iterations.

---

[5]For initial step size $\eta_1 \geq 1$ we encounter numerical problems in using the finite difference approximation. This is due to large value of the gradient of the penalizing term in our Shubert function outside the interval $[-10, 10]$.

Common Payoff Games of CALA

Initial Values		After 1000 Iterations	Final Function Value
x_1	Step(η_1)	x_{1000}	
3	0.7	1.502	-3.5876
3	0.7	4.46	-3.7498
3	0.5	3.46	-2.7111
8	0.7	-2.80	-2.7269
8	0.7	4.48	-3.7253
8	0.5	4.46	-3.7498
-6	0.7	4.47	-3.7341
-6	0.7	-3.76	-2.5996
-6	0.5	1.51	-3.587

Table 2.6.1. Simulation results while using stochastic approximation algorithm, with noise added to the Shubert function evaluations. $c_1 = 0.1$ in all these simulations.

Initial Values			After 8000 Iterations		Final Function Value
$\mu(0)$	$\sigma(0)$	Step(λ)	$\mu(8000)$	$\phi(\sigma(8000))$	
4	6	2×10^{-4}	2.534	0.01	-3.578
4	10	2×10^{-4}	0.4038	0.01	-12.87
8	3	3×10^{-4}	5.36	0.85	-8.5
8	5	2×10^{-4}	6.72	0.01	-12.87
-10	5	2×10^{-4}	-7.1	0.01	-8.5
-10	6	2×10^{-4}	-5.87	0.01	-12.87

Table 2.6.2. Simulation results for the CALA algorithm *with noise* added to the Shubert function evaluations.

(After 1000 iterations there is very little change in successive iterates). The table contains results for various initial points. For comparison, we reproduce the results (from Section 1.6.2) of the CALA algorithm for this problem in Table 2.6.2.

It may be observed that the stochastic approximation algorithm always converges to a *local minimum* in these simulations. Convergence to a global minimum may be possible if the algorithm is started in the region close to a global minimum. In contrast, by increasing the initial variance in the CALA algorithm we often converge to the global minimum or a better local minimum. Above results indicate that the initial search behavior of the CALA algorithm is better. This is generally due to the fact that we search in the space of probability distributions and thus can take occasional long jumps in the parameter space.

In our next example we choose a function of four variables to see how the performance of the two algorithms scale with the dimension of the problem. The problem is to maximize

$$f(x_1, x_2, x_3, x_4) = \frac{1}{2\sqrt{2\pi}} exp\left(-\frac{x_1^2 + x_2^2 + x_3^2 + x_4^2}{8}\right)$$

This is a simple function with a unique maximum at the origin. The stochastic approximation algorithm (with $\eta_1 = 10$ and $c_1 = 1$) started at $(2, 2, 2, 2)$ converges close to the minimum in about 1700 iterations. The CALA algorithm with initial values $\mu_i = 2$, $\sigma_i = 8$ and step-size $\lambda = 0.01$ also always converges close to the minimum in about 5500 iterations. This being a four dimensional problem, the Kiefer-Wolfowitz scheme needs 8 function evaluations per iteration while the CALA algorithm needs only two. Thus in terms of number of function evaluations the CALA algorithm does better. In our simulations on many problems we find this to be true.

Optimization algorithm when noisy gradient is available

Another stochastic approximation scheme is the Robbins–Munro algorithm or the LMS algorithm. Suppose we have access to a noisy version of the gradient. That is, suppose we can observe values of $\mathbf{r}(\mathbf{x})$ such that $E[\mathbf{r}(\mathbf{x})|\mathbf{x}] = \nabla f(\mathbf{x})$. (Note that $\mathbf{r}(\mathbf{x})$ is a vector of same dimension as $\mathbf{x}$). Then the Robbins–Munro algorithm is given by

$$\mathbf{x}_{k+1} = \mathbf{x}_k + \eta_k \mathbf{r}(\mathbf{x}) \qquad (2.6.56)$$

where the step-size satisfies $\sum_k \eta_k = \infty$; $\sum_k \eta_k^2 < \infty$. A typical choice for η_k is $\eta_k = \frac{\eta}{k}$, $\eta > 0$. Once again one can show that the algorithm converges to a local maximum of f under some assumptions. This algorithm, in general, is more robust and faster than the earlier scheme because it uses the gradient information (instead of being forced to estimate a noisy gradient using finite difference approximation).

The automata algorithm for a common payoff game of CALA is a useful optimization algorithm that does not need any gradient information. Let us call this as the CALA-team algorithm. Now, a natural question is: if a noisy gradient is available, can the automata algorithm use this information effectively? The answer to this question is in the affirmative. Suppose, as above, for any $\mathbf{x} \in \Re^M$, we can have access to $\mathbf{r}(\mathbf{x})$ such that $E[\mathbf{r}(\mathbf{x})|\mathbf{x}] = \nabla f(\mathbf{x})$. Now we employ a team CALA where we supply $\mathbf{r}(\mathbf{x})_i$, the i-th component of $\mathbf{r}(\mathbf{x})$, as β_i, the payoff or reinforcement to the i-th automaton in response to the action tuple $\mathbf{x}$. Recall that we need two reinforcements for the CALA team: β_i and β_i'. These are now given by

$$\beta_i = \mathbf{r}(\mathbf{x})_i, \quad \text{and} \quad \beta_i' = \mathbf{r}(\boldsymbol{\mu})_i$$

We use the above for β_i and β'_i in equations (2.2.3) and change the equations given by (2.2.4) to the following:

$$\mathcal{F}_1(\mu, \sigma, x, \beta, \beta') = \beta$$
$$\mathcal{F}_2(\mu, \sigma, x, \beta, \beta') = \left(\frac{\beta^2 - \beta'^2}{\phi(\sigma)}\right)\left[\left(\frac{x-\mu}{\phi(\sigma)}\right)^2 - 1\right] \qquad (2.6.57)$$

Using the same proof as in the original CALA algorithm we can show that now the algorithm converges to a local maximum of f (if $\mathbf{r}(\mathbf{x})$ is a noisy gradient of f).

We call this the *modified-CALA* algorithm. This would be an automata alternative to the classical LMS algorithm. In the next subsection we illustrate the performance of this algorithm for linear system identification under measurement noise.

2.7 Applications

In this section we briefly outline two applications of the automata game described in this chapter. In both cases we consider a common payoff game and thus both are examples of solving stochastic optimization problems using automata teams. The first application is in System Identification and here we would employ a team of CALA. The second application is in Concept Learning where we employ a team consisting of both FALA and CALA. We would be exploring some more applications in greater detail in Chapter 4.

2.7.1 System Identification

The system identification problem is that of finding the 'best' model from a parameterized class of chosen models to represent a given system based on training data of input-output pairs. For static systems, both the input and output spaces can be thought of as Euclidean spaces of appropriate dimension and identification is a function learning problem. Given a parameterized class of functions (e.g., a neural network of a specific architecture) we need to estimate the parameters (e.g., weights in the neural network) from the given training data (using, e.g., backpropagation algorithm or the CALA-team). From the earlier discussion it is easy to see that the CALA-team algorithm would be useful especially if the training data are noisy. The identification problem for dynamical systems also has the same flavor except that the class of models chosen would be dynamic. While this problem is extensively investigated for linear systems, for nonlinear systems the problem is rather difficult. The main difficulty is in choosing a good parameterized class of nonlinear dynamical systems using which one can efficiently learn a good approximation to the unknown system.

For any class of systems, once a parameterized model is chosen, the identification problem is one of identifying the parameter values to optimize a performance index. For static systems this could be the instantaneous squared error (between system and model outputs) while for dynamic systems we may want to average this squared error over a moving time window of, say, the last 100 time instants. In all these cases we can use the CALA-team algorithm as follows. We will have a team of as many automata as there are parameters. The choice of actions by each of the automata results in a specific parameter vector. The measured performance index with that parameter vector is what is supplied as the common payoff to all the automata. The algorithm can then converge to a local optimum of the performance index.

The main advantage of using the CALA team algorithm is that the learning algorithm itself is completely independent of the system models chosen and the performance index. (For example, the algorithm is as easily applied for learning nonlinear models as for linear models, mainly because it does not need any gradient information). Also, due to the probabilistic nature of the search effected by CALA it is often possible to go to global (or better local) optima than is possible with gradient following methods such as stochastic approximation.

In the remaining part of this section we describe how a CALA team can be used for identifying a linear system. We consider the case where the training data is noisy and, further, the order of the system may or may not be known. Since the class of models considered is linear it is possible to obtain gradient of the performance index also. Hence we show results obtained with both the CALA team algorithm as well as modified-CALA algorithm. Since this is a linear system identification problem, the classical LMS algorithm is very efficient for this problem. Hence we also compare our results with those obtained with LMS.

In our example problem, the specific system to be identified has the transfer function given by

$$H(z^{-1}) = \frac{0.05 - 0.4z^{-1}}{1.0 - 1.1314z^{-1} + 0.25z^{-2}} \qquad (2.7.58)$$

We need to estimate this transfer function using only input-output data. We explore both sufficient order and reduced order modeling. In the former case we assume that the order of the system is known and thus we use the class models given by

$$H(z^{-1}) = \frac{b_0 + b_1 z^{-1}}{1.0 + a_1 z^{-1} + a_2 z^{-2}}. \qquad (2.7.59)$$

This class of models is parameterized by four parameters, namely, b_0, b_1, a_1 and a_2. The identification here involves estimation of the four parameters to optimize the chosen performance measure.

Applications

In the case of reduced order modeling, we choose the model as a first order system given by

$$H(z^{-1}) = \frac{b}{1.0 - az^{-1}}. \qquad (2.7.60)$$

This class of models is parameterized by two parameters, namely, b and a. Here, identification involves estimation of the two parameters.

In the case of sufficient order modeling the optimization is over $\Re^4$ and in the case of reduced order modeling it is over $\Re^2$. Thus, for the former case we use a CALA team with four automata while in the latter case we use a team with two CALA. In each case we considered identification in the presence of noise. For simulating the noisy case, we added *iid* zero-mean Gaussian noise with $\sigma = 0.1$, to the output of the system given by (2.7.58).

The performance index chosen for the simulations here is the mean of the squared difference between the outputs of the system and the model, over the previous 100 time instants. The common reinforcement given to the CALA team is the normalized value of this performance index. (Here we want to minimize the expected value of the error while the CALA algorithm as presented earlier is for maximization. This is easily achieved by changing the sign in the update equations). During training, the input to the system is an *iid* sequence uniform over the interval $[0, 1]$.

As stated earlier, since we are considering linear models, we can also derive an analytical expression for the gradient of the performance index. Hence we used the modified CALA algorithm also (see eqn. 2.6.57). Here the reinforcement would be the corresponding gradient value. It may be noted here that in this case, different automata get different payoffs.

We present results showing how the error between system output and the model output decreases as learning proceeds. These are shown in Figs. 2.1–2.2. Fig. 2.1 shows the learning curves for CALA, modified-CALA and LMS algorithms for the case of sufficient order model while Fig. 2.2 does the same for the case of reduced order modeling. In both cases the training data is corrupted with noise. It is easy to see that the automata team learns well. In the case of sufficient order modeling, the asymptotic error becomes equal to the variance of the additive noise. In the case of reduced order modeling, even the best model, from the set of models considered, cannot achieve zero error. The residual error is a combination of that due to unmodelled dynamics and the noise variance. As can be seen from the figure, the CALA team algorithm is somewhat slower than the modified-CALA algorithm. However, the latter makes use of the gradient information and hence is applicable only for the linear system identification. The performance of CALA team is comparable to that of LMS. This performance of CALA team is good when we consider the fact that the same algorithm can be used for nonlinear system identification also.

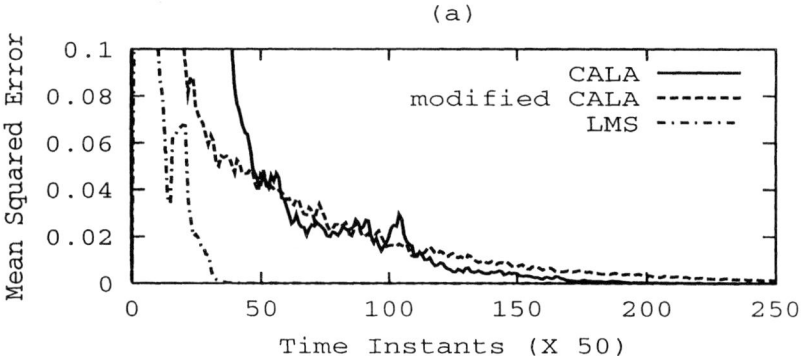

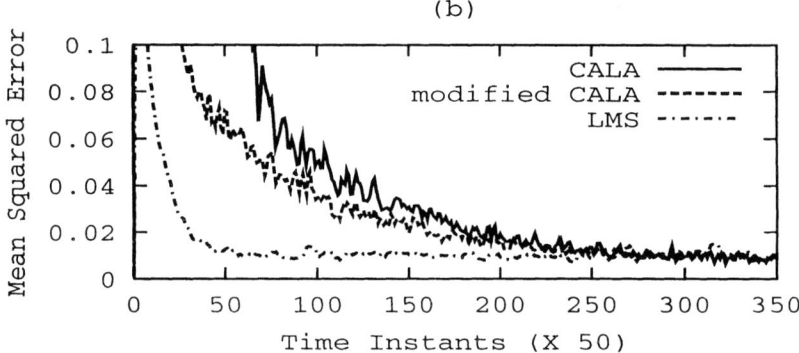

Figure 2.1. Performance of the CALA and Modified CALA algorithms on the Linear System Identification problem in the case of sufficient order modeling (a). without noise, and (b). with noise

2.7.2 Learning Conjunctive Concepts

In this section, we illustrate an application of the hybrid automata game described in this chapter to the problem of learning conjunctive concepts. A detailed description of this algorithm for concept learning can be found in [RS97] where we also discussed how such teams of FALA and CALA can efficiently learn certain classes of disjunctive concepts.

The problem is to learn a concept in the form of a conjunctive logic expression (over a finite set of attributes) given a set of positive and negative examples. Some of the attributes would be *nominal* while the others could be *linear*.

Each nominal attribute can assume only one of finitely many values and there is no algebraic structure on the set of values possible for a nominal attribute. Thus, while we can check whether the value of an attribute belongs

Applications

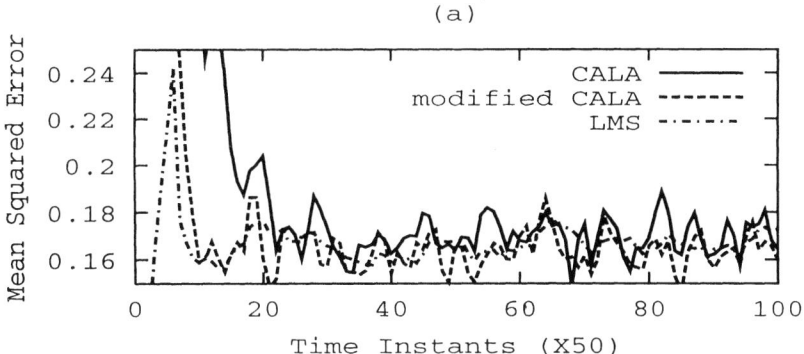

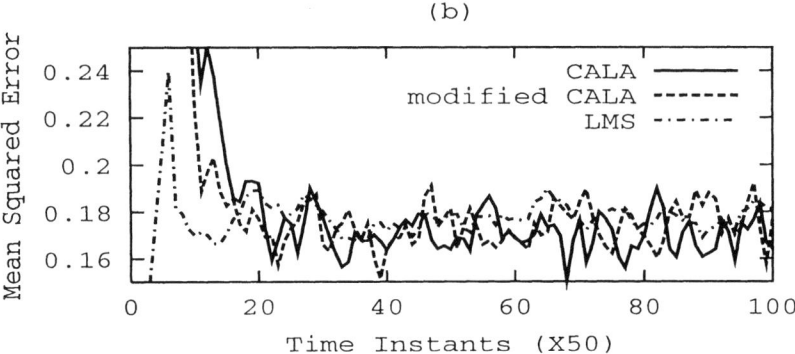

Figure 2.2. Performance of the CALA and Modified CALA algorithms on the Linear System Identification problem in the case of reduced order modeling (a). without noise, and (b). with noise

to some given subset of values, it is meaningless to add two values or to multiply an attribute value with a constant etc. Examples of nominal attributes could be color of an object, make of a car, nationality of a person etc. Each linear attribute takes values from a compact interval in $\Re$. Since values of linear attributes are real numbers, we can, if we wish, perform many algebraic operations on them.

All the training examples will be described by tuples of values for these attributes along with information regarding its classification, that is, whether it is a positive example or a negative example. In the problem we consider here, there may be classification noise in the training sample and thus what is given as a positive example, may not, in fact, be a positive example and so on. Like in any standard pattern recognition problem, given such a training sample of examples, the objective is to find the 'correct' concept by searching over the space of allowable concepts. Here we consider conjunctive logic expressions (defined below) as the space of possible concepts.

Let the attributes chosen for the domain be Y_i that take values from sets V_i, $i = 1, \ldots, N$.

DEFINITION 2.7.1 *A concept expressed as* $atom_1 \wedge \ldots \wedge atom_s$ *is called a* simple conjunctive concept *if each* $atom_i$ *is a logic expression of the form* $[Y_i \in v_i]$, $v_i \subset V_i$, *and, further, whenever* Y_i *is a linear attribute, then* $v_i \subset \Re$ *is a compact interval.*[6]

Now the learning problem is to search over the space of all possible simple conjunctive concepts (by making use of the training set of examples) to obtain the 'correct' concept. For this we have to first define the 'correct' concept.

Let $\mathcal{X}$ denote the set of all tuples of possible attribute values; that is, $\mathcal{X} = \Pi_{j=1}^{N} V_i$. It may be noted here that V_i is some finite set if Y_i is a nominal attribute and V_i is a closed interval on the real line if Y_i is a linear attribute. Let $\mathcal{Y} = \{0, 1\}$. Now, each example in the training set is a pair $(\mathbf{z}, y) \in \mathcal{X} \times \mathcal{Y}$. In an example $(\mathbf{z}, y)$, $y = 0$ indicates that $\mathbf{z}$ is a negative example and $y = 1$ indicates that $\mathbf{z}$ is a positive example. We assume that the training set examples are drawn in an independent and identically distributed (*iid*) fashion using an *unknown* probability distribution $\mathcal{P}_{xy}$ on $\mathcal{X} \times \mathcal{Y}$. Since this (unknown) distribution is over $\mathcal{X} \times \mathcal{Y}$, it amounts to saying that the the training sample is noisy. (For example, it is possible that the two pairs $(\mathbf{z}, y)$ and $(\mathbf{z}, y')$ both have positive probability even if $y \neq y'$). Let $\mathcal{H}$ denote the set of all possible simple conjunctive concepts. Each $h \in \mathcal{H}$ is a logic expression representing a simple conjunctive concept. Hence, as a notation, for any $\mathbf{z} \in \mathcal{X}$, $h(\mathbf{z})$ is 1 if $\mathbf{z}$ satisfies the concept h; otherwise $h(\mathbf{z})$ is zero. Note that any tuple $(v_1, \ldots, v_N)$ where $v_i \in V_i$ and v_i is a closed interval of real line if Y_i is a linear attribute, uniquely specifies a simple conjunctive concept. Hence $\mathcal{H}$ can be viewed as the set of all such tuples. For any $h \in \mathcal{H}$, define

$$R(h) = \text{Prob}[h \text{ correctly classifies a random example from } \mathcal{X}] \quad (2.7.61)$$

Now we can define the *correct* concept to be $h^* \in \mathcal{H}$ where R attains global maximum.

It may be noted that given a h, we cannot calculate $R(h)$ because we do not know the probability distribution $\mathcal{P}_{xy}$. We define another function $\hat{R}$ on $\mathcal{H}$ by

$$\begin{aligned} \hat{R}(h) &= 1 \quad \text{if } h(\mathbf{z}) = y \text{ for a random training example } (\mathbf{z}, y) \\ &= 0 \quad \text{otherwise} \end{aligned} \quad (2.7.62)$$

---

[6]This is called a simple conjunctive concept, because in a general conjunctive concept one may want v_i corresponding to linear attributes to be union of intervals as well.

Applications 95

Since training examples are drawn in an iid manner under the distribution $\mathcal{P}_{xy}$, it is easy to see $E[\hat{R}(h)] = R(h)$. For any h we can calculate $\hat{R}(h)$ using the training set. Thus the problem is to maximize the function $R(h)$ based on measurements of $\hat{R}(h)$.

Now, to use the automata team algorithm for this problem we need to represent each concept with a pair $(\boldsymbol{\alpha}, \mathbf{x})$ (where components of $\boldsymbol{\alpha}$ come from discrete sets and $\mathbf{x}$ is a real vector). Then, given an example $(\mathbf{z}, y)$ (where $\mathbf{z}$ is a tuple of attribute values and y is the classification as given in the training sample) we can supply $\hat{R}(h)$ as the common reinforcement for the choice of action tuple $(\boldsymbol{\alpha}, \mathbf{x})$ (which corresponds to concept h). Thus $F(\cdot, \cdot)$ (defined by (2.4.39)) would become the expected value of $\hat{R}$ and the team can learn h^*. Hence, in designing an automata team to solve the concept learning problem we need to answer two questions: (i). How do we formulate an automata team so that every tuple of actions of the team uniquely corresponds to an element of our concept space?, (ii). Since the automata team converges to optimal points of $F(.,.)$, are there optimal points other than h^*? (It is easy to see that h^* would be an optimal point).

Suppose the concept learning problem has N nominal attributes and M linear attributes. Every simple conjunctive concept is uniquely represented by a tuple $(v_1, \ldots, v_{N+M})$, $v_i \subset V_i$, $\forall i$. Further, for $N+1 \leq i \leq N+M$, the set v_i is an interval, say, $[c_i, d_i]$. Let n_i be the number of possible values for the ith nominal attribute. Then, given any conjunctive concept $(v_1, \ldots, v_{N+M})$, we can represent each v_i, for $1 \leq i \leq N$, by an n_i bit Boolean vector; and we can represent each v_i, for $N+1 \leq i \leq N+M$, by two real numbers. Hence, if we have an automata team with $2M$ CALA and $n_1 + \ldots + n_N$ FALA, where each FALA has two actions: {YES, NO}, then every tuple of actions of the team will represent a unique simple conjunctive concept.

The actions chosen by this team of automata will be a tuple of $n_1 + n_2 + \ldots + n_N + 2M$ elements. Let us denote this action tuple by $\mathbf{q} = (\boldsymbol{\alpha}, \mathbf{x})$ where $\boldsymbol{\alpha} = (\alpha_{ij}, j = 1, \ldots, n_i; i = 1, \ldots, N)$, with $\alpha_{ij} \in$ {YES, NO} being the action chosen by the FALA representing the j-th value of the i-th nominal attribute, and $\mathbf{x} = (x_{il}, l = 1, 2; i = 1, \ldots, M)$, with x_{il} being the action chosen by the CALA learning one end point(left or right depending on whether $l = 1$ or 2 respectively) of the i-th linear attribute. The action tuple $\mathbf{q}$ corresponds to the conjunctive concept $(v_1, \ldots, v_{N+M})$, $v_i \subset V_i$, $\forall i$, where

$$v_i = \begin{cases} \{z_{ij} | \forall j \, s.t. \, \alpha_{ij} = YES, 1 \leq j \leq n_i\} & i = 1, \ldots, N \\ [x_{i1}, x_{i2}] & i = N+1, \ldots, N+M \end{cases}$$
(2.7.63)

in which we have used the notation that V_i, the set of possible values for attribute Y_i, is given by $V_i = \{z_{i1}, \ldots z_{in_i}\}$.

We use this team of automata (consisting of both FALA and CALA) in a game with common payoff for learning the correct concept. At each instant, each of the automata chooses an action. Because of the way we designed the automata team, every tuple of actions chosen by the team corresponds to a simple conjunctive concept. We then generate the two reinforcements needed for the algorithm by checking whether or not the classification by the concept corresponding to the appropriate action tuple matches with that given in the training sample. For the reinforcement β (cf. equation 2.2.3) the action tuple would be the actions chosen by the automata; and for β' the action tuple would consist of the actions chosen in case of FALA and the means of the action probability distributions in case of CALA. We shall refer to this algorithm as CLearn.

The application of the automata team for concept learning illustrates well the general structure of how one employs an automata system to solve a stochastic optimization problem as explained in Section 2.1. The space over which we want to search is the set of all conjunctive concepts, $\mathcal{H}$. We designed an automata system (a common payoff game) so that choice of action by each of the automata results in choice of a point in $\mathcal{H}$. The natural parameterization for $\mathcal{H}$ needs some parameters to be Boolean while others are real. Thus the performance index (defined over $\mathcal{H}$) that we seek to optimize is defined over some discrete and some continuous variables. However, we are able to easily handle this stochastic optimization problem by employing some FALA and some CALA together.

From the results of Section 2.4, we know that the automata team converges to one of the optimal points of the payoff function of this game. Hence the next question to be answered is whether the optimal points of the payoff function are a good approximation to the correct concept h^*.

It is proved in [Raj96, RS97] that (under some mild conditions on the probability distribution $\mathcal{P}_{xy}$) the correct concept h^* would be an optimal point of the game and, further, any action tuple which does not correspond to h^*, would be an optimal point if and only if it corresponds to a simple conjunctive concept where the subset for one of the nominal attribute is null set. Since the correct concept will not contain null sets, (because otherwise there would not be even one positive example), this result means that we can effect correct learning.[7]

The above result is true under any general classification noise. Specifically, let ν_z denote the probability with which the classification label of an instance z is corrupted before being given to the learning system. The above result that

---

[7]For example, we can ensure correct learning by algorithm CLearn by making it automatically loop back till the concept learnt has no null sets. Though it cannot be proved that such a procedure always terminates, in our extensive empirical studies, algorithm CLearn always converged to correct concept and never needed such looping back.

h^* is the only optimal point (modulo null sets) is valid if $\nu_z < 0.5, \forall \mathbf{z}$ [Nag97]. Thus, this algorithm is effective even when the noise corrupting the examples is not uniform.

The algorithm CLearn is a strictly incremental learning algorithm. It need not store any examples or any other detailed statistics of the set of examples. This is a distinct advantage compared to other concept learning algorithms that can handle noisy examples. Further, our algorithm can be implemented in a distributed fashion. Though here we considered the problem of learning conjunctive concepts only, this technique is useful for learning richer concept classes also [Raj96].

Simulation Results

In this section, we present empirical studies of Algorithm CLearn on three examples. Two are problems taken from the Machine Learning Database maintained at UCI [MA94] and the third one is a synthetic problem.

For comparison purposes, we have implemented a version of the decision tree based algorithm[FI92] that can handle both nominal and linear attributes effectively. We refer to this as Algorithm DTree. This is among the best algorithms for handling both nominal and linear attributes and learning under classification noise.

Problem 1

We first consider the popular Iris Plants domain. There are three types of Iris plants, namely, Iris-setosa, Iris-versicolor and Iris-virginica. Each plant is characterized by four linear attributes *viz.*, petal width, petal length, sepal width and sepal length. It is known [DH73] that the class Iris-setosa is linearly separable from the other two and the other two are not linearly separable. We consider the nonlinear problem of classifying whether or not a given plant is Iris-virginica.

The training set here consists of 150 examples (taken from the Machine Learning databases at University of California, Irvine). we divide the data into two sets and use the first one as training set and the second as test test. The division is done carefully to keep the distribution of classes same in both sets. We study the performance of the algorithms by adding noise to the training set externally. The sequence of examples for algorithm CLearn is generated by uniform sampling from the training set.

We also train the algorithms with a set of examples (called duplicated training set) containing five copies of each example in the original training set. Hence, the duplicated training set has size five times that of the original set. This is done to illustrate the difference between incremental algorithms such as CLearn and non-incremental algorithms such as DTree. Since the duplicated training set contains no new information, the output of the algorithms would

Noise %	λ	CPU Time Secs.	Error rate %
0	0.005	5	2.6
5	0.005	7	4.0
10	0.005	8	4.0
20	0.005	11	4.0
30	0.005	16	5.3
40	0.004	28	5.3

Table 2.7.3. Performance of Algorithm CLearn on Iris Database

be the same. However, the time taken by non-incremental algorithms increases sharply because here the time depends on the total number of examples to be handled.

The results of simulation of Algorithm CLearn is given in Table 2.7.3. The columns in the table indicate % classification noise added to the training set, value of parameter λ in the algorithm, CPU time taken for learning and the final error rate on the test set after learning. Table 2.7.4 presents the results for Algorithm DTree. In addition to noise and error rate, Table 2.7.4 shows the size of the learnt decision tree in terms of the number of nodes in the learnt tree. In general, large trees could mean that the learnt tree has poor generalization and hence large error on the test set. In this table, the column CPU Time-I refers to the time taken by the algorithm on the original training set and CPU Time-II refers to the time taken on the duplicated training set. It may be noted that the final output of this algorithm is same on both training sets because the duplicated set contains no new examples. We indicate time taken for both case because, for nonincremental algorithms such as DTree the learning time increases sharply with the number of examples even if many examples may be redundant. It may be noted that there is no change in the execution time for algorithm CLearn, since at each instant the next example is drawn uniformly from the training set and that is why Table 2.7.3 contains only one CPU time.

Problem 2

We next consider the Wine recognition data from the UCI ML Database. These data are the results of a chemical analysis of wines grown in the same region but derived from three different cultivars. The analysis determined the quantities of 13 constituents found in each of the three types of wines and these are provided as the values of 13 linear attributes. We consider the problem of classifying the first type of wine from the other two.

The domain consists of 178 examples. We divide the data into two sets of 103 and 75 examples and use the first one as training set and the second as

Applications

Noise %	Nodes	CPU Time-I Secs.	CPU Time-II Secs.	Error rate %
0	4	0.5	2.0	5.3
5	8	0.5	3.0	8.0
10	17	1.0	9.0	8.0
20	23	2.0	10.0	24.0
30	31	2.5	12.0	32.0
40	46	2.5	13.0	48.0

Table 2.7.4. Performance of Algorithm DTree on Iris Database

Noise %	λ	CPU Time Secs.	Error rate %
0	0.004	4	2.6
5	0.004	6	4.0
10	0.004	8	6.6
20	0.004	10	6.6
30	0.004	13	8.0
40	0.004	18	12.0

Table 2.7.5. Performance of Algorithm CLearn on Wine Database

Noise %	Nodes	CPU Time-I Secs.	CPU Time-II Secs.	Error rate %
0	6	2.5	11.0	6.6
5	8	3.0	14.0	10.6
10	12	4.0	17.0	16.0
20	22	5.0	21.0	32.0
30	30	7.0	32.0	42.6
40	42	10.0	42.0	58.6

Table 2.7.6. Performance of Algorithm DTree on Wine database

test test. As in the previous problem, we train the algorithms with the duplicated training set having size five times that of the original set. We study the performance of the algorithms by adding noise to the training set externally.

The results of simulation of Algorithm CLearn are given in Table 2.7.5. Table 2.7.6 contains the results for Algorithm DTree. The column headings are as described in the previous problem.

Noise %	λ	CPU Time Secs.	Error rate %
0	0.005	10	4.0
5	0.005	13	5.0
10	0.005	17	5.0
20	0.005	20	6.0
30	0.005	23	8.0
40	0.004	28	11.0

Table 2.7.7. Performance of Algorithm CLearn on Problem 3:- Case 1

Problem 3

In this problem, we consider a synthetic domain to illustrate the ability of algorithm CLearn to learn in the presence of both nominal and linear attributes. Let the domain be characterized by 2 nominal and 2 linear attributes. Denote them by Y_1, Y_2, Y_3 and Y_4 respectively. The nominal attributes Y_1 and Y_2 take values in $\{A, B, C, D\}$ and the linear attributes Y_3 and Y_4 from $[0.0, 5.0]$. Let the target conjunctive concept be

$$[\{Y_1 \in \{A, C\}\} \wedge \{Y_2 \in \{B, D\}\} \wedge \{Y_3 \in [2.0, 4.0]\} \wedge \{Y_4 \in [2.0, 4.0]\}]$$

A fixed number of pre-classified examples is generated randomly according to a predefined probability distribution for training the algorithms. For testing purposes, we generate another set of examples with respect to the same probability distribution. The sets are generated in such a way that the number of positive and negative examples is approximately equal in each set.

The simulation results are presented below in a format similar to previous problems. However, instead of duplicating the training set, we present simulations on two training sets of 100 and 500 random examples respectively. In addition to showing the computational advantage of incremental learning, the results are used to study the generalization behavior of the algorithms.
Case 1: Training set size = 100; Test set size = 100;
See Table 2.7.7 and Table 2.7.8.
Case 2: Training set size = 500; Test set size = 100;
See Table 2.7.9 and Table 2.7.10.

Comments on Simulation Results

From the simulation results, it is clear that the automata team algorithm achieves good learning. The final error rate of the concept learnt using CLearn is always smaller than that learnt using DTree. This is particularly noticeable with noisy examples. The automata algorithm delivers acceptable level of performance even under 40% classification noise. However, the performance of

Applications

Noise %	Nodes	CPU Time Secs.	Error rate %
0	10	0.5	4.0
5	12	1.0	6.0
10	16	1.0	9.0
20	32	2.0	19.0
30	41	2.0	23.0
40	43	3.0	29.0

Table 2.7.8. Performance of Algorithm DTree on Problem 3:- Case 1

Noise %	λ	CPU Time Secs.	Error rate %
0	0.005	14	2.0
5	0.005	17	2.0
10	0.005	20	3.0
20	0.005	24	3.0
30	0.005	29	4.0
40	0.005	35	6.0

Table 2.7.9. Performance of Algorithm CLearn on Problem 3:- Case 2

Noise %	Nodes	CPU Time Secs.	Error rate %
0	15	2	3.0
5	17	5	4.0
10	24	18	8.0
20	39	28	14.0
30	55	44	22.0
40	61	58	28.0

Table 2.7.10. Performance of Algorithm DTree on Problem 3:- Case 2

DTree algorithm deteriorates rapidly with noise in the training examples. Such ability to tackle noise is the main strength of automata based methods. The time taken by the automata algorithm is more than that by DTree though the difference is not very large. A feature to be noted here is that the automata algorithm is strictly incremental while DTree is not. That is why, the time taken by DTree algorithm sharply increases when examples are duplicated. In general, for nonincremental algorithms, the time taken by the algorithm increases sharply with increasing number of examples, whereas for incremental learning algorithms such as the ones considered here, the time taken to learn does not depend so critically on the total number of examples. This is another distinct advantage of the automata algorithm.

2.8 Discussion

In this chapter we have presented one method of deploying together a number of automata for solving a problem. The model considered is a general stochastic game with unequal payoffs to different participants. Some of the (automata) players can have finite action sets while others can choose actions from real line. Using the general method of deriving an ODE approximation to the learning algorithm, some useful asymptotic properties of the learning algorithm are derived. For the special case of a common payoff game, it is shown that the algorithm achieves a close approximation to the optimal points of the game.

We have also considered the special case of games of only FALA. Use of L_{R-I} algorithm by each automaton ensures that the algorithm learns Nash equilibria. This is a decentralized learning technique and each automaton player is completely unaware of the others involved in the game. Similar results are available even when the automata use more general absolutely expedient algorithms [PST94]. Some details of such general nonlinear learning algorithms are given in the next chapter. In the special case of a common payoff game of FALA, L_{R-I} algorithm can learn only a mode of the reward matrix which is a local maximum. We have also explained pursuit algorithm for the common payoff game which results in both faster convergence and convergence to global maximum, but at the expense of some memory overhead. Also, unlike the L_{R-I} algorithm, the pursuit algorithm is not a decentralized technique.

Another special case considered is the common payoff game of only CALA. This algorithm finds local maxima of a function based on noisy function values and without needing any gradient information. The performance of CALA team is compared with stochastic approximation, a general technique for such optimization and the strengths of the CALA algorithm are brought out. In cases where a noisy gradient value is also available, a modification of the CALA team algorithm is presented which can use such information effectively. The general philosophy of CALA team algorithm is similar to the class of simultaneous perturbation stochastic approximation methods [Spa92, Spa98].

We have also illustrated the utility of all these algorithms on a few simple problems. We have shown how CALA team can be used for System Identification. Though we considered only linear models, the same algorithm is also useful for learning nonlinear models also because the algorithm per se does not make any assumptions on the model. For example, this algorithm is also effective in learning some recurrent neural network models called memory neuron networks [Mag98]. The second application considered, namely learning simple conjunctive concepts from (noisy) examples, illustrated another special feature of the automata algorithm. In this problem, the objective function to be maximized is defined on some discrete and some continuous variables. The automata game model with some FALA players and some CALA players is very effective at tackling such optimization problems.

The team of automata model discussed in this chapter can be used as a building block for designing more complex systems. The next chapter considers such complex learning systems synthesized by putting several teams together into networks of automata.

2.9 Supplementary Remarks

A single learning automaton is a simple method to learn one optimal action out of a given set. Hence in most realistic problems, it is often necessary to work with a collective of automata. It was realized very early that games of automata yield useful models. The first such model was that of fixed structure automata involved in a two person zero-sum game [KT63, Tse73]. Early studies of two-person games of FALA (or what were called variable structure automata) are in [CS69, VN74]. It was proved in [LN82] that automata algorithms can achieve the value of the game in 2-person zero sum games. FALA involved in a common payoff game were investigated in [TR82, NW83, TR84]. A good account of all these results can be found in the book by Narendra and Thathachar [NT89]. FALA involved in multiperson games with multiple payoffs were considered in [SPT94] where it is proved that the algorithm learns Nash equilibria. This is generalized to the case of a team of automata consisting of both FALA and CALA in [RS99]. The main convergence result proved in this chapter is from [SPT94] as generalized in [RS99]. A common payoff game of FALA using an estimator algorithm (which uses the same procedure as here for obtaining all estimates but uses a slightly different method for updating action probabilities) is first proposed in the context of pattern recognition in [TS87]. An FALA team using pursuit algorithm is discussed in [MT89, TS91]. Some of the recent results on games of FALA are in [BL99, ZBL99]. The CALA team was first investigated in [San94]. The comparison between CALA team and stochastic approximation is also found in [San94]. A good source for Stochastic approximation algorithms is [KY97, BB99]. Application of CALA team to system identification is from [Mag98]. The automata algorithm for concept learning discussed here is from [RS99]. Similar automata algorithms for learning many rich concept classes are discussed in [Raj91, Raj96, RS97].

Chapter 3

FEEDFORWARD NETWORKS

3.1 Introduction

Chapter 2 considered a number of Learning Automata (LA) involved in a general stochastic game. This is an example of several LA being employed to solve a learning problem. In the special case of a common payoff game, the LA are said to form a team because they share common interests. However, as seen in the previous chapter, each LA need not be 'aware' that it is part of a joint effort. Thus, the common payoff game could also be regarded as a decentralization of the learning problem as the LA participating in the game interact only with the environment.

Another way of looking at the common payoff game would be to highlight the computational advantage of a team of FALA over a single FALA. This advantage arises from the fact that while each LA is searching for the best action among its own r actions, the team arrives at the best combination of actions from all the N automata; thus effectively searching for the best among r^N choices. Computationally, only rN action probabilities are updated each instant in a team. On the other hand, r^N action probabilities would have to be updated if a single FALA were to be used. Moreover, the best action probability has to increase from a very low initial value (mostly $1/r^N$) to unity. Consequently, a single FALA would be very slow except for low values of r and N.

There is a price to be paid for this computational advantage. When algorithms like L_{R-I} are used, the team converges only to a local maximum of the expected reinforcement, whereas a single automaton would have converged to the global maximum. This drawback needs to be remedied.

Here we shall consider extensions of the ideas of Chapter 2 in two directions. One is to look at larger aggregations of LA so that more complex learning problems can be handled. This naturally leads to feedforward networks of learning units where each unit is a team of LA. The second extension is to consider a global learning algorithm which can work for teams as well as networks of LA so that the drawback of convergence to a local maximum is removed. As can be expected, global convergence would be at the expense of a somewhat longer convergence time and higher computational effort, but substantially lesser than those needed by a single LA.

In this chapter, results are first obtained for FALA and extensions to other LA are indicated later. Hence, unless otherwise stated, the term LA refers to FALA.

3.2 Networks of FALA

In several realistic learning problems, learning proceeds in stages. For instance, in hierarchical pattern classification where decision trees are used[1], the feature space is subdivided into sequentially smaller regions at successive nodes. A decision about classification has to be made at each node using the feature vector at that node. If a LA or a group of LA is employed at a node for this decision, one is naturally led to the concept of a feedforward network of a group of LA for the entire classification task. In such a model we essentially divide the feature space into regions such that in each region, a different classifier is appropriate. One example of such a network is the three layer network [Lip87] considered in Chapter 4 in detail (cf. Section 4.4).

Similar situations where one is required to learn some relevant regions occur in control problems also [Lew95]. In time-optimal control, for instance, it is necessary to divide the state space into two regions one corresponding to points where the maximum positive value of the control signal is appropriate and the other where the maximum negative value of control signal is appropriate. This is also called as bang-bang control. If the dynamics of the controlled plant are unknown, these regions have to be learnt during the operation of the system.

We have considered such learning situations briefly in Section 1.7. As explained there, one special feature of these problems is the crucial role played by the *context vector* (which could be the state vector or feature vector) in the decision to be made by each unit in the learning system.

In Section 1.7, we have seen how we can extend the the LA model to accommodate context vector by defining the Generalized Learning Automaton. (We will consider GLA again later in this chapter). Now we consider another approach, namely a network of FALA, where we use a team of LA to handle context vectors.

---

[1] See Section 4.5 for a brief introduction to decision tree classifiers

In this approach each node in the network corresponds to a learning unit which has several LA within it. The network is formed by a feedforward connection of these learning units. Each learning unit has its associated inputs and outputs. There would be two kinds of inputs to each unit. One is the context vector (which may be from the environment and/or outputs of units from earlier stages). The second type of input is the reinforcement from the environment indicating how well the entire learning system is performing. It may be noted that the two kinds of inputs are for the learning unit as a whole. The individual LA that make up the unit are the same FALA as in the earlier chapters and the only input received by these LA is the reinforcement feedback. The output of the learning unit is defined in a manner consistent with the objective of the learning system and could form part of the inputs to the units in succeeding stages. A network with these characteristics is described in detail in the next section.

3.3 The Learning Model

The learning model described in this section consists of a feedforward network interacting with an environment. The network could be regarded as a stochastic neural network organized from units of learning automata. The random environment, however, is more sophisticated than that considered so far, as it has to accommodate context vectors. This type of environment will be called Generalized Environment (G-Environment, for short) and will be described first.

3.3.1 G-Environment

The G-Environment is described by the tuple $\langle \mathcal{C}, A, B, \mathbf{F}, P_e \rangle$ where,

1 $\mathcal{C}$ is the set of context vectors (of the environment) which give the learning system information about the state of the environment. $\mathcal{C}$ is usually a compact subset of $\Re^n$, for some n.

2 $A = \{\alpha_1, \ldots, \alpha_r\}$ is the finite set of actions which can be input to the environment by the learning system.

3 B is the set of values that the reinforcement can take. Here we take B to be the closed interval $[0, 1]$; however, it can be any compact subset of $\Re$ in general. From now on we call the reinforcement supplied to an automata system by a G-environment as Scalar Reinforcement Signal or SRS. We denote the SRS by β.

4 $\mathbf{F} = \{F(\alpha, \mathbf{C}) : \alpha \in A, \mathbf{C} \in \mathcal{C}\}$ is a set of probability distributions over B. $F(\alpha, \mathbf{C})$ is the (unknown) distribution of the SRS when the state of the environment is $\mathbf{C}$ and the action input to the environment by the automata system is α.

5. P_e is a (unknown) probability distribution over $\mathcal{C}$. The state of the environment at any instant is a random realization of P_e. Thus, at any instant $k, k = 0, 1, 2\ldots$, and for any Borel set $\mathcal{B} \subset \mathcal{C}$,

$$\text{Prob}\{\mathbf{C}(k) \in \mathcal{B}\} = P_e(\mathcal{B}). \tag{3.3.1}$$

where $\mathbf{C}(k)$ is the context vector from the environment constituting the input to the learning system at instant k.

6. When the $F(\alpha, \mathbf{C})$'s are independent of time, the environment is said to be stationary. For every $\alpha \in A$ and $\mathbf{C} \in \mathcal{C}$ define

$$d(\alpha, \mathbf{C}) = E^{(\alpha, \mathbf{C})}[\beta] \tag{3.3.2}$$

where $E^{(\alpha, \mathbf{C})}$ denotes expectation with respect to the distribution $F(\alpha, \mathbf{C})$. The optimal action for $\mathbf{C}$ is defined to be that action which gives the highest value of $d(\alpha, \mathbf{C})$. There is an optimal action for each $\mathbf{C}$. These optimal actions are in general, different for different $\mathbf{C}$. The optimal action mapping is denoted by OA, which is a mapping from $\mathcal{C}$ to A, and it is such that

$$d(\text{OA}(\mathbf{C}), \mathbf{C}) = \max_{\alpha \in A} d(\alpha, \mathbf{C}). \tag{3.3.3}$$

In the above, it is assumed that P_e is independent of time. More complicated models, in which the present state of the environment depends on past actions/context vectors, are not considered here.

An Example

Here we illustrate the G-environment through a simple example. In this example, the set of context vectors, $\mathcal{C}$, is the unit square in $\Re^2$ as shown in Fig. 3.1. This space of context vectors is divided into regions I and II by straight lines OP and OQ. We take the set of actions as $A = \{\alpha_1, \alpha_2\}$. The set of possible values for SRS, B, is taken as binary. We take the distribution P_e from which context vectors are drawn as uniform. The distributions $F(\alpha, \mathbf{C})$ specify the probabilities of the SRS taking different values conditioned on the action and the context vector. In this example, these distributions are as follows.

$$\begin{aligned}
0.9 &= \text{Prob}[\beta(k) = 1 \mid \alpha(k) = \alpha_1, \mathbf{C}(k) \in \text{Region } I] \\
&= 1 - \text{Prob}[\beta(k) = 1 \mid \alpha(k) = \alpha_2, \mathbf{C}(k) \in \text{Region } I]
\end{aligned}$$

$$\begin{aligned}
0.9 &= \text{Prob}[\beta(k) = 1 \mid \alpha(k) = \alpha_2, \mathbf{C}(k) \in \text{Region } II] \\
&= 1 - \text{Prob}[\beta(k) = 1 \mid \alpha(k) = \alpha_1, \mathbf{C}(k) \in \text{Region } II] \tag{3.3.4}
\end{aligned}$$

The Learning Model

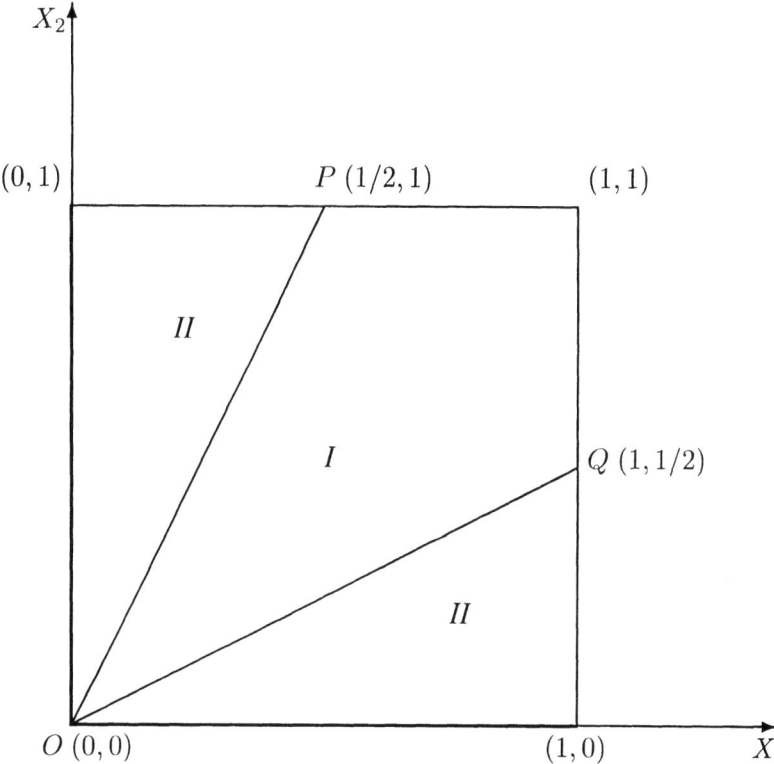

Figure 3.1. Example of the space of context vectors in a G-environment. See text for explanation

This completely specifies the distribution of SRS because the event $[\beta(k) = 0]$ is the complement of the event $[\beta(k) = 1]$.

Hence, in Region I,

$$E[\beta(k) \mid \alpha(k) = \alpha_1] = 0.9$$
$$E[\beta(k) \mid \alpha(k) = \alpha_2] = 0.1. \qquad (3.3.5)$$

In Region II, these numbers are reversed.

Thus, action α_1 is optimal in region I and action α_2 is optimal in region II. It is this optimal mapping which is to be learnt by the network.

3.3.2 The Network

The learning system which is in the form of a network attempts to infer or learn the best action associated with each context vector. As shown in Fig. 3.2, the network is a feedforward system of learning units. We will consider a simple form, where each unit is a team of LA playing a common payoff game.

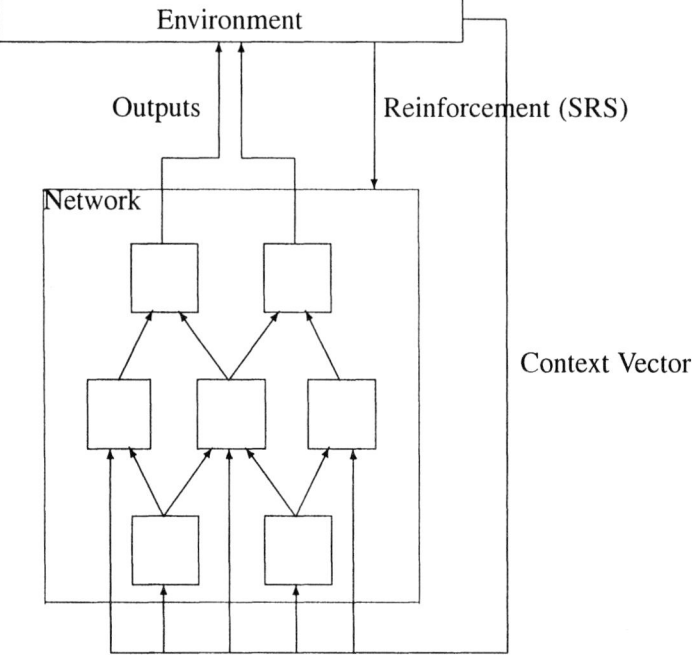

Figure 3.2. A Network of Learning Automata. Each of the boxes here corresponds to a unit which is a team of FALA.

This is sufficient for our purpose, although, in general, each unit could itself be more complex. These issues are discussed in Section 3.6.1.

A single unit in the network is described by the tuple $\langle \mathcal{X}, \mathcal{Y}, B, \mathbf{P}, \Gamma, \mathcal{T} \rangle$ where

1. $\mathcal{X}$ is the set of context vectors which are possible inputs to the unit.

2. $\mathcal{Y}$ is the set of outputs or actions of the unit.

3. B is the same as in the definition of the environment. It is the set of values the SRS can take.

4. $\mathbf{P}$ is the vector composed of the action probability vectors of all the learning automata in the unit. (Note that all automata here are FALA).

5. Γ is a vector of functions. The functions in Γ determine the output of the unit based on the actions selected by the FALA in the unit and the context vector input to the unit.

6. $\mathcal{T}$ is the learning algorithm which updates $\mathbf{P}$.

The Learning Model

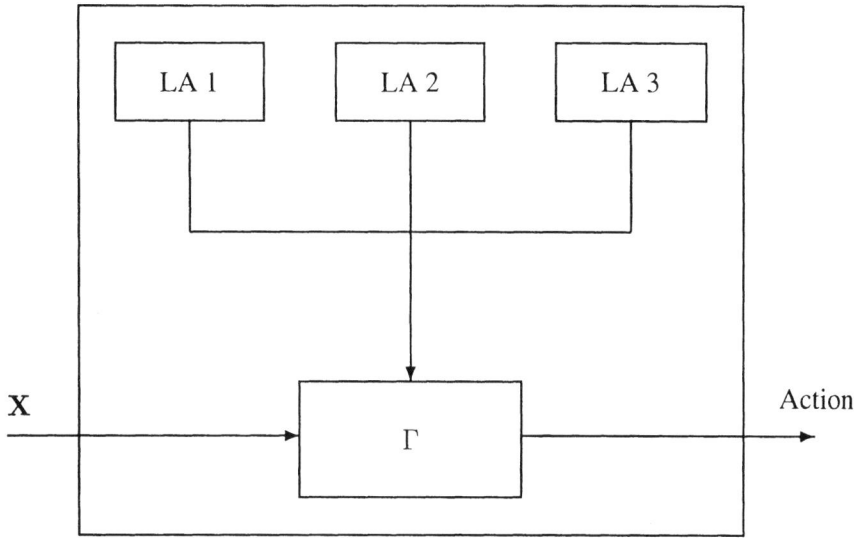

Figure 3.3. A Typical Unit in the Network

In the case of a single unit interacting with the environment, the context vector input to the unit and the context vector from the environment are the same. However, in a network of units, the context vector input to each unit could be composed of actions of other units and components of the context vector from the environment as in Fig. 3.2. This is the reason why we have used different symbols for set of context vectors of the environment and the set of context vectors possible for a unit. The output of the network is determined by the outputs of some prefixed units in the network. (It could be a vector composed of the outputs of the prefixed units). This output of the network is the action of the network which affects the environment. That is, it belongs to the set A, as in the definition of the G-environment. In the case of a single unit interacting with the environment, $\mathcal{X} = \mathcal{C}$ and $\mathcal{Y} = A$.

Each unit is composed of a number of learning automata as in Fig. 3.3 acting as a team. Each learning automaton in the unit chooses an action according to its action probability vector. The output of a unit could be defined in several ways depending on the manner of usage of the network. However, in every case, the output of the unit is determined based on the actions chosen by the automata in the unit and the context vector input to the unit.[2] As an example, consider a unit whose output is binary. (Such units would be useful, e.g., in 2-class pattern recognition problem). Based on the context vector $\mathbf{X}$ to the unit and the vector of actions $\boldsymbol{\alpha}$ chosen by the automata in the unit, we calculate the value of a function $\Gamma(\mathbf{X}, \boldsymbol{\alpha})$ and the output is taken as 1 if $\Gamma(\mathbf{X}, \boldsymbol{\alpha}) > 0$ and as 0 if $\Gamma(\mathbf{X}, \boldsymbol{\alpha}) \leq 0$. In case we want the unit to have m actions, m functions $\Gamma_j(\mathbf{X}, \boldsymbol{\alpha})$ $(j = 1, \ldots m)$ are defined and each function is associated with an action from a set of m possible actions. The ith action is output by the unit if,

$$\Gamma_i(\mathbf{X}, \boldsymbol{\alpha}) = \max_j \Gamma_j(\mathbf{X}, \boldsymbol{\alpha}). \tag{3.3.6}$$

The functions Γ_i that determine how the output is computed based on the context vector input and the actions chosen by the automata, are user specified and thus constitute part of the design of the network. We discuss below a simple example of network structure suitable for the example G-environment considered earlier. We will discuss a more complex pattern recognition example in Section 4.4.

Example

Consider the example G-environment discussed in the previous section. As shown in Fig. 3.1, the context vectors come from the unit square in $\Re^2$ and the optimal action mapping to be learnt is: in region I action α_1 is optimal while in region II α_2 is optimal. Suppose we decide to design the network so that it can represent regions expressible through a pair of lines. Then we can have a network consisting of two units. The set of context vectors for each unit would be same as the set of context vectors from the environment. We want each unit to represent a line. Since a line in $\Re^2$ needs three parameters, we make each unit to be a team of three FALA. Suppose we write the general equation for a line as $w_1 X_1 + w_2 X_2 + w_3 = 0$. The actions of the LA would be the possible values of the parameters, w_1, w_2, w_3. Since we are using FALA, we need finite action sets. Hence we would discretize the range of possible values for these parameters. Suppose, after discretization, all these parameters take values in the set $\{a^1, \ldots, a^r\}$. Then this set would be the action set of each LA (in either unit). Let $\boldsymbol{\alpha}^i = (\alpha^i_1, \alpha^i_2, \alpha^i_3)$, denote a general tuple of

---

[2] In Fig. 3.3, we have shown a box labeled Γ which determines the output or action of the unit based on the context vector, $\mathbf{X}$, and the actions chosen by all LA in the unit.

actions by LA in unit i, $i = 1, 2$. We can take the outputs of the units as binary and take the Γ function to be same for each of the two units and given by $\Gamma(\boldsymbol{\alpha}^i, \mathbf{X}) = \alpha_1^i X_1 + \alpha_2^i X_2 + \alpha_3^i$. Thus, each of the LA in a unit selects actions; this triple of actions determines a line in $\Re^2$ and the output of the unit indicates on which side of the line the current context vector is located. The final output of the network is α_1 if the outputs of the two units are different; and is α_2 if they are the same.[3] This output of the network can be thought of as coming from a third unit which has no LA in it and whose context vector input consists of the outputs of the other two units. It is easy to see that such a network structure is capable of representing the optimal action mapping in this example and hence can learn it if provided with a proper learning algorithm.

3.3.3 Network Operation

The network operates as an Associative Reinforcement Learning System [BA85, SB98] which learns an input-output mapping based on signals from the random environment. The sequence of events in such a system is as follows.

- At each instant, the environment generates a context vector which is input to the learning system.

- Based on its internal state and the context vector, the learning system (that is, the network) outputs an action.

- The environment then generates an evaluatory signal which indicates the appropriateness of the action to the particular context vector. This is the Scalar Reinforcement Signal (SRS) and a higher value of the SRS implies better performance of the system.

  Two of the main features of the SRS are that it is stochastic in nature with unknown distribution and its instantaneous value does not give any information about the relative performances of the other actions. The best action can be inferred only through repeated trials. Also each context vector has its own optimal action, and this mapping of context vector into its optimal action is what should be ideally learnt by the learning system. This is not always possible and what can be actually learnt by the learning system is discussed later.

- The learning system then updates its internal state depending on the context vector—action—SRS triple so as to improve its performance with respect

---

[3]Note that α_1, α_2 are the possible actions of the network as a whole. Thus, in this example, the set A in the definition of the G-environment corresponds to $A = \{\alpha_1, \alpha_2\}$. On the other hand, here we have used symbols α_j^i to denote the action chosen by the jth automaton in the ith unit and these actions are possible values of parameters that determine a line in $\Re^2$. These α_j^i should not be confused with α_1, α_2.

to an appropriate performance index. It may be noted that the internal state of the system consists of the action probability vectors of all the LA in all the units of the network.

The fundamental objective of the learning system is to find the best action associated with each context vector. As stated earlier, this amounts to finding the optimal action mapping OA defined by

$$d(\text{OA}(\mathbf{C}), \mathbf{C}) = \max_{\alpha \in A} d(\alpha, \mathbf{C}). \tag{3.3.7}$$

However, it may not always be feasible to learn such an optimal action mapping. Hence, we take the objective of learning to be solution of an optimization problem as explained below. We discuss why this is a good goal for the learning system and explain its relation with the problem of learning the optimal action mapping.

Let $\mathbf{p}_i$ be the action probability vector of the ith LA in the network.[4] Thus,

$$\mathbf{p}_i(k) = [p_{i1}(k), \ldots p_{ir_i}(k)]^T \tag{3.3.8}$$

where r_i is the number of actions of the ith automaton, is the action probability distribution of the ith FALA at instant k.

Let $\mathbf{P} = [\mathbf{p}_1^T, \mathbf{p}_2^T, \ldots, \mathbf{p}_N^T]^T$ be the supervector consisting of the action probability vectors of all LA in the network. (Here N denotes the total number of LA in the network). Since each $\mathbf{p}_i$ is a probability vector, it should satisfy

$$p_{ij} \geq 0 \quad 1 \leq j \leq r_i, \quad 1 \leq i \leq N$$
$$\sum_{j=1}^{r_i} p_{ij} = 1 \quad 1 \leq i \leq N \tag{3.3.9}$$

Now consider the optimization problem,

$$\text{maximize } f(\mathbf{P}) = E[\beta|\mathbf{P}] \tag{3.3.10}$$

subject to the constraints given by (3.3.9).

As should be clear from the previous two chapters, the general goal of any learning automata system is to maximize the reinforcement through modification of the action probability distributions. Thus, the above optimization problem is consistent with the general goal of an automata system.

In this chapter, we would be considering learning algorithms that would result in $\mathbf{P}(k)$ converging to a solution of this optimization problem. In the

---

[4]The index i for an LA comes through an arbitrary numbering of all LA in the network. Here we are not concerned with which LA is part of which unit etc.

The Learning Model

rest of this section, we discuss the relation between this optimization problem and the learning of optimal action mapping.

Suppose that the global maximum of $f(\mathbf{P})$ given by (3.3.10), subject to the constraints given by (3.3.9), is attained at a $\widetilde{\mathbf{P}} = (\widetilde{\mathbf{p}}_1^T \ \ldots \ \widetilde{\mathbf{p}}_N^T)^T$ which is such that each $\widetilde{\mathbf{p}}_i$ is a unit vector of dimension r_i. Any such $\widetilde{\mathbf{p}}_i$ is equivalent to an action of the ith LA and, hence, such a $\widetilde{\mathbf{P}}$ would be equivalent to (or a representation of) a tuple of actions of all automata in the network. Through the functions Γ_i in the units of the network, any such tuple of actions represents a mapping of context vectors to actions. The question then is whether the tuple of actions represented by $\widetilde{\mathbf{P}}$ corresponds to the optimal action mapping.

Consider a special case of the network where there are as many LA as the number of possible context vectors. We can then associate each LA with a context vector and the final action output by the network in response to any context vector input would be the action selected by the LA corresponding to that context vector. Then the optimal action mapping is nothing but a specific tuple of actions of all LA in the network. Hence, the $\widetilde{\mathbf{P}}$ which globally maximizes (3.3.10) would correspond to the optimal action mapping.

However, it is unrealistic to expect one LA for each context vector as the number of context vectors is large or is actually infinite in many problems of interest such as pattern classification. In the simple example considered earlier, the set of context vectors is uncountable though the network proposed has only six LA in all. In most realistic situations, it is necessary to consider the number of LA to be finite and not unduly large. In such situations, in general, all LA in the network affect, through their action choices, the output of the network in response to any context vector. When all LA participate in deciding the action for each context vector, the kind of mappings between context vectors and actions that can at all be represented by the network depend on the functions Γ_i which are part of the design of the network. Since the nature of the mapping OA is not known a priori, we do not know whether it is feasible for the network to learn this mapping. Even in such a case, the optimization problem defined by (3.3.10) is a good choice for goal of learning. However, now optimizing $f(\mathbf{P})$ may or may not correspond to learning the optimal association between context vectors and actions. One can say that maximizing $f(\mathbf{P})$ would result in the network learning the best possible approximation to the map OA, within the constraints of the chosen architecture of the network. This would be adequate as in most cases the context vector space could be divided into a finite number of regions each of which is associated with a single optimal action. In such an event, it is adequate for the network to have the flexibility to identify these regions. Obviously this ability depends on the complexity of the regions and the sophistication of the network architecture. In general, a given network would have some limitations on its performance and hence the optimization of (3.3.10) may not always lead to the ideal optimal action mapping. The bright side of the model is that by increasing the size of the network and designing the architecture in a proper manner one can approximate (to the required accuracy) any reasonable requirement for identification of relevant regions.

3.4 The Learning Algorithm

Let each LA in the network be numbered sequentially from 1 to N. (This number N would include the LA in all units in the network). Let $\mathbf{p}_i(k)$ be the action probability vector of the ith LA at instant k. Let this LA have r_i actions. Let A_i be the action set of this LA and let

$$A_i = \{a_{i1}, \ldots, a_{ir_i}\}. \qquad (3.4.11)$$

Denote by $\alpha_i(k)$ the action chosen by the ith LA at instant k. Then,

$$\text{Prob}\{\alpha_i(k) = a_{ij}\} = p_{ij}(k). \qquad (3.4.12)$$

The updating of $\mathbf{p}_i(k)$ is done according to the following learning algorithm which is same as the Linear Reward–Inaction ((L_{R-I})) algorithm. We update $p_{ij}, j = 1, \ldots, r_i, i = 1, \ldots N$, as

$$\begin{aligned} p_{ij}(k+1) &= p_{ij}(k) + \lambda \beta(k)(1 - p_{ij}(k)) && \text{if } \alpha_i(k) = a_{ij} \\ p_{ij}(k+1) &= (1 - \lambda \beta(k)) p_{ij}(k) && \text{if } \alpha_i(k) \neq a_{ij}. \end{aligned} \qquad (3.4.13)$$

In the above algorithm, $\beta(k) \in [0, 1]$ is the scalar reinforcement signal (SRS) from the environment at k. The learning parameter λ satisfies $0 < \lambda < 1$. This is fixed for all LA within a unit but could vary from one unit to another.

In general, the LA could use many algorithms other than L_{R-I} for updating the action probabilities. We discuss these in Section 3.6.2.

3.5 Analysis

In this section we show that if the action probability vectors of automata in the network are updated as given by (3.4.13), then the evolution of $\mathbf{P}(k)$ is such that it converges to a local maximum of $f(\mathbf{P})$ which, as explained earlier, is the goal of learning here. A special point that should be noted here is that this analysis is independent of details of the network structure such as which LA are part of which units, how the units are interconnected and the Γ_i functions that determine the overall output of the network based on the context vector input from the environment and the actions selected by different automata. Our objective is to maximize $f(\mathbf{P})$ which is the expected value of SRS conditioned on $\mathbf{P}$. The state of the network $\mathbf{P}$ contains the action probability vectors of all LA. Based on these probabilities, all LA select actions. Then, based on the Γ_i functions and the interconnections of units, the output of the network for

Analysis 117

the current context vector is determined. This output elicits a reinforcement β. The algorithm for updating the action probability vectors of LA is completely independent of the other architectural details of the system. We are going to show through our analysis that the algorithm would maximize the expected value of SRS. Thus this model of network of LA allows for a lot of flexibility in tackling different learning problems. We can choose the structure of our units, interconnections, the Γ_i functions etc. so that the network can capture a sufficiently rich class of mappings between context vectors and actions and then, using the same learning algorithm, learn the best mapping in this class.

As explained above, in this analysis, we are considering the case of a very general network architecture except for the assumption that the network is feedforward in nature. Thus, depending on the interconnections of units and the Γ_i functions, not all LA in the network may be required to select actions for determining the final output of the network in response to a specific context vector. We say that an LA is *activated* (in response to a context vector) if it selected an action. Only the activated LA would update their action probabilities using the algorithm given by (3.4.13). For an LA that is not activated at an instant k, the action probability vector remains unchanged.

Analysis of the operation of the network proceeds as follows [TP95a]. Basically we need to know how the action probabilities of the LA in the network evolve, especially in the long term. This is done (as usual!) by approximating the learning algorithm by an ODE so that for small values of the learning parameter, the long-term behavior of the solution of the ODE gives a very good approximation to the long term behavior of the action probabilities of LA. As in the earlier chapters, we follow the method outlined in Appendix A. Since the goal of learning is to solve the optimization problem (3.3.10), a connection is provided between the asymptotic behavior of the ODE (in terms of its equilibrium points and their stability properties) and maxima of $f(\mathbf{P}) = E[\beta \mid \mathbf{P}]$, the function being maximized.

It is useful to recall at this point the statement of the optimization problem and to state the necessary conditions for local maxima in this problem. The optimization problem is

$$\text{Maximize } f(\mathbf{P}) = E[\beta \mid \mathbf{P}] \tag{3.5.14}$$

subject to

$$p_{ij} \geq 0 \quad 1 \leq j \leq r_i, \quad 1 \leq i \leq N$$
$$\sum_{j=1}^{r_i} p_{ij} = 1 \quad 1 \leq i \leq N \tag{3.5.15}$$

where

$$\mathbf{P}^T = [\mathbf{p}_1^T, \mathbf{p}_2^T, \ldots \mathbf{p}_i^T, \ldots \mathbf{p}_N^T] \tag{3.5.16}$$

and $\mathbf{p}_i^T = [p_{i1}, p_{i2}, \ldots p_{ir_i}]$.

The region in which $\mathbf{P}$ is allowed to take values is the feasible region, denoted by $\mathcal{F}$. Thus,

$$\mathcal{F} = \{\mathbf{P} : p_{ij} \geq 0, \quad \sum_j p_{ij} = 1, \quad \forall i, j\}. \tag{3.5.17}$$

Conditions for a point $\mathbf{P}$ to be a solution of the above optimization problem are the so called Kuhn–Tucker conditions. The First Order Necessary Kuhn–Tucker conditions (FONKT) are as follows.

Necessary conditions for $\mathbf{P}$ to be a local maximum of $f(\mathbf{P})$ are that there exist η_{ij} and γ_i such that

$$\left. \begin{array}{ll} \frac{\partial f}{\partial p_{ij}}(\mathbf{P}) + \eta_{ij} + \gamma_i = 0 & (a) \\ \eta_{ij} p_{ij} = 0 & (b) \\ \eta_{ij} \geq 0 & (c) \\ p_{ij} \geq 0 & (d) \\ \sum_j p_{ij} = 1 & (e) \end{array} \right\} \forall i, j. \tag{3.5.18}$$

These are necessary conditions and form the counterpart of the condition in unconstrained optimization that the gradient must vanish. Second order necessary and second order sufficient conditions are also available, but will not be used here.

Now let us turn to the analysis of the learning algorithm. The first step is to obtain the approximating ODE. The learning algorithm specified by (3.4.13) can be rewritten in a vector form as

$$\mathbf{P}(k+1) = \mathbf{P}(k) + \lambda G(\mathbf{P}(k), \boldsymbol{\theta}(k)) \tag{3.5.19}$$

where $\boldsymbol{\theta}$ consists of the actions selected by all the automata in the network at k as well as the SRS at k and G represents the updating of action probabilities as given by (3.4.13). Both $\mathbf{P}$ and G are vectors of dimension $r_1 + \ldots + r_N$ where N is the total number of LA in the network. We denote the components of G function by G_{ij}, $j = 1, \ldots, r_i$, $i = 1, \ldots, N$. For the analysis here, we assume that the SRS, $\beta(k)$, is in $[0, 1]$ and that the context vector from the environment at each instant is random based on a probability distribution on the set of context vectors. Context vectors from the environment at different instants are stochastically independent. Since SRS is in $[0, 1]$, from (3.4.13), it is easily seen that each component of G is bounded.

For any $\mathbf{P} = (\mathbf{p}_1^T \ldots \mathbf{p}_N^T)^T$ where $\mathbf{p}_i$ is a r_i-dimensional probability vector, define, for $j = 1, \ldots, r_i$, $i = 1, \ldots, N$,

$$s_{ij}(\mathbf{P}) = E[G_{ij}(\mathbf{P}(k), \boldsymbol{\theta}(k)) \mid \mathbf{P}(k) = \mathbf{P}]. \tag{3.5.20}$$

Analysis

Now using Theorem A.1 in Appendix A, we can prove the following theorem that gives the approximating ODE for the learning algorithm of LA network.

THEOREM 3.1 *The approximating ODE for the learning algorithm specified by (3.4.13) for a network of LA is given by*

$$\frac{dz_{ij}}{dt} = s_{ij}(\mathbf{z}), \quad \mathbf{z}(0) = \mathbf{P}(0). \tag{3.5.21}$$

$j = 1, \ldots, r_i,\ i = 1, \ldots N.$

Proof: Comparing (3.5.19) with (A.1.1) of Appendix A, it is easily seen that our algorithm is in the same general form as the algorithm considered in that appendix. Now from (A.1.3) and equation (3.5.20) above, it is clear that this theorem directly follows from Theorem A.1 in Appendix A. Hence all we need to do to complete the proof is to verify assumptions (A1) to (A4) stated in Section A.2.1. For our algorithm, (A1) is easily seen to be true. The final output of the network depends only on the actions selected by all automata in the network and the context vector from the environment. Conditioned on current action probability vectors, actions selected by LA at the current instant are independent of the actions selected at the previous instant. The SRS depends on the output of the network and the current context vector which is independent of previous context vectors as per our assumption. Hence, we see that assumption (A2) is also satisfied. In our algorithm, all components of **P** and θ are bounded. Further, all G_{ij} functions are continuous and bounded. Hence, it is easily seen that assumption (A3) is also satisfied. Finally, since all G_{ij} are bounded, the variances are bounded as needed by assumption (A4). This completes proof of the theorem. ∎

The exact sense in which the solution of the ODE, $\mathbf{z}(t)$ approximates $\mathbf{P}(k)$ is explained in Appendix A (see equation (A.2.14)).

The next step in the analysis is to analyze the ODE for which we need to calculate the functions s_{ij}. In order to obtain these, the following definition and lemma are crucial.

DEFINITION 3.5.1 *The drift Δp_{ij} of the jth component of $\mathbf{p}_i$, the action probability vector of the ith automaton, is defined as*

$$\Delta p_{ij} = E[p_{ij}(k+1) - p_{ij}(k)|\mathbf{P}(k) = \mathbf{P}]. \tag{3.5.22}$$

∎

From the above definition and (3.5.20), we have

$$\Delta p_{ij}(k) = \lambda s_{ij}(\mathbf{P}(k)) \tag{3.5.23}$$

LEMMA 3.1 *The drift Δp_{ij} for algorithm (3.4.13) is*

$$\Delta p_{ij} = \lambda p_{ij} \sum_s p_{is} \left(\frac{\partial f}{\partial p_{ij}} - \frac{\partial f}{\partial p_{is}} \right) \quad (3.5.24)$$

where $f(\cdot)$ is as defined by (3.3.10).

Proof: Recall that we are considering a feedforward network of units where each unit is a team of LA. Now

$$\begin{aligned} f(\mathbf{P}) &= E[\beta \mid \mathbf{P}] \\ &= \int E[\beta \mid \mathbf{P}, \mathbf{C}] P_e(d\mathbf{C}) \end{aligned}$$

where the context vector from the environment, $\mathbf{C}$, is assumed to be drawn independently at each instant from some compact set according to the probability distribution $P_e(\cdot)$. Let $\varepsilon_i(k)$ be the event that the ith LA is activated at instant k and let $\varepsilon_i^c(k)$ be its complement. In the proofs to follow, we omit explicitly showing the dependence of various quantities on the time instant k whenever, there is no scope for confusion. Then,

$$\begin{aligned} f(\mathbf{P}) &= \int E[\beta \mid \mathbf{P}, \mathbf{C}, \varepsilon_i] \text{Prob}(\varepsilon_i \mid \mathbf{P}, \mathbf{C}) P_e(d\mathbf{C}) \\ &\quad + \int E[\beta \mid \mathbf{P}, \mathbf{C}, \varepsilon_i^c] \text{Prob}(\varepsilon_i^c \mid \mathbf{P}, \mathbf{C}) P_e(d\mathbf{C}) \\ &= \sum_s p_{is} \int E[\beta \mid \mathbf{P}, \mathbf{C}, \varepsilon_i, a_{is}] \text{Prob}(\varepsilon_i \mid \mathbf{P}, \mathbf{C}) P_e(d\mathbf{C}) \\ &\quad + \int E[\beta \mid \mathbf{P}, \mathbf{C}, \varepsilon_i^c] \text{Prob}(\varepsilon_i^c \mid \mathbf{P}, \mathbf{C}) P_e(d\mathbf{C}) \end{aligned}$$

by conditioning on the action chosen at k, $\alpha_i(k) = a_{is}$.

The second term is independent of $\mathbf{p}_i$ because of the feedforward property of the network (i.e., $\varepsilon_i(k)$ is independent of $\mathbf{p}_i(k)$) and the fact that the SRS $\beta(k)$ is independent of $\mathbf{p}_i(k)$ once $\alpha_i(k)$ is selected. In the first term, only p_{is} is a function of $\mathbf{p}_i$. Defining $f_{ij}(\mathbf{P}) \triangleq \frac{\partial f}{\partial p_{ij}}(\mathbf{P})$,

$$f_{ij}(\mathbf{P}) \triangleq \frac{\partial f}{\partial p_{ij}}(\mathbf{P}) = \int E(\beta \mid \mathbf{P}, \mathbf{C}, \varepsilon_i, a_{ij}) \text{Prob}(\varepsilon_i \mid \mathbf{P}, \mathbf{C}) P_e(d\mathbf{C}). \quad (3.5.25)$$

Now consider,

$$\Delta p_{ij} = E[p_{ij}(k+1) - p_{ij}(k) \mid \mathbf{P}(k) = \mathbf{P}].$$

Analysis 121

Conditioning on **C** and then on $\varepsilon_i, \varepsilon_i^c$ and noting that $p_{ij}(k+1) - p_{ij}(k) = 0$ under ε_i^c,

$$\Delta p_{ij} = \int E[p_{ij}(k+1) - p_{ij}(k) \mid \mathbf{P}, \mathbf{C}, \varepsilon_i] \text{Prob}(\varepsilon_i \mid \mathbf{P}, \mathbf{C}) P_e(d\mathbf{C}).$$

Conditioning further on $\alpha_i(k) = a_{is}$ and using the L_{R-I} learning algorithm (3.4.13),

$$\begin{aligned}
\Delta p_{ij} &= \sum_s p_{is} \int E[p_{ij}(k+1) - p_{ij}(k) \mid \mathbf{P}, \mathbf{C}, \varepsilon_i, a_{is}] \text{Prob}(\varepsilon_i \mid \mathbf{P}, \mathbf{C}) P_e(d\mathbf{C}) \\
&= \sum_{s \neq j} p_{is} p_{ij} \int E[-\lambda \beta \mid \mathbf{P}, \mathbf{C}, \varepsilon_i, a_{is}] \text{Prob}(\varepsilon_i \mid \mathbf{P}, \mathbf{C}) P_e(d\mathbf{C}) \\
&\quad + p_{ij}(1 - p_{ij}) \int E[\lambda \beta \mid \mathbf{P}, \mathbf{C}, \varepsilon_i, a_{ij}] \text{Prob}(\varepsilon_i \mid \mathbf{P}, \mathbf{C}) P_e(d\mathbf{C}) \\
&= \sum_{s \neq j} p_{is} p_{ij} \int E[-\lambda \beta \mid \mathbf{P}, \mathbf{C}, \varepsilon_i, a_{is}] \text{Prob}(\varepsilon_i \mid \mathbf{P}, \mathbf{C}) P_e(d\mathbf{C}) \\
&\quad + \sum_{s \neq j} p_{is} p_{ij} \int E[\lambda \beta \mid \mathbf{P}, \mathbf{C}, \varepsilon_i, a_{ij}] \text{Prob}(\varepsilon_i \mid \mathbf{P}, \mathbf{C}) P_e(d\mathbf{C}) \quad (3.5.26)
\end{aligned}$$

Simplifying using (3.5.25),

$$\begin{aligned}
\Delta p_{ij} &= \lambda p_{ij} \sum_{s \neq j} p_{is}(f_{ij} - f_{is}) \\
&= \lambda p_{ij} \sum_s p_{is}(f_{ij} - f_{is}) \quad (3.5.27)
\end{aligned}$$

which proves the lemma. ∎

REMARK 3.5.1 *It is seen that Δp_{is} is not directly a function of the time step k. Now, from (3.5.23), we have*

$$s_{ij}(\mathbf{P}) = p_{ij} \sum_s p_{is} \left(\frac{\partial f}{\partial p_{ij}} - \frac{\partial f}{\partial p_{is}} \right). \quad (3.5.28)$$

∎

The approximating ODE for our algorithm is given by (3.5.21) where s_{ij} are given by (3.5.28). The next step in the analysis is to characterize the solutions of the ODE.

From now on, the ODE will be represented in terms of **P** *rather than* **z**. *This is to facilitate comparison between the ODE and the optimization problem.*

This is an abuse of notation as **P** *is a discrete time variable and the same symbol will also be used to denote the continuous time variable for convenience. Thus, the ODE we consider is* $\frac{dp_{ij}}{dt} = s_{ij}(\mathbf{P})$.

As stated earlier, we want to prove that the ODE solutions would solve the optimization problem of maximizing $f(\mathbf{P})$. However, the ODE (3.5.21) (with s_{ij} given by (3.5.28)) is not a gradient ascent type. This is because there are constraints on **P**. We first show that $f(\mathbf{P}(t))$ increases monotonically along the ODE paths. In the following lemma, we use **q** as a dummy variable taking values in the same space as **P**.

LEMMA 3.2 *Along the paths of the ODE (3.5.21), $(df/dt)(\mathbf{q}) \geq 0$. Moreover, $(df/dt)(\mathbf{q}) = 0$ if and only if **q** is an equilibrium point of the ODE.*

Proof: In this proof, k is used as a subscript and has nothing to do with the k used in algorithm (3.4.13) as a discrete time step.
Using the chain rule of differentiation,

$$\frac{df}{dt}(\mathbf{q}) = \sum_i \sum_j \frac{\partial f}{\partial p_{ij}}(\mathbf{q}) s_{ij}(\mathbf{q})$$

$$= \sum_{i,j,k} q_{ij} q_{ik} \left[\frac{\partial f}{\partial p_{ij}}(\mathbf{q}) - \frac{\partial f}{\partial p_{ik}}(\mathbf{q}) \right] \frac{\partial f}{\partial p_{ij}}(\mathbf{q})$$

by substituting for $s_{ij}(.)$ using (3.5.28). By noting that there are precisely two terms in which $q_{ij} q_{ik}$ is the multiplying factor, the above expression for (df/dt) can be simplified.
For the remainder of the proof let $d_{ij} = (\partial f)/(\partial p_{ij})(\mathbf{q})$. Then,

$$\frac{df}{dt}(\mathbf{q}) = \sum_{i,j,k<j} q_{ij} q_{ik} [(d_{ij} - d_{ik}) d_{ij} + (d_{ik} - d_{ij}) d_{ik}]$$

$$= \sum_{i,j,k<j} q_{ij} q_{ik} [(d_{ij} - d_{ik})^2] \geq 0 \quad (3.5.29)$$

since the q_{ij}'s are all nonnegative, being probabilities. It is further seen that for $(df/dt)(\mathbf{q}) = 0$, each term in the above summation should be zero. Thus $(df/dt)(\mathbf{q}) = 0$ if and only if

$$q_{ij} q_{ik} (d_{ij} - d_{ik}) = 0 \quad \forall i, j, k \quad (3.5.30)$$

Summing the above equation over k, for fixed i and j,

$$\sum_k q_{ij} q_{ik} (d_{ij} - d_{ik}) = 0.$$

Analysis 123

From (3.5.28), the LHS of the above equation is the expression for (dq_{ij}/dt). Thus, **q** is an equilibrium point of the ODE. It is trivially true that $(df/dt)(\mathbf{q}) = 0$ if **q** is an equilibrium point of the ODE. Thus, $(df/dt)(\mathbf{q}) = 0$ if and only if **q** is an equilibrium point of the ODE. ∎

REMARK 3.5.2 *It is seen that $f(\mathbf{P}(t))$ is a strictly increasing function of t along the solutions of the ODE (except at equilibrium points of the ODE). This is a property similar to absolute expediency [NT89] but while absolute expediency holds at every step, the above lemma shows that a similar property holds for algorithm (3.4.13) with high probability for small enough learning parameter λ.*

The next lemma shows the relationship between the equilibrium points of the ODE and the points satisfying the FONKT conditions for the optimization problem.

LEMMA 3.3

1. *If **q** satisfies FONKT conditions given by (3.5.18) of the optimization problem (3.3.10) then **q** is an equilibrium point of the ODE (3.5.21).*

2. *If **q** is an equilibrium point of the ODE (3.5.21) and does not satisfy the FONKT conditions, then **q** is unstable.*

Proof: Here also we use k as a subscript and let $d_{ij} \triangleq (\partial f/\partial p_{ij})(\mathbf{q})$.

1. We show that if **q** satisfies FONKT conditions then

$$q_{ij}q_{ik}(d_{ij} - d_{ik}) = 0$$

for all i, j, k. This would then imply that **q** is an equilibrium point of the ODE.

If $q_{ij} = 0$ or $q_{ik} = 0$ then $q_{ij}q_{ik}(d_{ij} - d_{ik}) = 0$.

Let both q_{ij} and q_{ik} be strictly positive. Then the FONKT conditions (see (3.5.18)) imply $\eta_{ij}q_{ij} = \eta_{ik}q_{ik} = 0$. From this it is seen that $\eta_{ij} = \eta_{ik} = 0$. Then we get $d_{ij} + \gamma_i = d_{ik} + \gamma_i = 0$. From this we obtain $d_{ij} = d_{ik} = -\gamma_i$. Thus, again we get $q_{ij}q_{ik}(d_{ij} - d_{ik}) = 0$. This condition therefore holds for all i, j and k when **q** satisfies the FONKT conditions. Summing over k, for any fixed i, j

$$q_{ij} \sum_k q_{ik}(d_{ij} - d_{ik}) = 0.$$

From (3.5.28), it is seen that the above implies $s_{ij}(\mathbf{q}) = 0$ which is the same as $(dq_{ij}/dt) = 0$. Thus, as the above argument is for any i and j, **q** is an equilibrium point of the ODE.

2. Since **q** is an equilibrium point of the ODE, using the proof of Lemma 3.2,
$$q_{ij}q_{ik}(d_{ij} - d_{ik}) = 0$$
for all i,j,k. Consider the FONKT conditions given by (3.5.18). Since $\mathbf{q} \in \mathcal{F}$, the feasible region, conditions (3.5.18) (d), (e) are automatically satisfied. By choosing η_{ij} such that $\eta_{ij} = 0$ whenever $q_{ij} > 0$, (3.5.18)(b) is also satisfied. Let
$$\mathcal{I}_i = \{j : q_{ij} > 0\}.$$
Since $\mathbf{q} \in \mathcal{F}, \mathcal{I}_i \neq \phi$. For all $j \in \mathcal{I}_i$, $\eta_{ij} = 0$ so as to satisfy (3.5.18) (b). Thus, for all $j \in \mathcal{I}_i$, (3.5.18) (a) reduces to
$$d_{ij} + \gamma_i = 0 \quad . \tag{3.5.31}$$
For all $j \notin \mathcal{I}_i$, the η_{ij}'s can be calculated according to
$$\eta_{ij} = -\gamma_i - d_{ij}. \tag{3.5.32}$$
There are two ways the FONKT conditions can be violated. One is if at least one of the η_{ij}'s is strictly negative, violating (3.5.18) (c). The second is if the d_{ij}'s differ on $\mathcal{I}_i$, so that the γ_i cannot even be calculated. These two cases are considered separately.

Case 1: In this case, it is assumed that γ_i is well defined, but that there exists a $\eta_{ij} < 0$. Without loss of generality, let $\eta_{11} < 0$. It will be shown that **q** is unstable. Shift the equilibrium point **q** to the origin by the transformation (recall that **P** is the variable of the ODE)
$$\epsilon = \mathbf{P} - \mathbf{q}.$$
Then,
$$\frac{d\epsilon_{11}}{dt} = \frac{dp_{11}}{dt} = p_{11} \sum_j p_{1j} \left\{ \frac{\partial f}{\partial p_{11}}(\mathbf{P}) - \frac{\partial f}{\partial p_{1j}}(\mathbf{P}) \right\}.$$
For all $j \in \mathcal{I}_1$, $q_{1j} > 0$ and $q_{1j} = 0$ otherwise. Since η_{11} has been defined by (3.5.32), it has been assumed that $1 \notin \mathcal{I}_1$. Thus $q_{11} = 0$ and
$$\frac{d\epsilon_{11}}{dt} = \epsilon_{11} \sum_j p_{1j} \left\{ \frac{\partial f}{\partial p_{11}}(\mathbf{P}) - \frac{\partial f}{\partial p_{1j}}(\mathbf{P}) \right\}.$$
To prove instability we consider a linear approximation and hence only the constant contribution from the summation term of the above equation is needed. Thus, to a linear approximation, and noting that $q_{1j} = 0$ for all $j \notin \mathcal{I}_1$,
$$\frac{d\epsilon_{11}}{dt} = \epsilon_{11} \sum_{j \in \mathcal{I}_1} q_{1j}(d_{11} - d_{1j}).$$

Analysis 125

Substituting for d_{11} using (3.5.32) and for $d_{1j}, j \in \mathcal{I}_1$, using (3.5.31),

$$\frac{d\epsilon_{11}}{dt} = \epsilon_{11} \sum_{j \in \mathcal{I}_1} q_{1j}(-\gamma_1 - \eta_{11} + \gamma_1)$$
$$= -\eta_{11}\epsilon_{11}. \qquad (3.5.33)$$

As η_{11} has been assumed to be negative, $-\eta_{11}$ is positive. Thus the linearized differential eqn. (3.5.33) for ϵ_{11} is unstable and hence **q** is unstable for the ODE (3.5.21).

Case 2: In this case, it is assumed that γ_1 cannot be defined uniquely. Thus, there exist i, j in $\mathcal{I}_1$ such that $(\partial f/\partial p_{1i})(\mathbf{q}) \neq (\partial f/\partial p_{1j})(\mathbf{q})$. But, since both q_{1i} and q_{1j} are greater than zero, equation (3.5.30) implies that $(\partial f/\partial p_{1j})(\mathbf{q})$ and $(\partial f/\partial p_{1i})(\mathbf{q})$ have to be equal for **q** to be an equilibrium point of the ODE. Thus, this case cannot arise when **q** is an equilibrium point.

This completes the proof of the lemma.

∎

REMARK 3.5.3 *If **q** satisfies the FONKT conditions, the above lemma does not say anything about its stability or otherwise. For the algorithm to be useful, the equilibrium points which are stable should correspond to solutions of the optimization problem and vice-versa. It may not be possible to prove a result in such generality, but partial results have been obtained [Pha91]. These results indicate that the correspondence between the equilibrium points which are stable and the solutions of the optimization problem would be violated only in pathological cases. For illustrating the main arguments involved, here we concentrate on isolated maxima of the optimization problem and isolated equilibrium points of the ODE. The following lemma gives a straightforward connection between the two.* ∎

LEMMA 3.4

1. *If **q** is a locally asymptotically stable equilibrium point of the ODE (3.5.21), it is an isolated maximum of the optimization problem (3.3.10).*

2. *If **q** is an isolated maximum of the optimization problem, it is a stable equilibrium point of the ODE. Furthermore, if this equilibrium point is isolated, it is also locally asymptotically stable.*

Proof:

1. Let **q** be locally asymptotically stable. There exists a neighborhood Nb of **q** such that if $\mathbf{P}(0) \in \text{Nb}, \mathbf{P}(t) \to \mathbf{q}$. Consider any $\mathbf{u} \in \text{Nb} - \mathbf{q}$.

Let $P(0) = u$. As $P(t) \to q$, $(df/dt)(u) > 0$, else u would be an equilibrium point of the ODE and $p(t)$ would not converge to q. Thus, $f(u) < f(q)$, using Lemma 3.2. That is, for all $u \in Nb - q$, $f(u) < f(q)$. Thus q is a strict local maximum. If there is any other maximum of f in Nb−q, that point would satisfy the FONKT conditions and would thus be an equilibrium point of the ODE (3.5.21), by Lemma 3.3. This is not possible as q is locally asymptotically stable with at least Nb as its attractive bowl. Thus there is no maximum of f in Nb−q and therefore q is an isolated local maximum of f.

2. Let q be an isolated local maximum of the optimization problem. There is a neighborhood Nb of q such that $f(P) < f(q)$ for all $P \in Nb - q$. On Nb, define a Lyapunov function

$$V(P) = f(q) - f(P).$$

We have $V(q) = 0$ and $V > 0$ on Nb−q. Along the trajectories of the ODE (3.5.21),

$$\frac{dV}{dt} = -\frac{df}{dt} \le 0$$

by Lemma 3.2. Thus q is a (locally) stable attractor of the ODE [BS70].

If q is an isolated equilibrium point, $\frac{df}{dt} \ne 0$ in Nb−q and hence $\frac{dV}{dt} = -\frac{df}{dt} < 0$. Hence q is locally asymptotically stable.

∎

Actually the results derived in the lemmas can be extended further without assuming that the equilibrium points and maxima are isolated.

As $(df/dt) \le 0$, it is easily seen by application of LaSalle's Theorem [NA89] that the solutions of the ODE converge to the set ε where

$$\varepsilon = \left\{ P : \frac{df}{dt}(P) = 0 \right\}$$

which is exactly the set of equilibrium points of the ODE by Lemma 3.2. Thus, every solution of the ODE converges to the set ε.

Furthermore, one can show that ε is composed of a finite number of connected components, each of which is compact [Pha91]. Asymptotically $P(t) \to M$ as $t \to \infty$, where M is one such component of ε and for all $P \in M$, $f(P) = K > 0$, a constant.

The basic question is whether the solutions of the ODE converge to a local maximum of $f(P)$. Let us consider 3 cases that arise.

If M is locally asymptotically stable, then by an extension of Lemma 3.4, M is an isolated local maximum of $f(.)$. This is the most desirable situation.

If M is unstable, there is a sequence $\{\mathbf{P}_n\}$, $\mathbf{P}_n \to M$, and a neighborhood Nb of M such that the solution of the ODE exits Nb when the initial state is $\mathbf{P}_n$. Since $f(\mathbf{P}(t))$ increases monotonically (except over components of ε where it is constant), M is not a local maximum of f. This is again what is needed, since any unstable set of equilibrium points should not contain local maxima.

Suppose M is stable but not locally asymptotically stable. Then M need not necessarily be a local maximum. If M is not a local maximum, there could be a sequence of equilibrium points converging to M such that each of these equilibrium points has values all of which are either greater or less than K. This type of behavior is not expected as $f(.)$ is a polynomial in $\mathbf{P}$. An example of such a behavior is the function $t \sin 1/t$ near $t = 0$ where it is infinitely oscillatory. However, this is a transcendental function.

The proofs of some of the statements made above are along the same lines as those of lemmas stated earlier and can be found in [Pha91]. Summarizing the discussion we can say that barring some pathological cases the ODE (and hence the algorithm) converges to the local maxima of $f(.)$ and hence solves the optimization problem which is the goal of learning for the automata network.

3.6 Extensions

In the previous section we analyzed the asymptotic behavior of the network of LA model. We assumed a feedforward network of units each of which is a team of FALA. We also assumed that all LA update the their action probabilities using L_{R-I} algorithm. Both these assumptions are employed mainly to keep the analysis simple. In this section we briefly outline some of the possible extensions.

3.6.1 Other Network Structures

We have considered each unit to be a team of LA. One could consider a more complex structure for each unit. For instance, a unit consisting of a combination of hierarchy and team has been analyzed [TP95a]. In this structure, each unit consists of a tree hierarchy of subunits and each subunit is a team of LA. Use of such complex units could simplify the overall network structure in specific problems.

We have mentioned at the beginning of our analysis that in a general network structure, all the LA in the network may not be activated at each instant. This is specially true when the unit consists of a hierarchy as only one branch of the hierarchy is activated at a time. As already stated, the learning algorithm makes a distinction between the activated LA and the nonactive LA. Only the activated LA update their action probabilities using the learning algorithm.

For the LA that are not activated, the action probabilities are unchanged; i.e., $\mathbf{p}_i(k+1) = \mathbf{p}_i(k)$. This strategy appears reasonable as the nonactive LA do not contribute to the output.

One can generalize the network structure even further. For example, each unit could itself be a feedforward network of LA. For the results already derived to continue to hold good, the following is the condition needed on the unit and network structure [Pha91].

Let $E_i(k)$ denote the event that the ith LA is activated at instant k. The network should be such that for every k, given $\mathbf{p}_j(k)$, $j \neq i$, and $\mathbf{C}(k)$, $E_i(k)$ is independent of $\mathbf{p}_i(k)$. Formally, let

$$\mathbf{P}^{-i}(k) = (\mathbf{p}_1(k), \mathbf{p}_2(k), \ldots \mathbf{p}_{i-1}(k), \mathbf{p}_{i+1}(k), \ldots \mathbf{p}_N(k)). \quad (3.6.34)$$

Then,

$$\text{Prob}\{E_i(k)|\mathbf{P}(k), \mathbf{C}(k)\} = \text{Prob}\{E_i(k)|\mathbf{P}^{-i}(k), \mathbf{C}(k)\} \quad (3.6.35)$$

This condition implies that the probability of any LA being activated is independent of the action probability vector of that LA. It would also imply the LA in the network can be divided into groups which are ordered; each group getting inputs from groups 'before it' and the outputs going to groups 'after it'. The network could also be thought of as consisting of layers of units. Equation (3.6.35) defines a feedforward connectionist system or network of LA for which all the results obtained so far are applicable [PT96].

3.6.2 Other Learning Algorithms

The convergence results discussed so far have been established in relation to networks of FALA using the L_{R-I} algorithm. They can be extended to FALA using more general nonlinear algorithms; for instance, the absolutely expedient algorithms [PST94]. The most general conditions for absolute expediency for a single FALA were given by Aso and Kimura [AK76]. The corresponding result for networks has been established for a slightly stronger condition on the updating functions [PT96] as follows. Consider the learning algorithm described below.

If the ith LA is not activated at k,

$$\mathbf{p}_i(k+1) = \mathbf{p}_i(k)$$

If the ith LA is activated and $\alpha_i(k) = a_{ij}$, then,

$$\begin{aligned}
p_{ij}(k+1) &= p_{ij}(k) + \lambda \beta(k) \sum_{s \neq j} \phi_{ijs}(\mathbf{p}_i(k)) \\
&\quad - \lambda(1-\beta(k)) \sum_{s \neq j} \psi_{ijs}(\mathbf{p}_i(k)) \quad (3.6.36)
\end{aligned}$$

and for all $s \neq j$

$$\begin{aligned}p_{is}(k+1) &= p_{is}(k) - \lambda\beta(k)\phi_{ijs}(\mathbf{p}_i(k)) \\ &\quad + \lambda(1-\beta(k))\psi_{ijs}(\mathbf{p}_i(k)).\end{aligned}$$

In the above algorithm, $\phi_{ijs}(.)$ and $\psi_{ijs}(.)$ are nonnegative functions of $\mathbf{p}_i$ satisfying

$$\left.\begin{array}{l}\phi_{ijs}(\mathbf{p}_i) \leq p_{is} \quad \forall i, j, s \\ \sum_{s \neq j} \psi_{ijs}(\mathbf{p}_i) \leq p_{ij} \quad \forall i, j\end{array}\right\} \quad (3.6.37)$$

so that $\mathbf{p}_i$ remains a probability vector. Sufficient conditions to be satisfied by the functions for the convergence results of this chapter to hold good are,

$$\left.\begin{array}{l}p_{ij}\phi_{ijs}(\mathbf{p}_i) = p_{is}\phi_{isj}(\mathbf{p}_i) \\ p_{ij}\psi_{ijs}(\mathbf{p}_i) = p_{is}\psi_{isj}(\mathbf{p}_i)\end{array}\right\} \forall i, j, s. \quad (3.6.38)$$

The details can be seen in [PT96]. The L_{R-I} algorithm is a particular case of this algorithm obtained by setting

$$\left.\begin{array}{l}\phi_{ijs}(\mathbf{p}_i) = p_{is} \quad \forall i, j, s \\ \psi_{isj}(\mathbf{p}_i) \equiv 0 \quad \forall i, j, s\end{array}\right\}. \quad (3.6.39)$$

3.7 Convergence to the Global Maximum

The network of LA considered so far suffers from one limitation; with a learning algorithm such as L_{R-I}, it can only converge to a local maximum of the objective function $E[\beta \mid \mathbf{P}]$. This limitation exists even in the case of a single unit, namely a team of LA playing a common payoff game as seen in Chapter 2. In order to achieve convergence to the global maximum, one has to start the network with different initial conditions on the action probabilities and hope that at least in some of these cases the local and global maxima coincide and hence global convergence is attained.

An alternative way would be to change the learning algorithm. For example, it is known that [TS87, MT89] a team of LA using an estimator algorithm will rapidly converge to the global maximum. However, even in the case of a team, such an algorithm needs a large memory as it has to store a large number of reward probability estimates. This problem would become much worse in the case of a network and would not be practical.

In handling global optimization problems in general, Simulated Annealing [KGV83] has provided the impetus for a number of algorithms. The basic idea here is to make a perturbation of the variables and occasionally choose values which lead to a worse value of the objective function. This feature enables the search to move out of local optima.

If we apply the above idea to optimization with LA, we have to perturb the action probabilities as the objective function f is a function of $\mathbf{P}$ as in (3.3.10). Since these perturbations are random, it is possible that some components of p may go outside the simplex, thus violating (3.3.9). In order to avoid such an event, one can parameterize action probabilities in terms of real numbers and update these real numbers. We have discussed such automata models, called Parameterized Learning Automata (PLA), in Section 1.8. As explained there, the state of the PLA is a vector of real numbers and the action probabilities are obtained from it through a probability generating function. Let $\mathbf{u}_i = [u_{i1}, u_{i2}, \ldots u_{ir_i}]^T$ denote the internal state vector of the ith PLA and let $\mathbf{p}_i = [p_{i1}, p_{i2}, \ldots p_{ir_i}]^T$ denote its action probability vector. Then we can use the probability generating function given by

$$p_{is} = \frac{\exp(u_{is})}{\sum_{j=1}^{r_i} \exp(u_{ij})}. \qquad (3.7.40)$$

We have considered such probability generating functions in Section 1.8. The action probabilities $p_{is}(s = 1, \ldots r_i)$ will be within the simplex irrespective of the values that the real numbers $u_{ij}(j = 1, \ldots, r_i)$ take. It is u_{ij} which are updated using the environmental response.

In this section, global results for a network of PLA are developed using diffusion approximation techniques. The learning problem is posed as a constrained optimization problem as earlier. However, we present a learning algorithm for which we can prove that we achieve a close approximation to the global maximum [TP95b]. For the analysis, the suggested learning algorithm is approximated by a stochastic differential equation (SDE) using weak convergence techniques. The specific SDE that approximates our algorithm is called the Langevin equation and it is such that its invariant probability measure concentrates on the global maxima [APPZ85]. Hence the learning algorithm has the required property of converging to the global optimum.

3.7.1 The Network

The network of LA is the same as that described in section 3.3.2 except that each LA is a PLA and p_{ik} are generated through u_{ij} as in (3.7.40). The ith PLA is characterized by a vector of real numbers $\mathbf{u}_i$, which is called its internal state.

A single unit in the network is now described by the tuple $\langle \mathcal{X}, \mathcal{Y}, B, \mathbf{U}, \Gamma, \mathcal{T} \rangle$ where $\mathbf{U}$ is the super vector of $\mathbf{u}_i$, i varying over the LA in the unit. Other symbols remain the same. Now,

$$\text{Prob}\,[\alpha_i(k) = a_{ij}] = p_{ij}(k) = g_i(\mathbf{u}_i(k), a_{ij}) \qquad (3.7.41)$$

where g_i is the probability generating function for the ith PLA.

Since the primary variables now are u_{ij} rather than p_{ij}, the optimization problem (3.3.10) is slightly modified for the network of such PLA as

$$\text{Maximize } f(\mathbf{U}) = E[\beta|\mathbf{U}] \qquad (3.7.42)$$

subject to the constraints

$$|u_{ij}| \leq L, \quad \text{for } 1 \leq i \leq N, 1 \leq j \leq r_i \qquad (3.7.43)$$

where $L > 0$ is a positive constant. L can be chosen large enough so that there is not much difference in the value of the constrained and unconstrained global optimum. However, such a constraint is necessary as otherwise the learning algorithm may show unbounded behavior. For example, the global maximum of $[1 + \exp(x)]^{-1}$ is attained at $x = -\infty$ with a value of unity, but a close enough approximation can be obtained by maximizing over $|x| \leq L$, L large enough.

3.7.2 The Global Learning Algorithm

The proposed learning algorithm for achieving global convergence is as follows. The internal state of the ith PLA (which is activated) is updated as

$$\begin{aligned} u_{ij}(k+1) &= u_{ij}(k) + \lambda \beta(k) \frac{\partial \ln g_i}{\partial u_{ij}}(\mathbf{u}_i(k), \alpha_i(k)) \\ &\quad + \lambda h'(u_{ij}(k)) + \sqrt{\lambda} \zeta_{ij}(k). \end{aligned} \qquad (3.7.44)$$

For all PLA which are not activated at k,

$$u_{ij}(k+1) = u_{ij}(k) + \lambda h'(u_{ij}(k)) + \sqrt{\lambda} \zeta_{ij}(k) \qquad (3.7.45)$$

where, $h'(.)$ is the derivative of $h(.)$ given by

$$h(x) = \begin{cases} -K(x-L)^{2J} & x \geq L \\ 0 & |x| \leq L \\ -K(x+L)^{2J} & x \leq -L \end{cases} \qquad (3.7.46)$$

and

$\{\zeta_{ij}(k) : 1 \leq j \leq r_i, 1 \leq i \leq N\}$ is a set of *iid* random variables with zero mean and variance σ^2, σ and K are positive constants, $\lambda > 0$ is the learning parameter, J is a positive integer.

The second term on the RHS of (3.7.44) is a gradient following term. The term containing $h'(.)$ is introduced to keep the u_{ij} bounded. The last term containing ζ_{ij} is the random perturbation term to get the algorithm out of local optima which are not global optima. Since PLA in unactivated units do not contribute to the output of the network, there is no gradient term in (3.7.45).

The above algorithm is based on the REINFORCE algorithm [Wil92] which contains only the first two terms in (3.7.44). The REINFORCE algorithm has been shown to exhibit unbounded behavior and has only local convergence properties [Pha91]. The extra terms in (3.7.44) have been introduced to obtain global solutions which are bounded.

3.7.3 Analysis of the Global Algorithm

For each λ, (3.7.44), (3.7.45) generate a Markov process, $\{\mathbf{U}(k)\}$, where $\mathbf{U}$ is the super vector containing all $\mathbf{u}_i$. The evolution of this process depends of the step size λ. Denote this dependence explicitly by writing $\mathbf{U}^\lambda$. Let $U^\lambda(.)$ be the continuous time interpolation of $\mathbf{U}^\lambda(.)$ defined as

$$U^\lambda(t) = \mathbf{U}^\lambda(k) \quad \text{if} \quad t \in [k\lambda, (k+1)\lambda). \tag{3.7.47}$$

The following theorem gives the approximating stochastic differential equation (SDE). The following assumptions are needed to prove the theorem.

(A1) $\partial g_i / \partial u_{ij}$ is continuous for all i and j.

(A2) g_i is bounded away from zero on every compact $\mathbf{u}_i$ set.

THEOREM 3.2 *Under assumptions (A1) and (A2) above, the sequence $\{U^\lambda : \lambda > 0\}$ converges weakly [Bil68] as $\lambda \to 0$ to $\mathbf{z}$ where $\mathbf{z}$ satisfies the Langevin equation*

$$d\mathbf{z} = \nabla V(\mathbf{z})dt + \sigma d\mathcal{W}; \quad \mathbf{z}(0) = \mathbf{U}(0) \tag{3.7.48}$$

where

$$V(\mathbf{z}) = f(\mathbf{z}) + \sum_{i,j} h(u_{ij}) \tag{3.7.49}$$

and f is defined by (3.7.42). $\mathcal{W}$ is Brownian Motion of the appropriate dimension.

As part of the proof of the theorem, we first prove the following lemma.

LEMMA 3.5

$$\begin{aligned} H &\triangleq E\left[\beta(k)\frac{\partial \ln g_i}{\partial u_{ij}}\left(\mathbf{u}_i(k), \alpha_i(k)\right) \mid \mathbf{U}(k) = \mathbf{U}\right] \\ &= \frac{\partial f}{\partial u_{ij}}(\mathbf{U}). \end{aligned} \tag{3.7.50}$$

Proof: Let p_{ij} be the probability of choosing an action by the ith automaton of the network. This automaton is assumed to belong to a unit denoted by $\mathcal{U}$. $\mathbf{C}$ is the context vector from the environment. Context vectors from the environment arrive according to the probability distribution P_e.

Convergence to the Global Maximum 133

U is the vector composed of the state vectors of all the automata in the network. **X** is the context vector to unit $\mathcal{U}$ under consideration. To calculate H, condition first on the context vector **C** from the environment and then on **X**, the context vector to unit $\mathcal{U}$. The unit $\mathcal{U}$ can then be studied in isolation, since the effect of other units on $\mathcal{U}$ is only through **X**. As the system is a feedforward network, the units in the same and later layers do not affect the functioning of the unit $\mathcal{U}$. $Q(.|\mathbf{U},\mathbf{C})$ is the probability distribution of **X**, conditioned on **U** and **C**. Q is independent of u_i, since the unit $\mathcal{U}$ has no effect on **X**, once **U** and **C** are known. In fact, the conditioning need not be on the entire **U**, but only on those components of **U** which determine the action probability vectors of automata belonging to units in layers before unit $\mathcal{U}$. Conditioning on **C** and **X** (all conditioning is at the instant k)

$$H = \int\int E\left[\beta\frac{\partial \ln g_i}{\partial u_{ij}}\,\bigg|\,\mathbf{U},\mathbf{C},\mathbf{X}\right] Q(d\mathbf{X}|\mathbf{U},\mathbf{C})P_e(d\mathbf{C}). \quad (3.7.51)$$

Condition further on the actions of the automaton which has u_{ij} as one of its state vector components. The state vector of the automaton under consideration is $\mathbf{u}_i$. a_{iq} is the qth action corresponding to the probability $p_{iq} = g_i(\mathbf{u}_i, a_{iq})$. Thus

$$H = \int\int \sum_q p_{iq}\left\{E\left[\beta\frac{\partial \ln g_i}{\partial u_{ij}}\,\bigg|\,\mathbf{U},\mathbf{C},\mathbf{X},a_{iq}\right]\right\} Q(d\mathbf{X}|\mathbf{U},\mathbf{C})P_e(d\mathbf{C}) \quad (3.7.52)$$

$$= \int\int \sum_q \frac{\partial g_i}{\partial u_{ij}} E[\beta|\mathbf{U},\mathbf{C},\mathbf{X},a_{iq}]\, Q(d\mathbf{X}|\mathbf{U},\mathbf{C})P_e(d\mathbf{C}) \quad (3.7.53)$$

To prove (3.7.50), $\partial f/\partial u_{ij}$ needs to be calculated and compared with (3.7.53). To calculate $\partial f/\partial u_{ij}$, condition $f(\mathbf{U})$ on **C** and then **X**.

$$f(\mathbf{U}) = \int\int E[\beta|\mathbf{U},\mathbf{C},\mathbf{X}]Q(d\mathbf{X}|\mathbf{U},\mathbf{C})P_e(d\mathbf{C}). \quad (3.7.54)$$

Conditioning further on a_{iq}, the action chosen,

$$f(\mathbf{U}) = \int\int \sum_q p_{iq}E[\beta|\mathbf{U},\mathbf{C},\mathbf{X},a_{iq}]\, Q(d\mathbf{X}|\mathbf{U},\mathbf{C})P_e(d\mathbf{C}). \quad (3.7.55)$$

For all q, $E[\beta|\mathbf{U},\mathbf{C},\mathbf{X},a_{iq}]$ is independent of $\mathbf{u}_i$ because the automaton has already chosen its action. As mentioned before, $Q(.|\mathbf{U},\mathbf{C})$ is independent of $\mathbf{u}_i$. Therefore, differentiating (3.7.55) with respect to u_{ij}

$$\frac{\partial f}{\partial u_{ij}} = \sum_q \int\int \left\{E[\beta|\mathbf{U},\mathbf{C},\mathbf{X},a_{iq}]\frac{\partial g_i}{\partial u_{ij}}(\mathbf{u}_i,a_{iq})\right\} Q(d\mathbf{X}|\mathbf{U},\mathbf{C})P_e(d\mathbf{C}). \quad (3.7.56)$$

Comparing (3.7.53) and (3.7.56), we have

$$H = \frac{\partial f}{\partial u_{ij}}(\mathbf{U}) \qquad (3.7.57)$$

completing the proof. ∎

Proof of Theorem 3.2:

Algorithm (3.7.44), (3.7.45) can be rewritten in vector form as

$$\mathbf{U}^\lambda(k+1) = \mathbf{U}^\lambda(k) + \lambda \mathbf{G}(\mathbf{U}^\lambda(k), \boldsymbol{\theta}^\lambda(k)) + \sqrt{\lambda}\, \eta(\boldsymbol{\theta}^\lambda(k)). \qquad (3.7.58)$$

where $\boldsymbol{\theta}(k)$ is the vector composed of the SRS, the actions of the various units and the context vector from the environment and the ζ_{ij}'s, all at instant k.

Now, as shown in Section C.3.1 in Appendix C, the given algorithm can be approximated by the stochastic differential equation (SDE)

$$d\mathbf{z} = \mathbf{F}(\mathbf{z})dt + \sigma d\mathbf{W}; \quad \mathbf{z}(0) = \mathbf{U}(0) \qquad (3.7.59)$$

and $\mathbf{F}(\mathbf{z})$ is given by

$$\mathbf{F}(\mathbf{z}) = E^{\mathbf{z}}[\mathbf{G}(\mathbf{z}, \boldsymbol{\theta})]. \qquad (3.7.60)$$

where $E^{\mathbf{z}}$ denotes the expectation with respect to the appropriate invariant probability measure (see Appendix C for details). Using Lemma 3.5, it is seen that

$$F_{ij}(\mathbf{z}) = \frac{\partial f}{\partial z_{ij}}(\mathbf{z}) + h'(z_{ij}). \qquad (3.7.61)$$

That is,

$$\mathbf{F}(\mathbf{z}) = \nabla V(\mathbf{z}) \qquad (3.7.62)$$

where $V(\cdot)$ is defined in (3.7.49). Thus the SDE (3.7.59) reduces to

$$d\mathbf{z} = \nabla V(\mathbf{z})dt + \sigma d\mathbf{W}; \quad \mathbf{z}(0) = \mathbf{U}(0)$$

completing the proof.

REMARK 3.7.1 *The structure of the network does not explicitly affect (3.7.48). This is because the structure is reflected only via the function $f(\cdot)$. In fact, the above result would hold for almost any kind of feedforward network structure.*
∎

Global Maxima of $f(\cdot)$

Equation (3.7.48) is the Langevin equation and it is well known that the invariant probability measure of (3.7.48) concentrates on the global maxima of $V(\cdot)$ as $\sigma \to 0$ [APPZ85]. However, the goal of the algorithm (3.7.44), (3.7.45)

Convergence to the Global Maximum 135

is to maximize $f(.)$ as given by the constrained optimization problem (3.7.42), (3.7.43), not $V(.)$. Let V^* be the global maximum of $V(.)$. The following discussion relates the maxima of $V(.)$ to global solutions of the optimization problem given by (3.7.42), (3.7.43). Consider a point $\mathbf{U}^*$ such that

$$V(\mathbf{U}^*) = V^*. \tag{3.7.63}$$

Define the set FS as

$$FS = \{\mathbf{U} : |u_{ij}| \leq L \quad \forall i, j\}. \tag{3.7.64}$$

FS is the set over which f has to be optimized. Since $h(.)$ is zero over the closed interval $[-L, L]$,

$$V(\mathbf{U}) = f(\mathbf{U}) \quad \forall \mathbf{U} \in FS. \tag{3.7.65}$$

Let $\mathbf{U}^* \in FS$. Then, by (3.7.65)

$$f(\mathbf{U}^*) = V(\mathbf{U}^*) = V^* \tag{3.7.66}$$

and by the definition of the global maximum, $V(\mathbf{U}^*) \geq V(\mathbf{U})$ for all $\mathbf{U}$ and in particular for all $\mathbf{U} \in FS$. Thus for any $\mathbf{U} \in FS$

$$f(\mathbf{U}) = V(\mathbf{U}) \leq V(\mathbf{U}^*) = f(\mathbf{U}^*). \tag{3.7.67}$$

Thus $\mathbf{U}^*$ maximizes $f(.)$ globally over FS. There can be cases where $\mathbf{U}^* \notin FS$. Then, for all $\mathbf{U}$,

$$V(\mathbf{U}^*) \equiv f(\mathbf{U}^*) + \sum_{i,j} h(u_{ij}^*) \geq V(\mathbf{U}). \tag{3.7.68}$$

Consider any $\mathbf{U} \in FS$. Using (3.7.65) and (3.7.68) it is seen that

$$f(\mathbf{U}^*) \geq f(\mathbf{U}) - \sum_{i,j} h(u_{ij}^*). \tag{3.7.69}$$

Since $h(.)$ is non-positive, (3.7.69) implies

$$f(\mathbf{U}^*) > f(\mathbf{U}) \quad \forall \mathbf{U} \in FS. \tag{3.7.70}$$

Thus, the value of $f(.)$ at $\mathbf{U}^*$ is greater than the value of $f(.)$ at any point in FS. The original problem of maximizing $f(.)$ was converted to maximizing $f(.)$ over FS to avoid unbounded solutions. It is not a strict condition that the solution should lie within FS. It is acceptable if a point outside FS is located such that the value of $f(.)$ at that point is greater than or equal to the maximum of $f(.)$ over FS.

Since the complement of FS, FS^c, is an unbounded set, the behavior of the algorithm can become unbounded if there is a global maximum of $V(.)$ at infinity. The following lemma guarantees that this cannot happen. The set of points where $V(.)$ attains its global maximum is a bounded set. It is also compact since $V(.)$ is continuous.

LEMMA 3.6 *All the points where $V(.)$ attains its global maximum lie within a bounded set.* ∎

Proof: Since $f(\mathbf{U}) = E[\beta \,|\, \mathbf{U}]$ and the SRS is assumed to take values between 0 and 1, $0 \leq f \leq 1$. Let any one particular component of $\mathbf{U}$, say u_{kl}, be such that

$$u_{kl} > L + s \qquad (3.7.71)$$

where $s > 0$. Then, by simple algebraic calculations it can be shown that

$$V(\mathbf{U}) = f(\mathbf{U}) + \sum_{i,j} h(u_{ij}) \leq 1 - Ks^{2J}.$$

Thus, if $s > K^{(-1/2J)}$, it is seen that $V(\mathbf{U}) < 0$. A similar condition holds for $u_{kl} < -L - s$. Since $V(.)$ and $f(.)$ agree on FS and $f \geq 0$, the global maximum of $V(.)$ is nonnegative. Thus $V(.)$ cannot attain its maximum at $\mathbf{U}$ if even one component of $\mathbf{U}$ exceeds $L + K^{(-1/2J)}$ in absolute value. Define

$$\hat{\varepsilon} = \{\mathbf{U} : |u_{ij}| \leq L + K^{(-1/2J)} \forall (i,j)\}. \qquad (3.7.72)$$

Then all the points at which $V(.)$ attains its global maximum lie in the set $\hat{\varepsilon}$ which is bounded. ∎

3.8 Networks of GLA

The development outlined so far could be extended to feedforward networks where each unit is a Generalized Learning Automation (GLA) [PT95]. It may be recalled that a GLA uses the context vector in updating its state which in turn generates the action probability vector.

Consider the ith GLA in the network. Its state vector is $\mathbf{u}_i$ which has r_i components $u_{ij}(j = 1, \ldots, r_i)$. A learning algorithm which ensures boundedness of the solutions is as follows.

$$\begin{aligned} u_{ij}(k+1) &= u_{ij}(k) + \lambda \bigg\{ \beta(k) \frac{\partial \ln g_i}{\partial u_{ij}}(\mathbf{X}_i(k), h_i(\mathbf{u}_i(k)), y_i(k)) \\ &\quad + K_{ij}[h_{ij}(u_{ij}(k)) - u_{ij}(k)] \bigg\} \end{aligned} \qquad (3.8.73)$$

where
$$h_{ij}(\eta) = \begin{cases} L_{ij} & \eta \geq L_{ij} \\ \eta & |\eta| \leq L_{ij} \quad i = 1, \ldots, N \\ -L_{ij} & \eta \leq -L_{ij} \quad j = 1, \ldots, r_i \end{cases} \quad (3.8.74)$$

and
$$\mathbf{u}_i = [u_{i1}, u_{i2} \ldots u_{ir_i}]^T,$$
$$h_i(\mathbf{u}_i) = [h_{i1}(u_{i1}), h_{i2}(u_{i2}), \ldots h_{ir_i}(u_{ir_i})]^T,$$

g_i is the action probability generating function.

Analysis of the network of GLA can be carried out in a manner similar to that of FALA. Under similar assumptions about the nature of maxima it can be shown [PT95] that the network of GLA using the learning algorithm (3.8.73) converges to a constrained local maximum of $E[\beta \mid \mathbf{U}]$.

The convergence point can be changed from local maximum to global maximum by introducing a perturbation term and modifying the bounding term. The global algorithm is as follows.

$$u_{ij}(k+1) = u_{ij}(k) + \lambda \beta(k) \frac{\partial \ln g_i}{\partial u_{ij}} (\mathbf{X}_i(k), y_i(k), \mathbf{u}_i(k))$$
$$+ \lambda h'(u_{ij}(k)) + \sqrt{\lambda} \zeta_{ij}(k) \quad (3.8.75)$$

where
$$h(\eta) = \begin{cases} -K_{ij}(\eta - L_{ij})^{2J} & \text{for} \quad \eta \geq L_{ij} \\ 0 & \text{for} \quad |\eta| \leq L_{ij} \\ -K_{ij}(\eta + L_{ij})^{2J} & \text{for} \quad \eta \leq -L_{ij} \end{cases} \quad (3.8.76)$$

$K_{ij} \geq 0$, J, is a positive integer and h' is the derivative of h. ζ_{ij} is a sequence of i.i.d. random variables with zero mean and variance σ^2.

The details of the analysis can be found in [PT95, Pha91]. As in the earlier development, approximation of the algorithm by a stochastic differential equation plays a central role.

3.9 Discussion

We have indicated in this chapter how complex structures of LA can be constructed. Such structures could be feedforward networks of learning units each of which consists of many FALA. The unit could be a simple team of FALA playing a common payoff game or a sophisticated hierarchical organization of such teams or even a feedforward network by itself. The convergence results derived in this chapter hold good for all these structures.

Apart from networks of FALA, one could construct networks of PLA and GLA also. Moreover, for such networks, one could use the global learning algorithms which ensure convergence to the global maximum.

In situations where the action set of automata has to be an interval on the real line, one could use a network of CALA. Hybrid networks consisting of teams of FALA and CALA could also be considered along with local and global learning algorithms designed along the same lines as algorithms presented here. However, convergence results for such networks are not yet available.

This chapter has developed the theory needed for a general class of networks of LA. In specific applications, one has to particularize this theory to design specific network structures to suit the application on hand. This is illustrated for one particular application, namely pattern recognition, in Chapter 4. While the theory is general enough for any size of the network, in practice speed of learning reduces with increase in the number of LA in the network. We consider one general technique for improving the speed of LA systems in Chapter 5.

3.10 Supplementary Remarks

The idea of feedforward networks of units where each unit is a LA was first proposed in [Wil88]. Analysis of such networks with more sophisticated units consisting of hierarchies and teams of LA was made in [Pha91, TP95a]. Convergence properties of networks of GLA with local and global algorithms was studied in [Pha91, PT95]. Networks of PLA for learning global maxima were analyzed in [TP95b]. Generalization of units to feedforward connections of LA and to nonlinear learning algorithms was made in [PT96].

Chapter 4

LEARNING AUTOMATA FOR PATTERN CLASSIFICATION

4.1 Introduction

In the previous two chapters we have seen some methods of employing collectives of learning automata to tackle different learning problems. In Chapter 2 we have presented the model of a set of automata engaged in a general game situation. The automata involved can be FALA or CALA. We have provided learning algorithms so that the automata can converge to (close approximation of) the optimal points of the game. We have also illustrated how such a model can be used for maximizing a function under noisy measurements (with no gradient information available), in applications such as system identification, and learning conjunctive concepts from noisy examples. In Chapter 3 we have seen how we can build much more powerful models by combining such teams of automata into networks. The structure of such networks can be very general, thus allowing a lot of flexibility in designing automata solutions to specific learning problems. As has been mentioned, we can also utilize other automata models such as PLA and GLA in such structures. We have also shown that with PLA we can have learning algorithms (for networks of automata) that converge to the global maximum of expectation of reinforcement.

In this chapter, we consider one generic application area, namely, pattern recognition, in detail to show how all these ideas can be utilized in specific learning situations. We illustrate many ways of utilizing different multiautomata structures in pattern recognition problems. We begin with a general discussion on the pattern recognition problem. For simplicity of exposition we restrict our attention to 2-class problems. However, all the methods we discuss can be easily generalized to multi-class problems.

In a pattern recognition problem, one is interested in learning a decision rule for classification that is optimal under some criterion (such as probability of misclassification) using a given sample of noisy training patterns. The main strength of LA models is the flexibility they offer in constructing a parametric representation of the family of classification rules over which we wish to search for the best classifier. We illustrate this by considering different ways of utilizing all the LA models discussed so far for this problem.

4.2 Pattern Recognition

Classifying sensory information into some categories based on which appropriate action can be taken, is an important aspect of intelligent behavior. The problem of Pattern Recognition (PR) is an abstraction of such categorization tasks. Typical examples of PR problems are signature verification, speech recognition, character recognition, person identification based on finger prints etc. In a general PR problem, the input is the physical pattern (such as the image of a finger print, image of a document etc.) and the output is a class label representing a recognition decision [Nil65, DH73].

The nature of the input patterns as well as the classification labels to be output by the system depend on the specific application. For example, in OCR applications, the input pattern is a two dimensional (gray-scale) image of a character symbol and the output label (also called the class label or class) is the name (or some other representation) of the character present in the image. In speech recognition systems, the input pattern is a (sampled version of the) time-varying signal from a microphone and the output class label is the identity of the word(s) (or any other speech unit) that compose the speech utterance. In a fingerprint based identity verification system, input is the image of the fingerprint pattern of a person (along with some other claim of identity such as a name) and the output class label is binary specifying whether or not the person is who he claims to be. While the first two examples represent multiclass problems, the last example is a two class problem.

Any general PR system can be viewed as implementing a two step procedure: feature extraction and classification [DH73].

In the first step, the system extracts or determines some salient characteristics, called *features*, from the input pattern. In most cases the features are real numbers. Thus, though the input pattern may be in some arbitrary representation (such as an image), after feature extraction each pattern is represented by a vector of real numbers, called the *feature vector*. In some applications it may be desirable that not all features are numeric-valued. For example, in concept learning problems some of the features may be non-numeric. While it is possible to arbitrarily encode such features (e.g., color of an object) into some numbers, it may not be a satisfactory solution. In Chapter 2, we have shown how a team of LA can be used for learning conjunctive concepts where some of the features are real-valued while others are non-numeric (or nominal). Such teams of LA can be used for learning more complex classes of concepts also [Raj96]. However, in this chapter, we focus on the traditional PR problem where all features are assumed to be numerical valued.

After feature extraction, the classification step is to assign a class label to each feature vector. Since we are considering only 2-class problems here, the classifier is a function that maps the set of feature vectors to the set $\{0, 1\}$. Any such function is called a decision rule. A good classifier or a decision rule is one that maps a feature vector to a class label that most often corresponds to the class that the pattern represented by the feature vector belongs to.

The design of a PR system now involves two steps: design of a feature extractor and design of a classifier. Feature extraction is very much problem specific. For example, the features that we would like to measure from a fingerprint image so as to be able to make the correct identification decision would be different from those needed for correctly identifying a character from its image. However, after fixing the set of features, design of a classifier admits some general procedures. The techniques we discuss in this chapter are all meant for designing the classifier. Hence we start with formulating the problem of classifier design. For this, we shall be considering the Pattern Recognition problem in the statistical framework [DH73]. That is, we assume that the variations in the feature vectors corresponding to patterns of one class can be captured in terms of probability distributions which can be used for discriminating between feature vectors of different classes.

Let $\mathcal{X}$ be the feature space, that is, set of all feature vectors. If we are using n real valued features then $\mathcal{X}$ would be $\Re^n$, the n-dimensional real Euclidean space. Let $\mathcal{Y} = \{0, 1\}$ be the set of possible outputs of our classifier. Every classifier or decision rule is a function from $\mathcal{X}$ to $\mathcal{Y}$. Let $\mathcal{H}$ be the set of all possible classifiers or decision rules of interest. For a pattern or feature vector, $X \in \mathcal{X}$, let $y(X)$ denote its 'true' class label. It may be noted here that, in general, $y(X)$ would be a random variable. This is because the set of all possible feature vectors obtained from all patterns of one class is not necessarily disjoint with the set of feature vectors of the other class. Similar feature vectors can come from patterns of different classes (with different probabilities) and thus we can only talk about the probability of a feature vector belonging to one class or the other. Also note that this is more due to the specific set of features chosen in a given system than an inherent property of the underlying pattern spaces. For any $h \in \mathcal{H}$, $h(X)$ denotes the class label assigned to X by the classifier h. We can represent the goodness of the classifier h on pattern X using $L(h(X), y(X))$ where $L(\cdot, \cdot)$ is called a *loss function*. We assume here that $0 \leq L(x, y) \leq 1$. A simple loss function is given by: $L(x, y) = 0$ when $x = y$ and $L(x, y) = 1$ when $x \neq y$. This is called a $0 - 1$ loss function. Since $L(h(X), y(X))$ denotes the loss suffered by h on pattern X, an overall figure of merit for any classifier h can be given by a functional, $F(\cdot)$, defined on $\mathcal{H}$

by
$$F(h) = E\left(1 - L(h(X), y(X))\right), \quad h \in \mathcal{H} \tag{4.2.1}$$

where E denotes expectation with respect to the joint probability distribution of X and $y(X)$. A good classifier would have high value for F. If we are using the $0-1$ loss function, $F(h)$ would be the probability that $h(X) = y(X)$. That is,

$$\begin{aligned} F(h) &= E\, I_{\{h(X)=y(X)\}} \\ &= \text{Prob}[h(X) = y(X)], \end{aligned} \tag{4.2.2}$$

where I_A represents the indicator function of event A. Thus, a h which results in the maximum value of $F(h)$ would be a classifier with maximum probability of correct classification, or, equivalently, minimum probability of misclassification. Though $F(h)$ gives the probability of correct classification with h, only when we use $0-1$ loss function, the definition of $F(h)$ given by (4.2.1) is reasonable for many other types of loss functions. We note here that the criterion F defined by (4.2.1) is reasonable only if the values of loss function are in the interval $[0, 1]$. However, this general structure is applicable for any reasonable loss function. For example, if the loss function is positive but is not necessarily bounded above by unity, we could attempt to minimize expected value of loss function. We will discuss some of these later on in this section.

The next question is, how do we find a h to maximize $F(h)$? Define, $Q_0(X) = \text{Prob}[y(X) = 0|X]$ and $Q_1(X) = 1 - Q_0(X)$. These are called posterior probabilities (or posterior probability functions) of the classes. Define a classifier h by

$$\begin{aligned} h(X) &= 0 \text{ if } Q_0(X) > Q_1(X) \\ &= 1 \text{ if } Q_0(X) \le Q_1(X). \end{aligned} \tag{4.2.3}$$

It is easy to show that this h is optimal in the sense that it would achieve minimum probability of misclassification [DH73]. Let $f_0(X)$ denote the probability density function of feature vectors belonging to class 0 and similarly $f_1(X)$. These are called class conditional densities. Define $p_0 = \text{Prob}[y(X) = 0]$ and $p_1 = 1 - p_0$. These are referred to as prior probabilities. Since these are unconditional probabilities, p_0 denotes the probability that a random pattern belongs to class 0 and thus indicates the fraction of the total number of patterns that are in class 0. Now, using Bayes theorem of conditional probabilities, we have $Q_0(X) = f_0(X)p_0$ and similarly for Q_1. Define $g_B(X) = f_0(X)p_0 - f_1(X)p_1$. Now we can rewrite (4.2.3) as

$$\begin{aligned} h(X) &= 0 \text{ if } g_B(X) > 0 \\ &= 1 \text{ Otherwise.} \end{aligned} \tag{4.2.4}$$

The classifier given by (4.2.3) or equivalently by (4.2.4) is called the Bayes classifier. In a real problem it would be very difficult to actually derive the Bayes classifier because we do not know the class conditional densities or posterior probabilities. The main reason for this is that we generally have no information about the statistics of pattern classes (under the chosen feature vector representation). However, what we have are some examples or the so called training samples. The set of training samples is $\{(X_1, y_1), \ldots, (X_m, y_m)\}$ where each $X_i \in \mathcal{X}$ is a feature vector obtained from some pattern and y_i is the class label of that pattern. It may be noted here that the y_i is not obtained using some *special* classifier function on X_i. (If we have available such a function, we do not need to build a PR system for the problem!). It is obtained through a teacher (usually a human expert) classifying these training patterns. For example, in OCR application, we take some known characters (so that we can assign the y_i) and then obtain the corresponding feature vectors (X_i) by applying feature extraction programs. Hence, depending on the selected set of features, the training sample may be noisy in the sense that two arbitrarily close X_i may have different y_i as labels. In addition, in complicated PR problems the human expert acting as the teacher may himself make mistakes occasionally, thus introducing additional noise. Now the problem of classifier design is to build a procedure that infers or learns 'optimal' decision rules for classification using only these noisy training samples.

One approach to this problem is to assume that the form of the class conditional probability density functions is known. For example, we can assume that these are multidimensional Gaussians with unknown mean vectors and covariance matrices. Then, using some statistical estimation techniques, the unknown parameters of these density functions can be estimated from the given training samples [DH73]. Once the class conditional densities and prior probabilities are estimated, we can use the estimated quantities in the function g_B and the resulting classifier given by (4.2.4) would be an approximation to the Bayes classifier. While this method can be employed in some applications, there are some difficulties with this approach. The class of density functions that can be efficiently estimated is somewhat restricted. In addition, it is often difficult to relate errors in density estimation to probability of error in classification, which is a more important performance measure.

In cases where we cannot easily guess even the form of the density function or when it is difficult to do efficient density estimation for any other reason, we need an alternative method. A popular approach is to use what is known as a discriminant function. Here we consider a family of classifiers given by

$$h(X) = 0 \text{ if } g(W, X) > 0$$
$$= 1 \text{ Otherwise.} \qquad (4.2.5)$$

where $g(W, X)$ is called a discriminant function which is parameterized by a parameter vector W. By choosing different values for W we get different classifiers in this family. Since each classifier h in this family is specified by a value of parameter vector W, we can denote the decision made on X by a generic classifier in this family by $h(W, X)$ which is same as $h(X)$ in (4.2.5). The design of a classifier in this approach proceeds as follows. We first fix a form for the discriminant function g parameterized by a suitable W. Then we obtain the optimal classifier in the family given by (4.2.5) by searching for an optimal W with respect to some criterion.

Many choices are possible for the form of a discriminant function. If the feature vector has n components given by $X = (x_1, \ldots, x_n)$ then we can choose $W = (w_0, w_1, \ldots, w_n)$ and use a linear discriminant function given by $g(W, X) = w_0 + \sum_{i=1}^{n} w_i x_i$. Another choice is to use a neural network as a discriminant function when X would be input to the neural network and W would correspond to all the weights in the network.

Now the next question is to specify the criterion function used to define our 'optimal' classifier. The obvious choice would be to use the figure of merit F defined by (4.2.1). Since each classifier h in our case is specified by the parameter vector W, we can think of F as a function of W as

$$F(W) = E\left(1 - L(h(W, X), y(X))\right), \quad (4.2.6)$$

with L being a suitable loss function.

Obtaining the optimal parameter vector, say, W^*, that maximizes F defined by (4.2.6) is not a standard optimization problem. Recall that E in (4.2.6) denotes expectation with respect to the *unknown* probability distributions of the feature vectors of different classes. Hence given a W we cannot even calculate $F(W)$! We have to somehow use the given training patterns to solve this optimization problem. One way of doing this is to approximate the F in (4.2.6) by

$$\bar{F}(W) = \frac{1}{m} \sum_{i=1}^{m} \left(1 - L(h(W, X_i), y_i)\right), \quad (4.2.7)$$

where $\{(X_i, y_i), i = 1, \ldots, m\}$ is the set of training samples. If we can assume that the training samples are obtained in an *independent and identically distributed* (iid) manner and if the discriminant function g used in computing $h(W, X)$ and the loss function L are well behaved, then, using results from statistical learning theory [Vap97], it can be shown that for large m, the maximizer of $\bar{F}$ would be close to the maximizer of F. The requirement that training samples are *iid* is a formalization of our intuitive notion that the training samples should be *representative* of the kind of patterns that our PR system is likely to encounter in practice. The conditions on g and L needed to use the results of statistical learning theory are also fairly mild. Hence learning an optimal W

by maximizing the function defined by (4.2.7) is a very popular method used in designing pattern classifiers.

There is another (but essentially equivalent) way of finding W that maximizes F given by (4.2.6). Suppose we have an infinite sequence of *iid* samples $\{(X_i, y_i), i = 1, 2, \ldots\}$. Then we can find maximizer of F given by (4.2.6) using only observed values of $L(X_i, y_i)$ through stochastic approximation type algorithms [KY97, BB99]. However, in practice we have only a finite number of training samples. We can construct an infinite sequence out of this by uniformly sampling from the finite training set. In such a case, using stochastic approximation algorithms amounts to maximizing $\bar{F}$ given by (4.2.7) [Vap97].

The next question that needs to be addressed is: what kind of optimization algorithms would be suitable for finding the optimal parameter vector, W^*? A simple technique which is used often is a gradient ascent algorithm using the gradient of the criterion function with respect to W.[1] This would be an iterative algorithm to estimate the optimal parameter vector, W^*. Let $W(k)$ denote this estimate at iteration k. Then the algorithm is specified by

$$\begin{aligned} W(k+1) &= W(k) + \eta \, \nabla_W \bar{F}(W(k)) \\ &= W(k) - \eta \frac{1}{m} \sum_{i=1}^{m} \nabla_W L(h(W(k), X_i), y_i) \quad (4.2.8) \end{aligned}$$

where $\nabla_W \bar{F}$ and $\nabla_W L$ are gradients of the criterion function $\bar{F}$ and loss function L with respect to W respectively, and they are evaluated at the arguments shown. Here η is the step size for the optimization algorithm. The above is a simple deterministic algorithm that can find a local maximum of the criterion function $\bar{F}$.

As mentioned earlier, we could have also used a stochastic approximations based algorithm. One such algorithm is the Robbins–Munro algorithm [KY97, BB99] which has been briefly described earlier in Section 2.6.1.0. This algorithm, for our present case, would be

$$W(k+1) = W(k) - \eta_k \, \nabla_W L(h(W(k), X(k)), y(k)) \quad (4.2.9)$$

where, now, $(X(k), y(k))$ is the random example at iteration k. This is a stochastic algorithm because now $W(k+1)$ depends on the *random* example at iteration k. The step size η_k satisfies $\eta_k \geq 0$, $\sum \eta_k = \infty$ and $\sum \eta_k^2 < \infty$. This Robbins–Munro algorithm is an example of the general class of stochastic approximation algorithms [KY97, BB99] which can be used for finding a maximizer of F based on observed values of $L(h(W, X), y)$ on random (*iid*) samples (X, y).

---

[1] A crucial question here is whether the criterion function is differentiable with respect to W. We shall be considering this issue shortly.

To use either of the algorithms given by (4.2.8) and (4.2.9), we need $\bar{F}$ and hence L to be differentiable. If we use the $0-1$ loss function mentioned earlier, then L is not differentiable. We could use other loss functions such as $L(a,b) = (a-b)^2$. However, in addition to L being differentiable we also need $h(W, X)$ to be differentiable with respect to W. Since h represents a decision rule and hence is binary valued, it would not be differentiable. Hence to use such gradient based techniques, normally one uses a loss function given by $L(a,b) = (a-b)^2$ and defines $\bar{F}(W) = \frac{1}{m} \sum (1 - L(\bar{h}(W, X_i), y_i))$ where $\bar{h}$ is a continuous function of W. If we choose $\bar{h}$ to be a continuous function with range $[0, 1]$ with proper functional form, then the maximizer of this criterion function (using *iid* samples) can be shown to be a least mean square estimate of the posterior probability. That is, if W^* is the (global) maximum of this $\bar{F}$, then $\bar{h}(W^*, X)$ would be a good estimate of $Q_1(X)$ defined earlier. This is the approach that is followed in most neural network based algorithms for designing pattern classifiers.

However, the above method of obtaining W^* using a squared loss function and continuous $\bar{h}$ does not give us a classifier that minimizes the probability of misclassification [SW81]. As explained earlier, the maximizer of F defined by (4.2.6) (or its approximation, $\bar{F}$ given by (4.2.7)) with a $0-1$ loss function would correspond to a classifier with minimum probability of misclassification. From the viewpoint of the users of a PR system, often probability of misclassification is a more meaningful performance measure. If we want to obtain classifiers with minimum probability of misclassification, then we need to have techniques for maximizing F (given by (4.2.6) with $0-1$ loss function) without needing to calculate or estimate the gradient.

Such an alternative approach for optimization is provided by the Learning Automata models. As explained in Chapter 2, models such as a common payoff game of automata provide an interesting stochastic search algorithm for finding the optimum of a regression function without needing the gradient. Unlike stochastic approximation algorithms (see Section 2.6.1) we do not even need to explicitly estimate any gradient information. In this chapter we explain how the automata systems described in the previous two chapters can be used for designing pattern classifiers.

4.3 Common Payoff Game of Automata for Pattern Classification

The first LA model that we consider is the common payoff game of automata discussed in Chapter 2. In this section we see how the models of games of FALA or CALA can be used for learning the parameter vector that corresponds to a discriminant function with minimum probability of misclassification.

We pose the pattern classification problem as follows. Let $g(W, X)$, where X is the feature vector and W is the parameter vector, be the discriminant function. We classify a pattern using the classification rule given by (4.2.5). The form of $g(.,.)$ is assumed known (chosen by the designer). The optimal value for the parameter vector is to be determined by making use of a set of (possibly noisy) *iid* training samples. We call the set of training samples as the training set. We are interested in learning W that maximizes

$$F(W) = E\, I_{\{y(X)=h(W,X)\}} \qquad (4.3.10)$$

where E denotes expectation with respect to the joint distribution of X and $y(X)$, and h is a classifier (which is completely specified by the parameter vector W) as defined in (4.2.5). (Recall that $y(X)$ is the random variable denoting the class label for X). As explained in Section 4.2, $F(W)$ is the probability of correct classification with classifier W. F is defined over $\Re^N$ if there are N parameters and we are interested in finding a W that globally maximizes F.

In any realistic pattern recognition problem, the relevant probability distributions are unknown and hence the expectation in (4.3.10) cannot be evaluated. We need to find the maximizer of F using only the training set. The training set consists of (example patterns given as) pairs $(X, \bar{y}(X))$ where X is a feature vector and $\bar{y}(X)$ is the class label for X *as given* in the training set. Now define

$$\widetilde{F}(W) = E\, I_{\{\bar{y}(X)=h(W,X)\}}. \qquad (4.3.11)$$

Given the set of training patterns, we can evaluate $I_{\{\bar{y}(X)=h(W,X)\}}$ for any classifier W and any sample pattern X. This is what we can use for finding a maximizer of $\widetilde{F}$. Since the samples are *iid*, the value of $\widetilde{F}(W)$ will be same as the probability of correct classification with classifier W (that is, it will be same as $F(W)$) if $y(X) = \bar{y}(X)\ \forall X$, that is, if the training set is *noise free*.

When the training samples are noisy, in general, the class label given for a training pattern X, $\bar{y}(X)$, would not be same as $y(X)$. We can model all such noise as follows. The random variable $\bar{y}(X)$ is such that $\bar{y}(X) = y(X)$ with probability ρ and $\bar{y}(X) = 1 - y(X)$ with probability $1 - \rho$ where ρ is the level of noise. In general, ρ can be a function of X. Here we assume that ρ is a constant. We can think of ρ as the probability of correct classification by the 'teacher'. In this case, the $\widetilde{F}(W)$ defined by (4.3.11) will only give the probability that the classification of a random pattern by the system (that is, by the classifier W) agrees with that of the (noisy) teacher. But we want to actually maximize the probability of correct classification, $F(W)$. Now we have

$$\begin{aligned} F(W) &= \rho E\, I_{\{\bar{y}(X)=h(W,X)\}} + (1-\rho)E\,(1 - I_{\{\bar{y}(X)=h(W,X)\}}) \\ &= (2\rho - 1)E\, I_{\{\bar{y}(X)=h(W,X)\}} + (1-\rho) \\ &= (2\rho - 1)\,\widetilde{F}(W) + (1-\rho) \qquad (4.3.12) \end{aligned}$$

Thus, as long as $\rho > 0.5$, $\widetilde{F}(\cdot)$ and $F(\cdot)$ have the same maxima and hence it is sufficient to maximize $\widetilde{F}$.

In the above we have assumed a uniform classification noise. That is, the probability of the teacher correctly classifying a training set pattern is same for all patterns. Some of the automata algorithms discussed here can also handle the more general case where the probability of teacher correctly classifying X is $\rho(X)$, as long as $\rho(X) > 0.5$, $\forall X$, for certain classes of discriminant functions [Nag97]. However, in this chapter, we assume that ρ is a constant. Hence, as explained above, it is enough if we can find a W that maximizes $\widetilde{F}$.

The general procedure for using a team of LA for this problem is as follows. To learn a discriminant function with N parameters, we would need a team of N learning automata. The actions of the automata would be possible values for the respective parameters. The choice of action by each automaton in the team results in the random choice of a specific discriminant function. We classify the next training pattern with this discriminant function and the correctness or otherwise of this classification is supplied as the common reinforcement. In the next two subsections we describe teams of FALA and CALA for this problem.

4.3.1 Pattern Classification with FALA

As explained earlier, for learning the optimal classifier, we need to learn the optimal value of the parameter vector, $W = [w_1, \ldots, w_N] \in \Re^N$. Let $w_i \in V^i \subset \Re$. In any specific problem, knowledge of the parametric form chosen for the discriminant function and knowledge of the region in the feature space where the classes cluster, is to be utilized for deciding on the sets V^i. We use a team of N automata for learning optimal values of N parameters. In this section we are concerned with FALA and each automaton can have only finitely many actions. Hence we need to discretize the sets V^i to come up with action sets for the automata. For this, we partition each of the sets V^i into finitely many intervals V^i_j, $1 \leq j \leq r_i$. and choose one point, v^i_j, from each interval V^i_j, $1 \leq j \leq r_i$, $1 \leq i \leq N$. We take the action set[2] of ith automaton as $\{v^i_1, \ldots, v^i_{r_i}\}$. Thus, the actions of ith automaton are the possible values for the ith parameter, which are finitely many due to the process of discretization.

Now consider the following common payoff game played by these N automata. At each instant k, the ith automaton chooses an action $\alpha^i(k)$ independently and at random according to its action probabilities, $\mathbf{p}_i(k)$, $i = 1, \ldots, N$. Since actions of automata are possible values for parameters, this results in the choice of a specific parameter vector, say $W(k)$ by the automata team. The

---

[2]Recall that in Chapter 2, the action set of ith FALA is taken as $\{\alpha_{i1}, \ldots, \alpha_{ir_i}\}$. When the actions correspond to specific quantities of interest in the application domain, as is the case here, we employ different symbols to denote actions.

environment classifies the next sample pattern using this parameter vector, and the correctness or otherwise of this classification is supplied to the team as the common reinforcement, $\beta(k)$. Specifically

$$\begin{aligned}\beta(k) &= 1 \quad \text{if } h(W(k), X(k)) = \bar{y}(X(k)) \\ &= 0 \quad \text{otherwise}\end{aligned} \quad (4.3.13)$$

where $X(k)$ is the sample pattern at k and $\bar{y}(k)$ is its class label in the training set.

It is easy to see from equations (4.3.11) and (4.3.13) that the expected value of the common payoff to the team at k is equal to $\widetilde{F}(W(k))$ where $W(k)$ is the parameter vector chosen by the team at k. Now it follows from equations (4.3.11), (4.3.12) and (4.3.13) that the optimal tuple of actions for the team (corresponding to the maximum element in the reward matrix) is the optimal parameter vector that maximizes the probability of correct classification. Hence all we need is a learning algorithm for the team which ensures convergence to such optimal actions. For this we can use the algorithms discussed in Chapter 2 and we discuss two such algorithms in the next two subsections.

Before proceeding further, it should be noted that this method (in the best case) would only converge to the classifier that is optimal from among the *finitely* many classifiers in the set $\prod_{i=1}^{N} V^i$. This can only be an approximation to *the* optimal classifier due to the inherent loss of resolution in the process of discretization. While this approximation can be improved by finer discretization, it can result in a large number of actions for each automaton and consequently, slow convergence. One can also improve the precision in the learnt classifier by progressively finer discretization. That is, we can first learn a rough interval for the parameter and then can choose the V^i set as this interval and further subdivide it and so on. However, the method is most effective in practice mainly in two cases: when there is sufficient knowledge available regarding the unknown parameters so as to make the sets V^i small enough intervals or when it is sufficient to learn the parameter values to a small degree of precision. Since we impose no restrictions on the form of the discriminant function $g(.,.)$, we may be able to choose a convenient parameterization for the discriminant function so as to have some knowledge of the sets V^i. In Section 4.3.2 we will employ a team of CALA for solving this problem where no discretization of parameter ranges would be necessary.

L_{R-I} Algorithm for the team

As discussed in Section 2.5, we can use L_{R-I} algorithm to update the action probability vector of each automaton using the common payoff given by (4.3.13).

This will be a decentralized learning technique. No automaton needs to know the actions selected by other automata or their action probabilities. In fact each automaton is not even aware that it is part of a team because it is updating its action probabilities as if it were interacting alone with the environment.

Recall from Chapter 2 (cf. Section 2.5.1), that in a common payoff game, if each automaton uses an L_{R-I} algorithm with sufficiently small step-size, then the team will converge with arbitrarily high probability to a set of actions that is a *mode* of the reward matrix. The mode is a Nash equilibrium in the common payoff game. From the point of view of optimization, it amounts to a local maximum. If the reward matrix of the game is unimodal then the automata team using the L_{R-I} algorithm will converge to the optimal classifier. For example, it is easy to show that if the class conditional densities are normal (with both classes having the same covariance matrix) and if the discriminant function is linear, then the game matrix would be unimodal. Another example where the automata team with L_{R-I} algorithm is similarly effective is that of learning simple conjunctive concepts [SRR93, RS97] as discussed in Section 2.7.2. However, in general, with this algorithm the team can converge only to a local maximum of $F(.)$ and, depending on the specific application, it may or may not be acceptable.

Pursuit Algorithm for the Team

As discussed in Section 2.5.2, in a common payoff game of FALA, by using pursuit algorithm the team converges to the global maximum of the reward matrix. In the present case, this implies convergence to the global maximum of F (with discretized parameters). Thus the automata team would learn the optimal parameter vector even if the reward matrix is multi-modal.

It may be noted that the pursuit algorithm is not decentralized unlike the L_{R-I} algorithm. To maintain the estimated reward probability matrix, we need to know the actions chosen by all the automata at that instant. As noted in Section 2.5.2, though the algorithm is computationally not very intensive, the memory overhead becomes severe as the dimensionality of the problem increases, due to the need to store the estimated reward probability matrix. In Chapter 6 we would consider another application of a common payoff game of FALA using pursuit algorithm, where we illustrate how the memory overhead can be considerably reduced if we are mainly interested in some 'good' parameter tuples rather than only in the global optimum. However, in general, this memory overhead is the price to be paid for achieving convergence to global maximum using the pursuit algorithm.

4.3.2 Pattern Classification with CALA

We can use a team of N continuous action learning automata for learning an N-dimensional parameter vector. Since the action set of a CALA is the real line, we need not discretize the parameter space. The model now is a common payoff game of CALA which is discussed in Section 2.6. Each automaton will be using a normal distribution for the action probability distribution. The N automata will independently choose actions resulting in the choice of a parameter vector by the team. As in the previous section, we classify the next sample pattern with the classifier specified by the chosen parameter vector and supply a binary reinforcement to the team depending on whether the classification agrees with that of the teacher or not.[3] Each of the automata use the algorithm described in the previous chapter to update the action probability distribution. This is once again a completely decentralized learning scheme for the team.

From the results of the previous chapter, it is easy to see that such a CALA team algorithm would converge to a local maximum of the function F defined on $\Re^N$.

4.3.3 Simulations

In this section we present results obtained with the automata team models on some pattern classification problems. We present results with FALA team using the Pursuit algorithm and with CALA team. For the FALA team, we need to discretize the parameters and the pursuit algorithm finds the global maximum. In the case of CALA, convergence to only local maxima is assured but we need not discretize the parameters. As will be seen below, by proper choice of initial variance in the CALA algorithm, we can obtain very good performance.

We present simulation results on only one problem. More details on empirical performance of these algorithms can be found in [TS87, ST99]. Let $f_i(X), i = 0, 1$ denote the two class conditional densities. The prior probabilities of the two classes are assumed equal.

Example 1: The class conditional densities for the two classes are given by Gaussian distributions:

$$\begin{aligned} f_0(X) &= N(\mathbf{m}_1, \Sigma_1) \\ f_1(X) &= N(\mathbf{m}_2, \Sigma_2) \end{aligned}$$

$$\begin{aligned} \text{where } \mathbf{m}_1 &= [2.0, \ 2.0]^T \\ \mathbf{m}_2 &= [4.0, \ 4.0]^T \\ \Sigma_1 &= \begin{bmatrix} 1.0 & -0.25 \\ -0.25 & 1.0 \end{bmatrix} \end{aligned}$$

---

[3]It may be recalled that for the CALA team, at each instant we generate two reinforcements: one for the parameter corresponding to the tuple of actions chosen and one for the parameter vector corresponding to the means of the action probability distributions. Both these are generated as per equation (4.3.13).

$$\Sigma_2 = \begin{bmatrix} 1.5 & -0.25 \\ -0.25 & 1.5 \end{bmatrix}$$

For this problem, a quadratic discriminant function is considered. The form of the discriminant function is

$$g(W, X) = \left[mx_2 + x_1 - (m^2 + 1)\left(x_0 - \frac{a}{(1+m^2)^{1/2}}\right) \right]^2 \frac{1}{1+m^2}$$
$$- \left[x_1 - \left(x_0 + \frac{a}{(1+m^2)^{1/2}}\right) \right]^2$$
$$- \left[x_2 - m\left(x_0 + \frac{a}{(1+m^2)^{1/2}}\right) \right]^2$$

where $W = (m, x_0, a)$ is the parameter vector and $X = (x_1, x_2)$ is the feature vector. This is a parabola described by three parameters m, x_0 and a as shown in Fig. 4.1. As mentioned earlier, in our method we can choose any form for the discriminant function. With the specific form chosen, it is easier to guess the ranges of the parameters based on some knowledge of the regions of feature space where patterns of the two classes cluster. It would be considerably more difficult to guess the parameter ranges if we had chosen a general quadratic expression for our discriminant function. It may be noted that the above discriminant function is nonlinear in its parameters.

The parameters of the optimal discriminant function for this problem are: $m = 1.0$ $x_0 = 3.0$ $a = 10.0$. A sketch of the optimal discriminant function is shown in Fig. 4.2.

Learning Automata Team with Pursuit Algorithm

For the simulation, 300 samples of each class were generated. At each instant one of the patterns from this set was selected at random and given to the learning system. The learning parameter, λ, was set at 0.1. A team of three automata was used. The ranges for the parameters, m, x_0 and a were taken to be [0.5, 1.5], [2, 6] and [1, 10] respectively. The range of each parameter was discretized into five levels. In seven out of ten experiments the team converged to the optimal parameters. In the other three runs, while two of the automata converged to their optimal actions, one of the automata converged to the next best action. The average number of iterations needed for convergence was 1970.

CALA Team

Since there are three parameters, we used a team of three continuous action set Learning Automata. As earlier, 300 sample patterns were generated from each class. Also a test set consisting of 100 samples was generated from the

Automata Network for Pattern Recognition 153

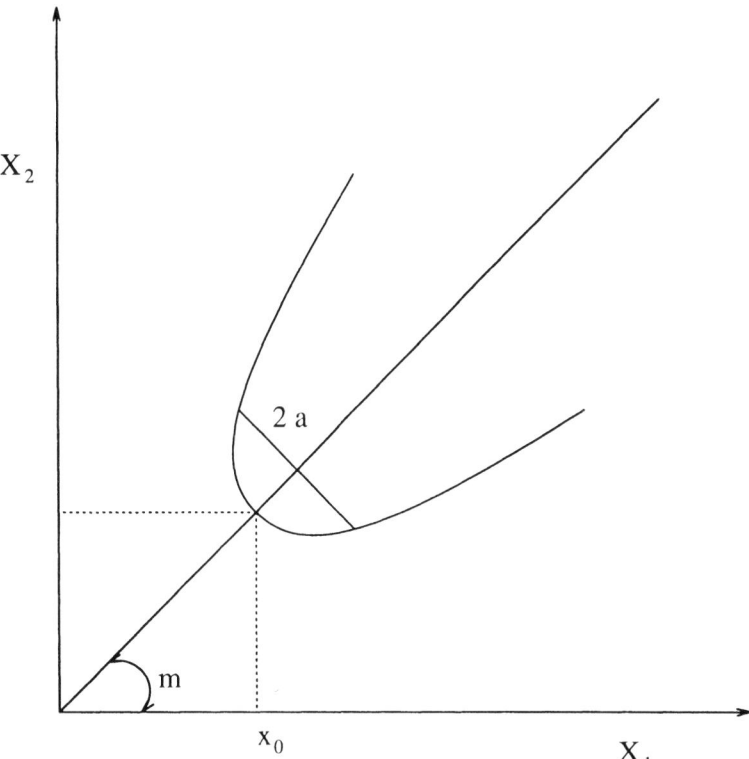

Figure 4.1. The form of the discriminant function used in Example 1. It is a parabola specified by three parameters: m is the slope of the axis, x_0 is the abscissa of the point of intersection of the parabola with its axis and a is half the length of its directrix.

two classes. In this problem it is difficult to analytically compute the minimum probability of misclassification. The number of misclassifications on the generated test set of patterns, with the parameter values set to those corresponding to the optimal discriminant function (mentioned above), was found to be 11 (out of a total of 100 test patterns). The results of the simulations are provided in Table 4.3.1, for 3 different initial values. As can be seen from the results presented, the probability of misclassification is between 30% and 40% with the initial values of the parameters. However, after learning, the probability of misclassification is close to the best expected. Also, it may be noted that, from many different starting points, the team converges to a classifier whose probability of correct classification is close to global maximum.

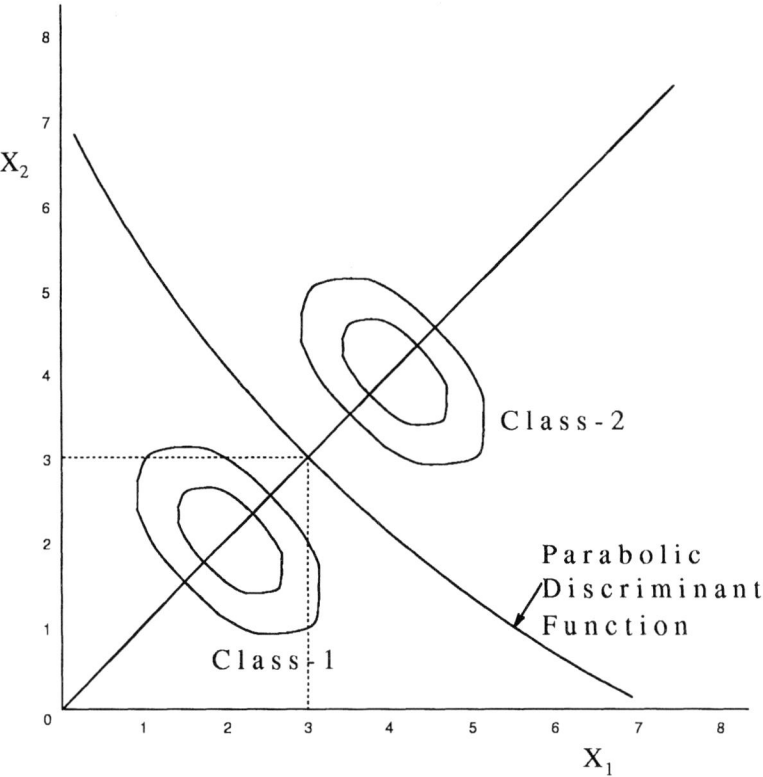

Figure 4.2. Class conditional densities and the form of the Optimal discriminant function in Example 1.

Initial Values								Final
$\mu_1(0)$	$\mu_2(0)$	$\mu_3(0)$	$\sigma_1(0)$	$\sigma_2(0)$	$\sigma_3(0)$	% Error	# Iterations	% Error
1	0.4	6	6	6	4	32	3700	11
5	2	12	6	6	4	45	6400	12
4	2	12	6	6	4	42	3400	12

Table 4.3.1. Results obtained with CALA team on Example 1.

4.4 Automata Network for Pattern Recognition

In the previous section, we have considered a common payoff game of automata and shown how it can be utilized for pattern classification. In that approach, we have to first decide on a parametric representation for the discriminant function, $g(W, X)$. The algorithm itself is independent of what this

Automata Network for Pattern Recognition

function is or how it is represented. For example, we could have represented it as an artificial neural network with parameters being the weights and then the algorithm would be converging to the 'optimal' set of weights and thus could be regarded as an alternative to the backpropagation algorithm. In all team models, the actions of each automaton always represent possible values for a parameter irrespective of the form chosen for the discriminant function.

In a general multiautomata system, actions of different automata may be representing different types of quantities. In this section we illustrate this in the context of pattern recognition using the automata network model discussed in Chapter 3. Though we can construct such networks with all varieties of automata, in this section we consider networks with only FALA.

In a two class PR problem, we are interested in finding a surface that appropriately divides the feature space into two regions. Such a surface, in general, is well approximated by a piecewise linear function [Lip87] and can be implemented using linear threshold units in a three layer feedforward network. In this network, the first layer units represent linear discriminant functions. That is, each unit is characterized by a hyperplane and for any vector as input, the output of the unit is binary based on which side of the hyperplane that vector falls. Units in the second layer perform *AND* operation on the outputs of *some selected* first layer units. Thus the first two layers together can represent convex sets with piecewise linear boundaries. The final layer performs an *OR* operation on the outputs of the second layer units. Hence this network, with appropriate choice of the internal parameters of the units and connections, can represent any subset of the feature space that is expressed as a union of convex sets with piecewise linear boundaries. Note that such a subset can itself be nonconvex. Such a network can well approximate the desired region of feature space (corresponding to one class) in most pattern recognition problems.

We now describe how we can configure a network of FALA (of the kind described in Section 3.3) to represent such a classifier. Our LA network would also have three layers. Each unit in the network would be a team of FALA.

Let the first layer consist of M units and let the second layer have L units. That means we can learn utmost M distinct hyperplanes and L distinct convex pieces. The final layer consists of a single unit.

As before let $X(k) = [x_1(k), \ldots, x_N(k)]^T \in \Re^N$ be the feature vector. Denote by U_i, $1 \leq i \leq M$, the units in the first layer, each of which should learn a N-dimensional hyperplane. A hyperplane in $\Re^N$ can be represented by (N+1) parameters. Hence we will represent each unit, U_i, $1 \leq i \leq M$, by an (N+1)-member team of automata, $\tilde{A}_{ij}$, $0 \leq j \leq N$. The actions of automaton $\tilde{A}_{ij}$ are the possible values of the jth parameter of the ith hyperplane being learnt. Since we are using finite action set automata here, as in Section 4.3.1, we discretize the ranges of parameters for making up the action sets of automata. Let A_{ij} be the set of actions[4] of automaton $\tilde{A}_{ij}$ whose elements will be denoted by a_{ijs}, $1 \leq s \leq r_{ij}$, $0 \leq j \leq N$, $1 \leq i \leq M$. Let $\mathbf{p}_{ij}(k)$ be the action probability vector of $\tilde{A}_{ij}$ with components p_{ijs}. Thus we have

$$\text{Prob}[\alpha_{ij}(k) = a_{ijs}] = p_{ijs}(k)$$

where $\alpha_{ij}(k)$ is the action chosen by the automaton $\tilde{A}_{ij}$ at time k. The output of the first layer unit U_i at k is $y_i^1(k)$ where

$$\begin{aligned} y_i^1(k) &= 1 \quad \text{if } \alpha_{i0}(k) + \sum_{j=1}^{N} \alpha_{ij}(k) x_j(k) > 0 \\ &= 0 \quad \text{otherwise} \end{aligned} \quad (4.4.14)$$

where $X(k) = [x_1(k) \ldots x_N(k)]^T$ is the feature vector of the sample pattern at iteration k.

Let V_i be the ith second layer unit that has connections with $n(i)$ first layer units, $1 \leq i \leq L$. The $n(i)$ first layer units that are connected to V_i are prefixed. (We could, for example, choose $n(i)$ equal to M). Thus V_i can learn a convex set bounded by *utmost* $n(i)$ hyperplanes. The unit V_i is composed of a team of $n(i)$ automata $\tilde{B}_{ij}$, $1 \leq j \leq n(i)$, each of which has two actions: 0 and 1. The action probability distribution of $\tilde{B}_{ij}$ at k can be represented by a single real number $q_{ij}(k)$ with

$$\text{Prob}[z_{ij}(k) = 1] = 1 - \text{Prob}[z_{ij}(k) = 0] = q_{ij}(k)$$

where $z_{ij}(k)$ is the action selected by $\tilde{B}_{ij}$ at k. Let $y_i^2(k)$ be the output of the second layer unit V_i at instant k. $y_i^2(k)$ is the AND of the outputs of all those first layer units which are connected to V_i and are activated, i.e., for which $z_{ij}(k) = 1$. More formally,

$$\begin{aligned} y_i^2(k) &= 1 \text{ if } y_j^1(k) = 1 \text{ whenever } z_{ij}(k) = 1,\ 1 \leq j \leq n(i), \\ &= 0 \quad \text{otherwise.} \end{aligned} \quad (4.4.15)$$

The third layer contains only one unit whose output is a boolean OR of all the outputs of the second layer units. This unit contains no automata. The output of this third layer unit is the output of the network.

---

[4]Recall that in Chapter 3, we have denoted the action set of ith automaton by A_i and the individual actions of this automaton by a_{ij}. In that chapter, for the analysis of the algorithms, we did not need to know which automata constitute which unit. However, in a specific application as in this chapter, our notation should allow for associating automata with units. Thus the notation for set of actions contains two subscripts – one to denote the unit and other to denote the automaton inside that unit. Similarly, the notation for individual actions now contains three subscripts.

This network of automata functions as follows. At each instant k, all the automata in all the first and second layer units choose an action at random based on their current action probability vectors. That is, in each U_i, each of the automata $\tilde{A}_{ij}$ chooses an action $\alpha_{ij}(k)$ from the set A_{ij} at random based on the probability vector $\mathbf{p}_{ij}(k)$. This results in a specific parameter vector and hence a specific hyperplane being chosen by each U_i. Then, based on the next feature vector, $X(k)$, each unit U_i calculates its output $y_i^1(k)$ using equation (4.4.14). In each second layer unit V_i, all the $n(i)$ automata $\tilde{B}_{ij}$ choose an action $z_{ij}(k)$ at random based on the probability $q_{ij}(k)$. Using these $z_{ij}(k)$ and the outputs of first layer units, each V_i would calculate its output $y_i^2(k)$ using equation (4.4.15). Using the outputs of the second layer units, the unit in the final layer will calculate its output which is 1 if any $y_i^2(k)$ is 1; and 0 otherwise. Let $Y(k)$ denote the output of the final layer unit which is also the output of the network. $Y(k) = 1$ denotes that the pattern is classified as Class-1 and $Y(k) = 0$ denotes that the pattern is Class-0. For this classification, the environment supplies a reinforcement $\beta(k)$ as

$$\begin{aligned} \beta(k) &= 1 \quad \text{if } Y(k) = \bar{y}(X(k)) \\ &= 0 \quad \text{otherwise} \end{aligned} \tag{4.4.16}$$

where $\bar{y}(X(k))$ is the classification supplied for the current pattern $X(k)$ in the training set. $\beta(k)$ is supplied as the common reinforcement to all the automata in all the units and then all the automata update their action probability vectors using the L_{R-I} algorithm as below.

For each i, j, $0 \leq j \leq N$, $1 \leq i \leq M$, the probability vectors $\mathbf{p}_{ij}$ are updated as

$$\begin{aligned} p_{ijs}(k+1) &= p_{ijs}(k) + \lambda\beta(k)(1 - p_{ijs}(k)), \quad \text{if } \alpha_{ij}(k) = a_{ijs} \\ &= p_{ijs}(k)(1 - \lambda\beta(k)), \quad \text{otherwise} \end{aligned} \tag{4.4.17}$$

For each i, j, $1 \leq j \leq n(i)$, $1 \leq i \leq L$, the probabilities q_{ij} are updated as

$$\begin{aligned} q_{ij}(k+1) &= q_{ij}(k) + \lambda\beta(k)(1 - q_{ij}(k)), \quad \text{if } z_{ij}(k) = 1, \\ &= q_{ij}(k)(1 - \lambda\beta(k)), \quad \text{otherwise} \end{aligned} \tag{4.4.18}$$

Let $\mathbf{P}(k)$ denote the internal state of the network. This includes the action probabilities of all the automata in all the units. That is, it includes all $\mathbf{p}_{ij}$ and all q_{ij}. Define

$$f(\mathbf{p}) = E[\beta(k) \mid \mathbf{P}(k) = \mathbf{p}] \tag{4.4.19}$$

From the results presented in Chapter 3, it is easy to see that for this network of teams of automata, the learning algorithm given by (4.4.17) and (4.4.18) will make $\mathbf{P}(k)$ converge to a local maximum of $f(.)$.

It is clear from the earlier discussion in Section 4.3 and from equation (4.4.16), that maximizing the expected reinforcement will result in maximizing the probability of correct classification. Thus this network will learn a classifier which locally maximizes the probability of correct classification.

In this network model, actions of automata in the first layer units are possible values for parameters that define hyperplanes. Actions of automata in the second layer units represent boolean decisions on which hyperplanes to choose to make the appropriate convex regions. As explained in Chapter 3, the learning algorithm as well as the proof that the learning algorithm converges to local maxima of the function f defined by (4.4.19) are independent of the specific architecture of the network. As illustrated here, this flexibility can be exploited to design network structures to capture interesting classes of functions. All that we need to ensure is that the reinforcement, β, is generated suitably so that maximizing f is what is needed from the point of view of the application.

4.4.1 Simulations

We consider two examples here to illustrate the three layer network presented above. Both these examples are from [TP95a]. The first one is an artificial problem while the second one is on a real data set. For the first one we consider a problem where the region in the feature space in which the optimal decision is Class 0, is a convex set with linear boundaries. We consider a 2-dimensional feature space because it is easier to visualize the problem.

Example 2: The feature vectors from the environment are uniform over the set $[0, 1] \times [0, 1]$. The discriminant function to be learnt is shown in Fig. 4.3. Referring to the figure, the optimal decision in region A is Class 0 and that in region B is Class 1. In region A,

$$\text{Prob}[X \in \text{Class } 0] = 1 - \text{Prob}[X \in \text{Class } 1] = 0.9,$$

and in region B

$$\text{Prob}[X \in \text{Class } 0] = 1 - \text{Prob}[X \in \text{Class } 1] = 0.1.$$

It may be noted that this example is essentially the same as the example of a G-environment given in Section 3.3.1. The optimal discriminant function (to be learnt) in this problem is

$$[2x_1 - x_2 > 0] \text{ AND } [-x_1 + 2x_2 > 0]$$

where $X = (x_1 \ x_2)^T$ is the feature vector.

Since we need to learn only one convex set, the network is made up of two first layer units, U_1 and U_2, and one fixed second layer unit. The second layer

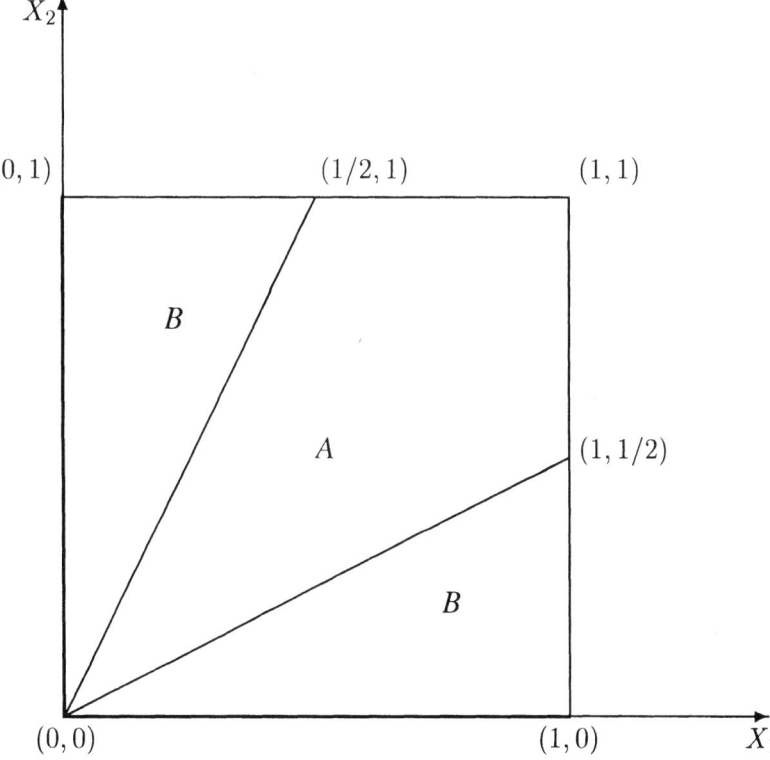

Figure 4.3. The class regions in Example 2

unit performs AND operation on the outputs of the first layer units. Each first layer unit has two automata. Using only two parameters for a hyperplane in $\Re^2$ means that we are considering only those that pass through origin. Since this is sufficient for our purposes here, we use only two rather than three automata in each unit. Each of the automata has four actions which are the possible values of the parameters to represent the hyperplanes. All four automata have the same action set given by $\{-2\ -1\ 1\ 2\}$. In this problem, there are two sets of choices of actions by the four automata given by (-1, 2, 2, -1) and (2, -1, -1, 2) at which the global optimum is attained. (Recall that a tuple of actions of the team corresponds to a parameter vector). The learning parameter, λ, is fixed at 0.005 for all automata. The initial action probability distribution is uniform. That is, initial probability of each of the four actions is 0.25. The number of samples generated is 500 and at each instant one pattern chosen randomly from this set is presented to the network. Twenty simulation runs were conducted and the network converged to one of the two sets of optimal actions in every run. The average number of iterations needed for the probability of the optimal action to be greater than 0.98 for each automaton, is 10,922 steps. (Since the computations per iteration are very simple, the actual time taken is only a few seconds on a PC).

Example 3: In this example, a 2-Class version of the Iris data [DH73] was considered. This is the same data that was used in one of the examples in Section 2.7.2. As mentioned there, the data was obtained from the Machine Learning databases maintained at University of California, Irvine. Recall that the problem is one of determining the species of a plant based on four features, namely, petal length, petal width, sepal length and sepal width. We need to classify each such feature vector into one of three classes, namely, iris-setosa, iris-versicolor and iris-virginica. As mentioned in Chapter 2, setosa is linearly separable from the other two; but virginica is not linearly separable from versicolor. Since we are considering only 2-Class problems here, setosa was ignored and we consider classifying versicolor versus virginica. The data used was 50 samples of each class with the correct classification.

The network consisted of 9 first layer units and 3 second layer units. Each first layer unit has 5 automata (since this is a 4-feature problem). Each automaton had 9 actions which were {-4,-3,-2,-1,0,1,2,3,4}. Uniform initial conditions were used. The learning parameters were 0.005 in the first layer and 0.002 in the second layer.

In this problem we do not know which are the optimal actions of the automata and hence we have to measure the performance based on the classification error on the data after learning. Hence we use part of the data for training and the rest for testing.

For a comparison of the performance achieved by the automata network, we also simulated a standard feedforward neural network where we used backpropagation with momentum term (BPM) for the learning algorithm. The network has four input nodes (to take in the feature vector) and one output node. We tried two and three hidden layers. For the two hidden layer network we used 9 and 3 units in the hidden layers. For the three hidden layer network we used 8 nodes in each hidden layer. Initial weights for the network were generated randomly. In the learning algorithm the learning parameter for the momentum term was set at 0.9 and various values of the learning parameter for the gradient term were considered and the best results are reported here.

Simulations were conducted for perfect data (0% noise) and noisy cases. Noise was introduced by changing the known classification of the feature vector at each instant by a fixed probability. Noise levels of 20% and 40% were considered. With 40% noise, the probability of each training sample having a wrong class label is 0.4.

The results obtained are summarized in Table 4.4.2. These are averages over 10 runs. The error reported in the table for the backpropagation algorithm is the root mean square error while that for the automata network is the probability

Algorithm	Structure	Noise(%)	Error	Steps
BPM	9 3 1	0	2.0	66,600
BPM	9 3 1	20	–	No Convergence
BPM	9 3 1	40	–	No Convergence
BPM	8 8 8 1	0	2.0	65,800
BPM	8 8 8 1	20	–	No Convergence
BPM	8 8 8 1	40	–	No Convergence
L_{R-I}	9 3 1	0	0.1	78,000
L_{R-I}	9 3 1	20	0.1	143,000
L_{R-I}	9 3 1	40	0.15	200,000

Table 4.4.2. Simulation Results for Iris data. The entry in the fourth column refers to RMS error for BPM and probability of misclassification for L_{R-I}.

of misclassification. While they cannot be directly compared, the performance was about the same at the values reported.

The results show that in the noise-free case, the backpropagation with momentum converges about 20% faster. However, this algorithm fails to converge even when only 20% noise is added. The learning automata network continues to converge even with 40% noise and there is only slight degradation of performance with noise. This problem thus empirically illustrates the fact that as long as the noise level is less than 50%, the automata models can still learn proper classifiers.

4.4.2 Network of Automata for Learning Global Maximum

In the three layer network of automata considered above, all automata use the L_{R-I} algorithm. As shown in Chapter 3, with this algorithm, the network can learn only a local maximum of the function f defined by (4.4.19). In Section 3.7, we have presented another learning algorithm that ensures convergence to global maximum. For this we employ PLA instead of FALA in the network. In this subsection we illustrate this global algorithm for the case of the three layer network for pattern recognition.

The units in the network now would be composed of teams of PLA. Recall that a PLA would have an internal state vector which is updated through the learning algorithm. The probabilities of different actions are calculated using the internal state through a probability generating function.

The complete learning algorithm for this three layer network of automata is as follows. We would be following the same general notation as earlier. Under this notation, p_{ijs} is the probability of sth action, a_{ijs}, of automaton $\tilde{A}_{ij}$ which is the jth automaton in U_i, the ith first layer unit and so on. The functioning of the network is same as before. However, the learning algorithm now updates the internal state of each automaton and the actual action probabilities are calculated using the probability generating function. Let $\mathbf{u}_{ij}$ be the state vector of automaton $\tilde{A}_{ij}$ with components u_{ijs}. Similarly, let $\mathbf{v}_{ij}$ be the state vector of the second layer automaton $\tilde{B}_{ij}$ with components v_{ij0} and v_{ij1}. Let $g^1_{ij}(.,.)$ be the probability generating function for automaton $\tilde{A}_{ij}$ and let $g^2_{ij}(.,.)$ be the probability generating function for automaton $\tilde{B}_{ij}$. (Recall that automata $\tilde{A}_{ij}$ are part of first layer units and automata $\tilde{B}_{ij}$ are part of second layer units. Thus the superscript on the probability generating function indicates the layer). Following the model as discussed in Section 3.7, the various action probabilities, p_{ijs} and q_{ij} are now given by

$$p_{ijs} = g^1_{ij}(\mathbf{u}_{ij}, a_{ijs}) = \frac{exp(u_{ijs})}{\sum_s exp(u_{ijs})}$$
$$q_{ij} = g^2_{ij}(\mathbf{v}_{ij}, 1) = \frac{exp(v_{ij1})}{exp(v_{ij1}) + exp(v_{ij0})} \quad (4.4.20)$$

The algorithm given below specifies how the various state vectors should be updated. Unlike in equations (4.4.17) and (4.4.18), there is a single updating equation for all the components of the state vector. $\beta(k)$ is the reinforcement obtained at k, which is calculated as before by equation (4.4.16).
For each i, j, $0 \leq j \leq N$, $1 \leq i \leq M$, the state vectors $\mathbf{u}_{ij}$ are updated as

$$u_{ijs}(k+1) = u_{ijs}(k) + \lambda\beta(k)\frac{\partial \ln g^1_{ij}}{\partial u_{ijs}} + \lambda h'(u_{ijs}(k)) + \sqrt{\lambda}\zeta_{ij}(k) \quad (4.4.21)$$

For each i, j, $1 \leq j \leq n(i)$, $1 \leq i \leq L$, the state vectors $\mathbf{v}_{ij}$ are updated as

$$v_{ijs}(k+1) = v_{ijs}(k) + \lambda\beta(k)\frac{\partial \ln g^2_{ij}}{\partial v_{ijs}} + \lambda h'(v_{ijs}(k)) + \sqrt{\lambda}\zeta_{ij}(k) \quad (4.4.22)$$

where

- The function $g^1_{ij}(.,.)$ and its partial derivatives are evaluated at the current state vector and the current action of $\tilde{A}_{ij}$, $(\mathbf{u}_{ij}(k), \alpha_{ij}(k))$. Similarly, the function $g^2_{ij}(.,.)$ and its partial derivatives are evaluated at $(\mathbf{v}_{ij}(k), z_{ij}(k))$.

- $h'(.)$ is the derivative of $h(.)$ which is defined by

$$\begin{aligned} h(x) &= -K(x-L)^{2J} \quad \text{for } x \geq L \\ &= 0 \quad \text{for } |x| \leq L \\ &= -K(x+L)^{2J} \quad \text{for } x \leq -L \end{aligned} \quad (4.4.23)$$

where K and L are positive real numbers and J is a positive integer, all of which are parameters of the algorithm.

- $\{\zeta_{ij}(k)\}$ is a sequence of *iid* random variables (which are also independent of all the action probabilities, actions chosen etc.) with zero mean and variance σ^2. σ is a parameter of the algorithm.

It is easily seen that this algorithm is the same as the algorithm given in Section 3.7.2. From that chapter, we recall the following features of the algorithm. In the updating equations given by (4.4.21) and (4.4.22), the $h'(\cdot)$ term on the right hand side is essentially a *projection* term which ensures that the algorithm exhibits bounded behavior and the ζ_{ij} term adds a *random walk* to the updating. The $\beta(k)$ term on the RHS is essentially the same updating as the L_{R-I} given earlier. To see this, it may be noted that

$$\frac{1}{g_{ij}^1(\mathbf{u}_{ij}, a_{ijs})}[\frac{\partial g_{ij}^1}{\partial u_{ijs}}(\mathbf{u}_{ij}, a_{ijs})] = 1 - p_{ijs}$$

$$\frac{1}{g_{ij}^1(\mathbf{u}_{ij}, a_{ijs})}[\frac{\partial g_{ij}^1}{\partial u_{ijs'}}(\mathbf{u}_{ij}, a_{ijs})] = -p_{ijs'} \qquad (4.4.24)$$

It is proved in Chapter 3 (see Theorem 3.2 in Section 3.7.3) that the asymptotic properties of this algorithm are obtained by deriving an approximating stochastic differential equation (SDE) which turns out to be of the Langevin type. As discussed in Section 3.7.3, by analyzing this SDE, it can be concluded that the algorithm essentially converges to the global maximum of the f function given by (4.4.19). Hence, in our current context, we can conclude that this three layer network of PLA, under the above learning algorithm, would converge to a classifier with minimum probability of misclassification.

Simulations with the Global Algorithm

Here we briefly give results of simulations with this algorithm on one example considered earlier in Section 4.4.1, namely, Example 2.

In that example, we have seen that the global maximum is attained at two parameter vectors (-1,2,2,-1) and (2,-1,-1,2). One of the local maxima in that problem is given by (1,1,1,1). We have seen that the L_{R-I} algorithm converges to the global maximum when started with uniform initial conditions, that is, equal initial probabilities to all actions in all automata. Here we pick the initial conditions such that the effective probability of the parameter vector corresponding to the local maximum is greater than 0.98 and the rest of the probability is distributed among the other parameter vectors. With this much of bias, the L_{R-I} algorithm always converged to the local maximum.

We tried the algorithm presented above with these initial conditions. The parameters in the $h(.)$ function are set as $L=3.0$, $K=1.0$, and $J=2$. The learning parameter is set at 0.05. The value for σ is initially 10 and was reduced as

$$\sigma(k+1) = 0.999\sigma(k), \quad 0 \le k \le 5000.$$

and was kept constant thereafter. (Here $\sigma(k)$ is the value of σ used at iteration k).

As earlier, the training set had 500 samples. Twenty simulations were done and each time the algorithm converged to one of the two global maxima. The average number of iterations needed for convergence was 33,425. This, of course, does not compare favorably with the time taken by L_{R-I}. In this algorithm, the computation time per iteration is also more than that of L_{R-I}. The extra time seems to be mainly due to the fact that the action probabilities are to be computed at each instant and not directly stored. The extra terms in the algorithm (the term for bounding the iterates and the random term) do not seem to slow down the algorithm much. A different choice of the probability generating function may result in a faster algorithm. However, the higher computational effort and lower rates of convergence appear to be the price to be paid for convergence to global maximum in all such algorithms.

4.5 Decision Tree Classifiers

Decision tree is another popular structure for a pattern classifier. Consider a two class problem with N real-valued features. A decision tree in this case is a binary tree with the following structure. Each non-leaf node of the tree is associated with a hyperplane in the feature space. Each leaf node has a class label attached to it. Such a decision tree is used for classifying a pattern or, equivalently, a feature vector, X as follows. We start with the root of the tree. Suppose the hyperplane associated with the root is H_1. Then, depending on which side of H_1 is X, we go to the left or right child of the root node. This process is repeated at every node so encountered till we reach a leaf node. That is, at every node we find out which side of the associated hyperplane this feature factor falls and appropriately go to the left or right child of that node. By this process we would traverse a path in the tree. Finally, when we reach a leaf node, we assign to X the class label attached to that leaf node.

Fig. 4.4 illustrates a decision tree classifier for a 2-class pattern recognition problem with $\Re^2$ as the feature space. The left panel in the figure shows the class regions. The shaded region corresponds to, say, class C_1 and the rest to class C_0. The right panel of the figure shows a decision tree which can exactly represent the class regions in this example. In the decision tree, the non-leaf nodes are labeled with the hyperplanes as H_1, H_2 etc. and these hyperplanes (which are straight lines in $\Re^2$) are also shown in the feature space depicted in the left panel of the figure. The root of the tree is labeled H_1. As can be seen from the figure, all feature vectors falling on one side of H_1 belong to class C_1. That is why the right child of the root node is a leaf labeled C_1. The feature vectors falling on the other side of H_1 are now tested with hyperplane H_2. Once again the ones (from among those that come to this node) that fall on one side of H_2 are in class C_0. The rest are tested with hyperplane H_3 which

Decision Tree Classifiers

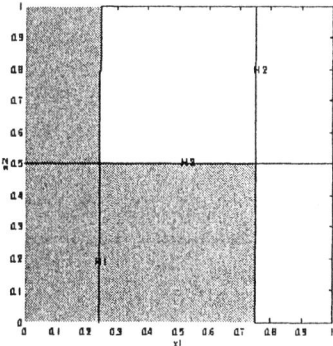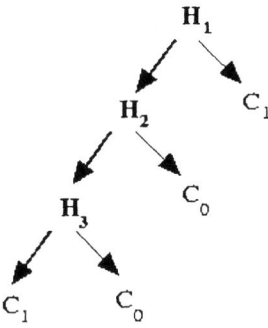

Figure 4.4. An example of a decision tree representation for a classifier on a 2-class problem

completes the classification. In this example, each of the hyperplanes is parallel to one of the coordinate axes. Such decision trees are called univariate or axis-parallel decision trees. Decision trees with general hyperplanes are called multivariate or oblique decision trees. In an axis-parallel decision tree, each hyperplane is represented by an equation of the type $X_i - \theta = 0$. Learning of such hyperplanes involves deciding on the component of feature vector to use and getting a 'good' value for θ. These trees are called univariate because here each hyperplane depends only on one of the feature variables. In an oblique decision tree, each hyperplane would be represented by $W^T X - \theta = 0$. With an n-dimensional feature space, learning of each hyperplane involves learning the n-dimensional vector W and the scalar θ.

As can be seen from the example depicted in Fig. 4.4, each path in the tree from the root to a leaf node represents a region in the feature space which is a convex set bounded by hyperplanes. Since multiple leaf nodes can have the same class label, a decision tree represents each class region as a union of such polyhedral sets in the feature space. Thus, a decision tree represents class boundaries by piece-wise linear surfaces.

From the above discussion, one can see that the three layer network considered in Section 4.4 is structurally equivalent to a decision tree in a 2-class pattern recognition problem. However, there are some differences. In the three layer network as defined in Section 4.4, the number of first layer units (and hence the number of hyperplanes) is fixed a priori by the designer. On the other hand, in the decision tree structure as defined above (and as used in pattern recognition problems) there is no such restriction on the number of non-leaf nodes in the tree. Most decision tree learning algorithms construct the tree in a top down fashion and hence the final number of non-leaf nodes in the tree is also learnt and is dependent on the training set of examples. The structure of a decision tree as defined above is the same irrespective of whether we are considering a 2-class problem or a multi-class problem. For example, the decision tree in Fig. 4.4 can be converted to a three class classifier by, e.g., changing the label on the right child of node labeled H_3 as C_2. This corresponds to taking an appropriate rectangular portion on the top side of the feature space as another class. In contrast, the structure of the three layer network defined in Section 4.4 is applicable only to 2-class problems. However, this does not mean that the three layer network cannot be used in multi-class problems because a multi-class classifier can always be designed using a collection of 2-class classifiers [DH73].

Algorithms for learning decision trees include ID3 [Qui83, Qui86], CART [BFOS84], OC1 [MKS94], APDT [SS99] etc. All such algorithms construct the decision tree in a top down fashion as follows. Suppose S is the set of training patterns. The first step is to learn the hyperplane at the root of the tree. For this some objective function that measures the 'goodness' of any hyperplane is defined and some optimization technique is used to find the 'best' hyperplane. Normally, the criterion function used to rate the 'goodness' of any hyperplane depends on the training set of patterns, S, at this node. (See the references cited above for details). Suppose H_1 is the 'best' hyperplane obtained in this process. We now label the root with this hyperplane. Then we divide the set S into two parts using H_1: those that are on one side of H_1 and hence would go to left child of this node (say S_l) and those that are on the other side of H_1 and hence would go to right child of this node (say, S_r). Now the problem is reduced to two subproblems: learning a tree for the sets of samples S_l and S_r. Same procedure is applied for both these problems recursively. This recursion finally stops at a node where the set of patterns coming to that node satisfies some stopping criterion (such as sufficiently high fraction of patterns in the set are all of the same class).

We now explain how learning automata algorithms can be used for constructing a decision tree in a top down fashion. As in all the earlier sections, we restrict our attention to a 2-class problem.

4.5.1 Learning Decision Trees using GLA and CALA

As explained earlier, the main computation in constructing a decision tree is the following. Given a set of sample patterns S at a node, we have to learn the 'best' hyperplane at that node (or if a stopping criterion is satisfied then label that node appropriately as a leaf). As seen in Section 1.7.2, a GLA can be used to learn a hyperplane classifier. So, we use a two action GLA to learn the hyperplane at any node. If the feature vector is n-dimensional then the internal state of the GLA, $\mathbf{u}$, would be $(n+1)$-dimensional. Let $\mathbf{u} = (u_0, u_1 \ldots u_n)^T$ denote the internal state vector of GLA and let $X = (X_1, \ldots, X_n)^T$ denote the feature vector. The feature vector would be the context vector input to the GLA. Let $\bar{X} = (1, X_1, \ldots, X_n)^T$ denote the so called augmented feature

Decision Tree Classifiers

vector. Let the two actions of GLA be y_1 and y_2. We use the probability generating function given by (1.7.39) explained in Section 1.7. The action probabilities of GLA are given by

$$g(X, y_1, \mathbf{u}) = 1 - g(X, y_2, \mathbf{u}) = \frac{1}{1 + \exp(-\mathbf{u}^T \bar{X})} \quad (4.5.25)$$

. From the above, we have $g(X, y_1, \mathbf{u}) \geq g(X, y_2, \mathbf{u})$ if and only if $1 \geq \exp(-\mathbf{u}^T X)$ which is same as $\mathbf{u}^T X \geq 0$. Thus, when the context vector is on one side of the hyperplane, the probability of one action would be high and the probability of the other action would be high when it falls on the other side of the hyperplane. It is in this sense that the GLA is like a hyperplane classifier.

The learning process can be thought of as follows. At each instant we take the next sample pattern as the context vector. The GLA would use this and the current value of its internal state to randomly choose an action. After that, the GLA gets a reinforcement (for the choice of action) and this is used to update the internal state.

All we need to do now is to devise the reinforcement signal appropriately. The reinforcement should give a noisy indication of the goodness of the hyperplane chosen by the GLA. As we have done in the earlier sections of this chapter, we can supply a reinforcement of 1 or 0 depending on whether or not the action chosen by GLA corresponds to the class label of the pattern in the training set. That way, we can expect the GLA to learn a hyperplane that minimizes probability of misclassification. However, in a decision tree classifier, the hyperplane that is being learnt at a node is not the whole classifier; it is only one step in the process of classification. The hyperplane at a node should aid the overall classification in the sense that it should split the set of samples (at that node) into two subsets each of which are 'easier' to classify. Thus, we should not expect the 'best' hyperplane at a node to be necessarily the one that gives minimum classification error (for the patterns reaching that node) by itself. For example, a hyperplane at a node would be good if it can split the patterns at that node into two parts each of which are linearly separable. This is because, if we pick this hyperplane at this node, at the next level in the tree the classification would be properly completed. Such a hyperplane may have very high misclassification rate because it may divide the set of samples into two parts so that both parts contain equal number of patterns of either class (though both parts are sets of patterns that are linearly separable). In view of this, we use a kind of look-ahead strategy to generate the reinforcement as explained below.

In addition to the GLA, we would have two teams of CALA. Each team would have $n + 1$ automata. (Recall that the feature vector is n-dimensional). We call these teams left-CALA team and right-CALA team. We view the actions of CALA as values of parameters to specify a hyperplane.

The system functions as follows. Like in any automata system, each iteration consists of the automata choosing actions, getting reinforcement and updating their states. First, using the current sample pattern (which is chosen with a uniform distribution from the set S) as the context vector, the GLA chooses an action. If the action is y_1, we activate the left-CALA team; otherwise we activate the right-CALA team. Now all automata in the activated CALA team select an action and this results in the team selecting a hyperplane. Now we generate reinforcements as explained below and the GLA and all CALA in the activated team update their state. For the GLA we use the algorithm in Section 1.7.1 and for the CALA team we use the algorithm from Section 2.2.1. Note that the CALA in the non-activated team do not update their state.

Recall that the CALA algorithm needs two reinforcements from the environment at each iteration. We generate these as follows. Let a denote the tuple of actions chosen by the CALA team and let μ denote the vector whose components are the means of action probability distributions of the CALA. We classify the next sample pattern with (the hyperplane represented by) a and generate a 1 − 0 reinforcement based on whether or not the classification agrees with that given in the training sample. Call this reinforcement β. Then we classify the same sample with μ and generate another binary reinforcement, say, β'. The reinforcements, β and β' are supplied as common payoffs to all the members of the activated CALA team. Each CALA uses these two in the learning algorithm as explained in Section 2.2.1. The GLA uses β' as the reinforcement to update its internal state.

This cycle goes on till one of two conditions are met: a prefixed-fixed number of iterations are over or the classification errors made over a prefixed-fixed number of consecutive iterations is below a threshold. We refer to these two as condition 1 and condition 2 respectively and the remaining part of the algorithm depends on the condition on which the learning process as above is terminated. This is explained below.

Let $\mathcal{N}$ denote the current node of the decision tree for which we are learning a hyperplane using the set of samples S. Let $\mathbf{u}^*$ denote the internal state vector of GLA at termination of the learning process as above. Let μ_l^* denote the vector whose i^{th} component is the mean of the action probability distribution of the ith CALA in the left-CALA team when the above learning process is terminated and let μ_r^* denote the corresponding quantity for the right-CALA team.

Suppose the learning process with GLA and the two teams of CALA was terminated by condition 1. (This means we have not reached the required level of misclassification rate and hence have to continue the tree under this node for more levels). Then we associate the $\mathbf{u}^*$ as the hyperplane with the current node $\mathcal{N}$. We then split the set S into two sets S_l and S_r using the hyperplane $\mathbf{u}^*$. (That is, $S_l = \{X \in S \mid X^T \mathbf{u}^* \geq 0\}$ and $S_r = S - S_l$). We recursively call

Decision Tree Classifiers 169

the same algorithm once with S_l and once with S_r, to construct the subtrees under the left and right child of the current node $\mathcal{N}$. In these recursive calls, we initialize the internal state of the GLA in the algorithm that is learning with S_l, to $\boldsymbol{\mu}_l^*$ which is the hyperplane learnt by the left-CALA team at the current stage. Similarly, for the recursive call with S_r, we initialize the internal state of GLA with $\boldsymbol{\mu}_r^*$.

Suppose the learning process is terminated with condition 2. (This means we have reached the required level of misclassification at this node and hence construction of the tree under this node can be completed). We once again associate $\mathbf{u}^*$ as the hyperplane with node $\mathcal{N}$ and split S into S_l and S_r based on $\mathbf{u}^*$. We create nodes $\mathcal{N}_l$ and $\mathcal{N}_r$ as the left child and right child of $\mathcal{N}$. We associate the hyperplane given by $\boldsymbol{\mu}_l^*$ with $\mathcal{N}_l$ and associate hyperplane given by $\boldsymbol{\mu}_r^*$ with $\mathcal{N}_r$. The children of these nodes would be leaves of the tree. We split S_l into S_{ll} and S_{lr} using the hyperplane $\boldsymbol{\mu}_l^*$. The left child of $\mathcal{N}_l$, which would be a leaf, is given the class label which is same as that of the majority of patterns in S_{ll} and the right child of $\mathcal{N}_l$ is given the class label similarly based on the set S_{lr}. In a similar way we create the two children (which would be leaves) for the node $\mathcal{N}_r$.

The above recursive procedure would be started with the full set of training examples. When this top level call returns, we would have a complete decision tree. As can be seen, the main learning unit is a GLA plus two teams of CALA. Together, this unit is trying to learn a three node tree structure. By keeping on splitting the examples and calling this learning algorithm recursively, we construct the complete tree. There are some extra features that can be added to the basic algorithm as above to improve its efficiency. For example, if the number of sample patterns at a node is below some threshold, then we can treat that node as a leaf and label it with a class based on the predominant class in the set of samples at that node. Recall that in the GLA learning algorithm described in Section 1.7.1, we have a projection term that tries to keep components of $\mathbf{u}$ bounded. (See equation (1.7.44)). The constants used there, namely, the L_i, can be chosen so that we keep the hyperplane represented by the internal state of the GLA within a bounding hyper-rectangle (in the feature space) of the sample patterns. A discussion of such added features to the basic algorithm described above can be found in [Pra03].

Fig. 4.5 shows the results obtained with this algorithm on two example problems. Both are 2-feature 2-class problems. In the figure these problems are referred to as the XOR problem and the Triangle problem. The figure shows the class regions in the feature space. In both cases the shaded region represents one class and the rest is the other class. In both cases the tree consists of only three non-leaf nodes. The hyperplanes learnt by GLA and the two teams of CALA are shown superimposed on the class regions. As can be seen, the algorithm learns a very good classifier. It may be noted that in the triangle problem,

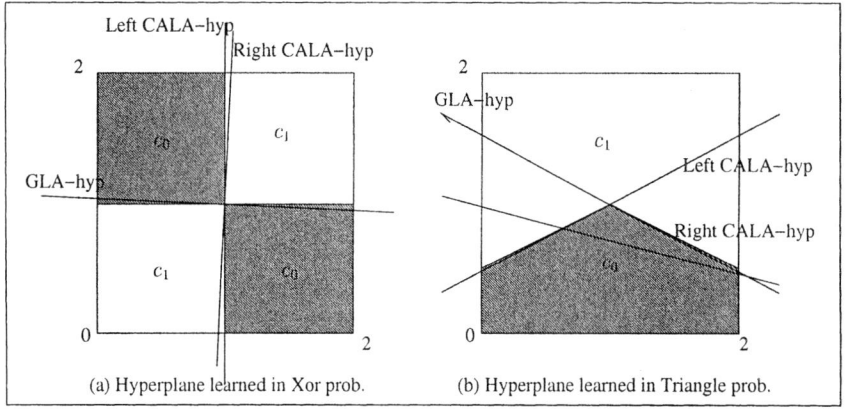

Figure 4.5. The decision trees learnt with the automata method on two 2-class problems

given the hyperplane learnt by GLA, all patterns going to right-CALA team are of the same class and hence the hyperplane learnt by this CALA team is of no consequence. (As a matter of fact both the children of this node are labeled with the same class). For both the problems we have used 200 training patterns generated with a uniform distribution over the rectangle $[0, 2] \times [0, 2]$ in $\Re^2$. After classifying these patterns using the class regions as shown, we have added 10% classification noise. The results shown are obtained with these noisy samples. This algorithm performed well on many other example problems also. For example, on the Iris data, (with 10% classification noise added), this method achieved an accuracy of 93%. The reader is referred to [Pra03] for more details regarding the performance of this algorithm.

4.5.2 Learning Piece-wise Linear Functions

An interesting aspect of a decision tree is that the same structure can be used to represent certain piece-wise linear functions also. In a decision tree, the path from the root to a leaf represents a convex region bounded by hyperplanes. When used as a classifier, we label each leaf node with a class and all feature vectors falling in the region represented by the path to this leaf node would be classified into that class. Instead of assigning a class label to a leaf we can associate with each leaf some real valued function. Now this tree represents a real-valued function defined on $\Re^n$ as follows. Given any $X \in \Re^n$, we traverse the tree as before to reach a leaf. Then we take the value of the function represented by the tree (at this X) to be the value of the function associated with the leaf, at the argument given by X. Such trees are sometimes called regression trees or model trees and learning such representations for functions is useful in applications such as datamining [WF00]. The usual decision tree

Decision Tree Classifiers

used for classification can be considered as a special case of such model trees where the function associated with each leaf is a constant.

Usually, the function associated with each leaf is of the form $W^T X$, where X is the input vector and W is a parameter vector. Then a model tree essentially divides the domain of the function into some polyhedral sets and represents the function by a linear model within each such region.

Model trees are useful in function learning problems. In a function learning problem, we are given a set of samples $\{(X_1, z_1), \ldots, (X_l, z_l)\}$ where X_i are some points in the domain and z_i are the (possibly noise-corrupted) values of the target function. Learning a model tree from these examples amounts to dividing the domain into polyhedral regions and approximating the target function with a linear model in each region. Such an approximation would be good for many piece-wise linear functions.

The automata algorithm for decision trees based on GLA and two teams of CALA is easily adopted for learning model trees also. Given a set of examples at a node, the main step in the algorithm proceeds in a similar fashion as earlier. We use the next example as the context vector for the GLA and choose one of the two actions. This results in one of the CALA teams being activated. Now all the CALA in the activated team choose actions. The actions are viewed as values for the parameters in a linear function of the input vector. Suppose the example vectors X_i are n-dimensional. We would have $n + 1$ members in the CALA team. Let $X(k) = (X_1(k) \ldots X_n(k))^T$ denote the example at iteration k and let $a_i(k)$ denote the actions chosen by the CALA, $i = 0, \ldots, n$. Let $z(k) \in \Re$ denote the desired function value for this example as given in the training set. Denote the reinforcement (for this choice of actions by the team) by $\beta(k)$, which is generated as follows. We calculate $\hat{z}(k) = a_0 + a_1 X_1(k) + \cdots + a_n X_n(k)$ and set $\beta(k) = \frac{1}{1+(z(k)-\hat{z}(k))^2}$. (As in the decision tree algorithm explained in the previous subsection, we generate two reinforcements for the CALA team one of which is also given as reinforcement to the GLA). Now all the CALA and GLA update their states and the cycle repeats.

The above routine for learning a hyperplane at a node using GLA plus two teams of CALA can be used recursively to construct a model tree by following essentially the same procedure as in the case of learning a decision tree. After termination of the above learning routine there are essentially two cases. First case is when the two linear functions learnt by the two CALA teams give good enough approximation to the desired function (as measured, e.g., by the value of $(\hat{z} - z)^2$ summed over all examples). Then we label the current node by the hyperplane learnt by the GLA and make both its children as leaves. The left child would be labeled by the linear function learnt by the left-CALA team and the right child would be labeled by the right-CALA team. The second case occurs when the approximation as given by the linear functions learnt by the

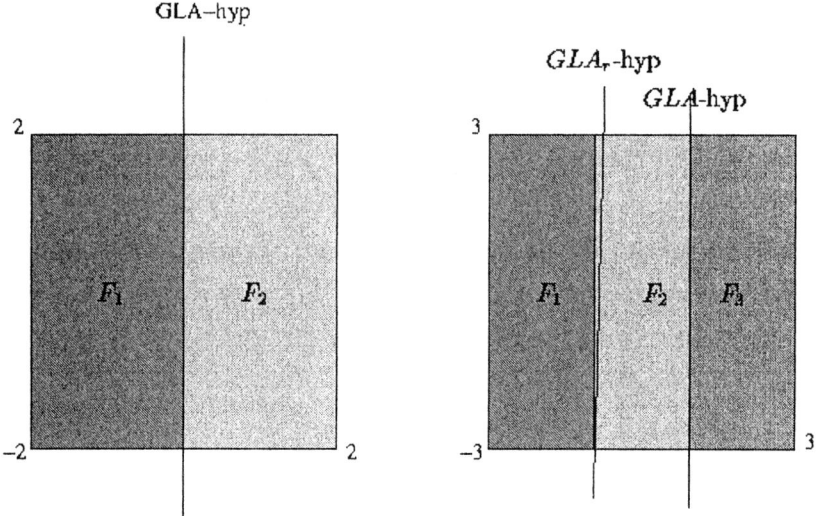

Figure 4.6. Learning of piecewise linear functions using the automata Algorithm

CALA teams are not good enough. In such a case, we first label the current node with the hyperplane learnt by the GLA. Then we split the set of examples at this node into two sets as earlier and recursively call the same routine to learn the left and right subtrees of the current node.

This algorithm is found to work reasonably well on some examples. Full details of the algorithm are available in [Pra03]. We describe below results obtained with this algorithm on two examples.

Fig. 4.6 shows the results obtained on two problems of learning piece-wise linear functions. The figure shows the domains of the function. In each shaded region (marked F_1, F_2 etc. in the figure) the target function is a linear function and these linear functions are different in different regions.

In the first problem (represented by the left panel in Fig. 4.6), the target function is $f_1(\mathbf{x})$ in region F_1 and is $f_2(\mathbf{x})$ in region F_2, where f_1 and f_2 are given by

$$\begin{aligned} f_1(\mathbf{x}) &= 10x_1 + 3x_2 + 2.5 \\ f_2(\mathbf{x}) &= 4.2x_1 - 5x_2 + 1.5 \end{aligned} \qquad (4.5.26)$$

In the second problem (represented by the right panel in the figure) the target function is given by f_i in region F_i, $i = 1, 2, 3$, where f_1 and f_2 are given by (4.5.26) and f_3 is given by (4.5.27) below.

$$f_3(\mathbf{x}) = 6.8x_1 - 1.5x_2 - 4 \qquad (4.5.27)$$

Discussion 173

In fig. 4.6, we show the hyperplanes learnt by the GLA superimposed on the target function domains. As can be seen, the algorithm has learnt a very good approximation to the underlying region structure of the target function. *It may be noted that the algorithm does not know that there are two regions in the first problem and three in the second problem.* The basic learning unit consisting of GLA plus two CALA teams is run again and again recursively till the required level of approximation error is reached. Thus, in the second problem, after learning the hyperplane at the first or root node, the algorithm automatically decided to further grow the tree under one of the children. The approximating linear functions learnt in the first problem are:

$$\hat{f}_1(\mathbf{x}) = 10.33x_1 + 3.16x_2 + 2.49$$
$$\hat{f}_2(\mathbf{x}) = 4.45x_1 - 5.25x_2 + 1.57 \qquad (4.5.28)$$

Comparing the above with (4.5.26), we see that the learning achieved is quite good. In the second problem, the functions learnt are:

$$\hat{f}_1(\mathbf{x}) = 9.76x_1 + 3.11x_2 + 1.5$$
$$\hat{f}_2(\mathbf{x}) = 4.5x_1 - 5.18x_2 + 1.75$$
$$\hat{f}_3(\mathbf{x}) = 6.7x_1 - 1.52x_2 - 3.36 \qquad (4.5.29)$$

Comparing this with (4.5.26) and (4.5.27), we see that even in this problem the learning has been quite effective.

4.6 Discussion

In this chapter we have considered one generic application area, namely, pattern recognition, to illustrate the effectiveness of learning automata. The main intention is to show how the various automata models and the multiautomata structures discussed in the earlier chapters can be deployed in applications. Towards this end, this chapter has described a number of ways in which automata can be utilized for pattern classification.

In a simple problem where the number of features is not too large and the form of discriminant function can be chosen by the designer, the problem of learning the optimal parameters of the discriminant function can be handled by a team of automata (playing a common payoff game), a structure introduced in Chapter 2. If the parameters can be discretized then a team of FALA can be used. With a team of FALA, we can employ the L_{R-I} algorithm if convergence to a local optimum is good enough; otherwise estimator algorithms such as the pursuit algorithm can be used to attain global optimum (provided the consequent memory overhead is affordable). If one does not wish to discretize the parameters, one can employ a team of CALA. In all these cases, the form of the discriminant function can be very general. In particular, the algorithms need no change to handle discriminant functions that are nonlinear in their parameters also. As seen earlier in this chapter, the automata team delivers very satisfactory performance.

For more complex problems, we can employ feedforward networks of automata discussed in Chapter 3. We have shown how a three layer network of automata can be used for learning a classifier that represents each class region in the feature space as a union of convex polyhedral sets. Such piece-wise linear approximation of the boundary between classes would be effective in many complicated pattern recognition problems. By using the L_{R-I} algorithm for updating the action probability vectors of all automata, we can get convergence to a local optimum and in many cases that is good enough. We have also shown how we can use PLA along with the global learning algorithm discussed in Section 3.7 so that the system converges to the globally optimal point. In the case of global algorithms, longer learning times are needed. More general network structures can also be used in a similar way.

This network of automata structure also illustrates the usefulness of one of the important characteristics of automata models, namely, the fact that the actions of automata can represent any quantities of interest. In the three layer network, actions of automata in the first layer units represent values of parameters to specify some hyperplanes while the actions of automata in the second layer units are essentially logical decisions regarding which hyperplanes to combine together. However, using the same common reinforcement signal, all automata update their action probability distributions and the system learns the classifier as desired. Another example of such generality of automata models is the team of automata algorithm for learning conjunctive concepts which was discussed in Section 2.7.2. As is easy to see, this is also a pattern recognition problem but where not all features are numerical. When we include nominal features also, the natural parameterization of the class of discriminant functions of interest results in some parameters that are real-valued while others take some discrete values. The team of automata that includes both FALA and CALA can handle this problem. It may be noted that overall structure of the algorithm for concept learning is same as that of the common payoff game considered in this chapter for learning discriminant functions in the case where all features are real-valued.

We have also discussed how automata can be used for learning decision trees, which are popularly used in many pattern recognition applications. The automata based method for decision tree learning illustrated the use of the GLA model. As seen in Section 4.5, the basic learning unit consisting of a GLA and two teams of CALA is very effective for learning decision trees as well as for learning piece-wise linear functions from examples using the regression tree or model tree to represent the learnt function.

Discussion

While we illustrated a variety of automata structures for pattern recognition, the underlying design philosophy of all these is the same. We essentially design a multiautomata system in such a way that when each automaton in the system chooses an action (using its current action probability distribution), this tuple of actions represents a classifier in the set of classifiers of interest. We then classify the next sample pattern with this chosen classifier and the result of this classification determines the reinforcement supplied to the automata system. In all algorithms discussed in this chapter, we employed a very simple binary reinforcement signal which simply indicates whether or not the classification by the classifier corresponding to the chosen tuple of actions, agrees with that given in the training sample. As explained in Section 4.3, if the automata system maximizes the expected value of this reinforcement, then the system would be learning a classifier with minimum probability of misclassification even when the examples are corrupted by uniform classification noise which could go up to 50%. In the examples presented here, the various automata algorithms are tested using such noise corrupted training samples and the results show the ability of these algorithms to tolerate classification noise.

In all the methods illustrated in this chapter, the collection of automata tends to minimize the probability of error in classification. This is a better performance measure than other criteria such as mean square error which are often used in most of the other methods, mainly for convenience in the learning algorithm. It may be noted, however, that by suitably defining the reinforcement (using some other loss function), the automata algorithms can be used for learning the optimal classifier with respect to other criteria as well.

In most of the algorithms presented here, the actions of automata correspond to possible values of parameters. (The only exception being the GLA in the decision tree learning algorithm). This may look somewhat unnatural especially in pattern recognition problems with real-valued features where it is now necessary to discretize the parameters if we want to use FALA. Making the automata actions correspond to class labels seems, at first sight, to be a better option. In the decision tree learning algorithm discussed in this chapter, the actions of the GLA do indeed correspond to something akin to class labels. If we want to identify actions of automata with class labels, then GLA are needed so that we can use the feature vector as context vector to associate different class labels with different feature vectors. The example of a GLA presented in Section 1.7.2 is, in fact, a 3-class 2-feature pattern recognition problem. Thus, that example illustrates how a GLA can be used in some PR problems. However, in general, in many PR applications it is difficult to learn the classifier using only such GLA models. In the example considered in Section 1.7.2, we could properly design a GLA algorithm only because of the knowledge that the three classes are separable by two hyperplanes. For a general PR problem, it is more effective to use a team of automata, a feedforward network of automata or the decision tree method.

Pattern Recognition constitutes only one of the many possible application areas for automata methods. In Chapter 6, we would be briefly indicating many other areas where these methods are effective. Before that, in the next chapter we discuss parallel operation of automata which can greatly improve the speed of learning and hence is useful in many applications including pattern recognition.

4.7 Supplementary Remarks

An overview of application of automata in pattern recognition can be found in [TS01]. Pattern recognition has been one of the general areas on which applications of automata models are illustrated. The team of FALA algorithm for pattern recognition is discussed in [TS87]. The team of CALA algorithm for PR is from [San94]. The ability of automata algorithms to tolerate many types of noise is discussed in [Nag97]. For example, the team of CALA algorithm can learn optimal hyperplane classifiers even when the examples are corrupted with non-uniform classification noise (where the probability of the class label given in the example being wrong can depend on the feature vector of the example). The three layer network for pattern recognition is from [Pha91]. This network structure is useful in many other situations also. For example, it can be used in pruning decision trees. Most standard decision tree algorithms construct the tree in a top down fashion and hence may learn a tree that does not generalize well to new examples. One way to mitigate the problem is to prune the learnt tree. Most standard pruning methods can only replace some subtrees with leaves. An interesting automata based algorithm for pruning decision trees is presented in [SS99]. Here we use the three layer network where the first layer units are fixed and there is a first layer unit corresponding to every hyperplane in the decision tree. Learning is employed only for the second layer units. This amounts to reorganizing the learnt hyperplanes in the decision tree so as to improve classification accuracy. The algorithm is efficient and results in good pruning of decision trees learnt with any method. The decision tree and model tree algorithms presented here are from [Pra03]. As mentioned earlier, a single GLA can also be used as a classifier. One of the first suggestions for allowing a context vector input to an automaton was motivated by pattern recognition applications [BA85]. Details of GLA methods for pattern recognition can be found in [Jav94].

Chapter 5

PARALLEL OPERATION OF LEARNING AUTOMATA

5.1 Introduction

A decisive aspect of any learning system is its rate of learning or equivalently, speed of convergence. It is decisive because most learning systems operate in slowly varying environments and the learning process should be completed well before significant changes take place in the environment; otherwise, learning is ineffective. In the case of learning automata algorithms, speed of convergence can be increased by increasing the value of the learning parameter, λ. For example, it is easy to see that in the L_{R-I} algorithm, the amount by which an action probability is changed, is directly proportional to λ. However, increasing λ results in reduced accuracy in terms of probability of convergence to the best action. To take an extreme example, if we make $\lambda = 1$ in the L_{R-I} algorithm, then it converges in a single step to the first action that resulted in $\beta = 1$, though it may not be the optimal action. The result that L_{R-I} algorithm is ϵ-optimal only says that if λ is sufficiently small then with probability arbitrarily close to unity, the algorithm converges to the optimal action. As we have seen in Chapters 2 and 3, all convergence results hold only when λ is sufficiently small. Small value of λ implies slow rate of convergence. The problem therefore is to increase speed of convergence without reducing accuracy. This is not possible in the models considered so far, as a single step-size parameter controls both speed as well as accuracy. In this chapter, we discuss a method, which is a way of parallelizing LA algorithms, that would help in increasing speed of convergence (without sacrificing accuracy) for any LA system.

Parallel operation is known to increase the speed of convergence in general. In order to conceive of parallel operation of learning automata, one has to change the sequential nature of the models considered earlier. Any single learning automaton discussed so far, generates one action at a time, interacts with the environment and obtains the reinforcement signal. If we have a number of learning automata acting in parallel in the place of a single automaton, each of these LA generates its own action and gets a corresponding reinforcement signal from the environment simultaneously. Such a model would be appropriate in applications such as pattern recognition considered in the previous chapter. In these applications, different action tuples of the automata system correspond to different classifiers and for any choice of action tuple, the reinforcement is generated by testing the corresponding classifier using the next training sample. Parallel operation is feasible because one could test several parameter vectors (or classifiers) simultaneously. On the other hand, there are applications such as routing in communication networks [NT89, Ch.9] where the action corresponds to a route, and hence simultaneous choice of several actions is not feasible unless one is studying a simulation of the network. The method we present here involves choice of more than one action at each instant and the algorithms we present here result in considerable increase in speed whenever such parallel operation is feasible.

The basic idea in such a parallel operation of LA is that since the environmental responses are stochastic, a decision based on several responses would have less error (or variance) than a decision based on a single response. The updating of the action probability vector would thus be more accurate and facilitate faster convergence.

5.2 Parallel Operation of FALA

Consider first the operation of n FALA in parallel in the place of a single one. These FALA could be said to form a module which replaces a single LA considered earlier. We will consider the operation of the module with an extended L_{R-I} algorithm. The schematic is shown in Fig. 5.1. The main features are the following:

- The action set is $\{\alpha_1, \alpha_2, \cdots, \alpha_r\}$.

- The action probability vector $(\mathbf{p}(k))$ is common to all the n LA of the module.

- Each LA (say τ^{th} LA, $\tau = 1, 2, \ldots n$) in the module selects an action $(\alpha^\tau(k))$ based on the common action probability vector (independently of all other LA) and obtains its own reinforcement signal $(\beta^\tau(k))$.

- The action probability vector is updated depending on the actions selected by all the members and the corresponding reinforcement signals obtained. A fuser combines all the available information.

It is assumed that $\beta^\tau(k) \in [0, 1]$, $\tau = 1, \ldots n$. At each instant the fuser computes the following:

Parallel Operation of FALA

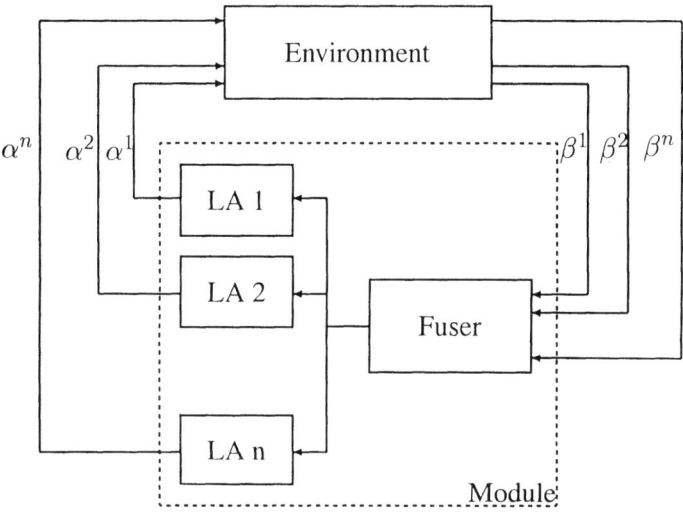

Figure 5.1. A Module of Learning Automata.

- the total response to any action α_i at k:

$$q_i(k) \triangleq \sum_{\tau=1}^{n} \beta^\tau(k) I\{\alpha^\tau(k) = \alpha_i\} \qquad (5.2.1)$$

where $I\{A\}$ is the indicator function taking values 1 or 0 depending on whether A is true or false respectively.

- the total response at the instant k:

$$q(k) \triangleq \sum_{\tau=1}^{n} \beta^\tau(k) = \sum_{i=1}^{r} q_i(k) \qquad (5.2.2)$$

After this, the action probability distribution is updated using the following learning algorithm. In this algorithm, $\lambda \in (0, 1]$ is the learning parameter and $\tilde{\lambda} = \frac{\lambda}{n}$ is called its normalized value. In updating $\mathbf{p}(k)$, $\tilde{\lambda}$ controls the step-size. The learning algorithm is:

$$p_i(k+1) = p_i(k) + \tilde{\lambda}(q_i(k) - q(k)p_i(k)) \quad \forall\, i = 1, \dots r. \qquad (5.2.3)$$

We note the following to get an intuitive understanding of this algorithm.

- Since the action choice is stochastic, the action probability $p_i(k)$ is also the expected fraction of choices of α_i at the instant k. Thus the quantity $q_i(k)/q(k)$ could be considered as a figure of merit of the performance of α_i at k. The updating term moves $p_i(k)$ towards $q_i(k)/q(k)$.

- It can be seen that the updating ensures that $\mathbf{p}(k+1)$ remains a probability vector if $\mathbf{p}(k)$ is. Computationally, $\mathbf{p}(k)$ is updated once and not necessarily by each LA; the updated value $\mathbf{p}(k+1)$ is shared by all the LA in the module.

5.2.1 Analysis

It is shown in the sequel that when the optimal action is unique, algorithm (5.2.3) is ϵ-optimal. Let

$$d_i = E[\beta^\tau(k)|\alpha^\tau(k) = \alpha_i] \qquad \tau \in \{1, 2, \ldots, n\}. \tag{5.2.4}$$

The environment is assumed to be stationary and hence d_i are constant. Define

$$\Delta p_i(k) = E[p_i(k+1) - p_i(k) \mid \mathbf{p}(k)]. \tag{5.2.5}$$

From the algorithm, (omitting k for convenience and using $\tilde{\lambda} = \frac{\lambda}{n}$)

$$\Delta p_i = \left(\frac{\lambda}{n}\right) E[q_i - qp_i \mid \mathbf{p}] \tag{5.2.6}$$

Substituting for q_i and q from (5.2.1), (5.2.2),

$$\Delta p_i = \left(\frac{\lambda}{n}\right) \sum_{\tau=1}^{n} E[\beta^\tau(I\{\alpha^\tau = \alpha_i\} - p_i) \mid \mathbf{p}] \tag{5.2.7}$$

Since $\beta^\tau(k)$ depends only on $\alpha^\tau(k)$ and not on $\alpha^s(k)$, $s \neq \tau$,

$$E[\beta^\tau(I\{\alpha^\tau = \alpha_i\} - p_i) \mid \mathbf{p}] = p_i(1 - p_i)d_i - p_i \sum_{s \neq i} p_s d_s \tag{5.2.8}$$

Simplifying (5.2.8) using $\sum_{s \neq i} p_s = 1 - p_i$, and substituting in (5.2.7),

$$\Delta p_i = \lambda \sum_{s=1}^{r} p_i p_s (d_i - d_s) \tag{5.2.9}$$

Let $d_l = \max_i d_i (1 \leq i \leq r)$. Then, from (5.2.9),

$$\Delta p_l \geq 0, \qquad \forall \mathbf{p} \in S_r, \forall k, \tag{5.2.10}$$

where S_r is the simplex,

$$S_r \triangleq \{\mathbf{p} \in [0,1]^r : \sum_{i=1}^{r} p_i = 1\}.$$

Moreover, if d_l is unique, then

$$\Delta p_l > 0 \qquad (5.2.11)$$

whenever $\mathbf{p} \in S_r^0$, the open simplex. The following result is a natural consequence.

PROPOSITION 5.1 *Let d_l be unique. Then $p_l(k)$ converges to 0 or 1 with probability one.*

Proof: Since $\Delta p_l(k) \geq 0$, $p_l(k)$ is a submartingale. As d_l is unique, $\Delta p_l = 0$ only when $p_l = 0$ or $p_l = 1$ in (5.2.9). Convergence of $p_l(k)$ to 0 or 1 w.p.1 follows from the martingale convergence theorem [KT75]. ∎

5.2.2 ϵ-optimality

In this subsection, sufficiency conditions are derived on $\tilde{\lambda}$ such that the algorithm is ϵ-optimal in all stationary random environments. The concept of subregular functions [NT89] is used for this purpose. A continuous function $\phi: S_r \to \Re$ is said to be subregular if

$$E[\phi(\mathbf{p}(k+1))|\mathbf{p}(k) = \mathbf{p}] \geq \phi(\mathbf{p}), \forall \mathbf{p} \in S_r. \qquad (5.2.12)$$

Also, let

$$\Gamma_i(\mathbf{p}) \triangleq \Pr\{p_i(\infty) = 1|\mathbf{p}(0) = \mathbf{p}\}. \qquad (5.2.13)$$

PROPOSITION 5.2 *Let $\phi_i(\mathbf{p})$ be subregular with, $\phi_i(\mathbf{e_i}) = 1$ and $\phi_i(\mathbf{e_j}) = 0$, $\forall j \neq i$ ($\mathbf{e_j}$ is the unit vector with 1 in the jth component). Then $\Gamma_i(\mathbf{p}) \geq \phi_i(\mathbf{p})$.*

Proof: See [NT89, Ch.5]. ∎

The Γ_i defined by (5.2.13) is the probability that the automaton converges to i^{th} action. Hence, to show ϵ-optimality, we need to show that Γ_l can be made arbitrarily close to 1 by choosing the learning step-size appropriately. To show this we would be using an appropriately constructed subregular function. Let

$$\phi_i(\mathbf{p}) \triangleq \frac{1 - e^{-x_i p_i}}{1 - e^{-x_i}}, \quad x_i > 0. \qquad (5.2.14)$$

Then $\phi_i(\mathbf{p})$ satisfies the boundary conditions of Proposition 5.2. Now, using (5.2.3) it can be shown that

$$E[\phi_l(\mathbf{p}(k+1))|\mathbf{p}(k)] - \phi_l(\mathbf{p}(k))$$
$$= \frac{1}{1 - e^{-x_l}}(e^{-x_l p_l(k)} - E[e^{-x_l p_l(k+1)}|\mathbf{p}(k)])$$
$$= \frac{e^{-x_l p_l}}{1 - e^{-x_l}}(1 - (A_l + B_l)^n) \qquad (5.2.15)$$

where, for any $\tau \in \{1, 2, \ldots, n\}$

$$A_l \triangleq p_l E[\exp(-x_l \tilde{\lambda} \beta^\tau (1 - p_l)|\mathbf{p}, \alpha^\tau = \alpha_l], \qquad (5.2.16)$$
$$B_l \triangleq \sum_{i \neq l} p_i E[\exp(-x_l \tilde{\lambda} \beta^\tau p_l|\mathbf{p}, \alpha^\tau = \alpha_i]. \qquad (5.2.17)$$

Hence, to ensure subregularity of $\phi_l(\mathbf{p})$ it is sufficient if we have $A_l + B_l \leq 1$. The following inequalities are used while obtaining sufficiency conditions:

i) for $y \geq 0$, $e^y \leq 1 + y + \dfrac{y^2}{2} e^y$; $\qquad (5.2.18)$

ii) for $y \geq 0$, $e^{-y} \leq 1 - y + \dfrac{y^2}{2}$. $\qquad (5.2.19)$

Let $d_m = \max_{i \neq l} d_i$, $1 \leq i \leq r$. That is, α_m is the second best action. Let $\theta \triangleq d_l - d_m$.

LEMMA 5.1 *Let $\theta > 0$. Then $\phi_l(p)$ is subregular whenever*
$x_l \tilde{\lambda} \leq ((1 + \theta)^2 + 4\theta)^{1/2} - (1 + \theta))$.

Proof: From (5.2.16) and (5.2.17) and the inequalities given above,

$$A_l + B_l \leq 1 - x_l \tilde{\lambda} p_l \sum_{j=1}^{r} p_j(d_l - d_j) + \frac{(x_l \tilde{\lambda}(1 - p_l))^2 p_l}{2}$$
$$+ \frac{(x_l \tilde{\lambda} p_l)^2 (1 - p_l)}{2} e^{x_l \tilde{\lambda}}$$
$$\leq 1 - x_l \tilde{\lambda} p_l (1 - p_l) \left(\theta - \frac{(x_l \tilde{\lambda})}{2} e^{x_l \tilde{\lambda}}\right). \qquad (5.2.20)$$

For the quantity on the right hand side to be smaller than unity, we set $x_l \tilde{\lambda} \in (0, 1)$. Using the inequality,

$$e^x \leq \left(\frac{2 + x}{2 - x}\right); \quad x \in (0, 1) \qquad (5.2.21)$$

(5.2.20) simplifies to

$$A_l + B_l \leq 1 - x_l\tilde{\lambda}p_l(1 - p_l)\left(\theta - \frac{(x_l\tilde{\lambda})}{2}\frac{2 + x_l\tilde{\lambda}}{2 - x_l\tilde{\lambda}}\right). \quad (5.2.22)$$

The lemma follows by enforcing $\left(\theta - \frac{(x_l\tilde{\lambda})}{2}\frac{2+x_l\tilde{\lambda}}{2-x_l\tilde{\lambda}}\right) > 0$. ∎

The main result of this section is now proved.

THEOREM 5.3 *The proposed algorithm is ϵ-optimal in all stationary random environments with $\theta > 0$.*

Proof: It is enough to prove that for any given $\delta \in (0, 1)$, it is possible to have $\Gamma_l(\mathbf{p}) \geq (1 - \delta)$, by a suitable choice of $\tilde{\lambda}$.

Now, by definition $\phi_l(\mathbf{p}) > (1 - e^{-x_l p_l})$ where $p_l = p_l(0)$. Choosing $x_l = \ln(\frac{1}{\delta})/p_l(0)$, $\phi_l(\mathbf{p}) > (1 - \delta)$ is ensured. By Lemma 5.1, subregularity of $\phi_l(\mathbf{p})$ is assured whenever $\tilde{\lambda} \in (0, \tilde{\lambda}_0)$, where

$$\tilde{\lambda}_0 = \left(\frac{p_l(0)}{\ln\left(\frac{1}{\delta}\right)}\right)\left(((1 + \theta)^2 + 4\theta)^{1/2} - (1 + \theta)\right). \quad (5.2.23)$$

Choosing $\tilde{\lambda} \in (0, \tilde{\lambda}_0)$, and using Proposition 5.2, the result follows. ∎

From the above, it is seen that the accuracy of the learning process is controlled by the parameter $\tilde{\lambda}$. In the following subsection it is shown that a large value of λ indicates large speed. Since λ and $\tilde{\lambda}$ are related by $\tilde{\lambda} = \lambda/n$, n & λ can be varied to control speed of convergence while keeping $\tilde{\lambda}$ small enough for the required level of accuracy. The issues involved here are stated more formally in the corollaries below.

COROLLARY 5.4 *If the optimal action is unique, given any $\delta \in (0, 1)$, $n \geq 1$ and $\theta_0 \in (0, 1)$, there exists $\lambda_0 \in (0, 1)$ such that $\forall\, \lambda \in (0, \lambda_0)$, we have $\Pr\{\lim_{k \to \infty} p_l(k) = 1\} \geq (1 - \delta)$ in all stationary random environments with $\theta \geq \theta_0$.* ∎

COROLLARY 5.5 *If the optimal action is unique, given any $\delta \in (0, 1)$, $\lambda \in (0, 1)$, $\theta_0 \in (0, 1)$ there exists $n_0 \geq 1$ such that $\forall\, n > n_0$, $\Pr\{\lim_{k \to \infty} p_l(k) = 1\} \geq (1 - \delta)$ in all random environments with $\theta \geq \theta_0$.* ∎

These corollaries follow from (5.2.23), and the fact that $0 \leq \lambda \leq 1$. They indicate the freedom in selection of parameters. While Corollary 5.4 gives sufficiency conditions on the step size of the algorithm for a given module size and given accuracy, Corollary 5.5 gives sufficiency conditions on the module size for a given step size. Using (5.2.23), sufficiency conditions on λ_0 and n_0

are given by

$$\lambda_0 = \min\left(1, \left(\frac{np_l(0)}{\ln\left(\frac{1}{\delta}\right)}\right)\left(((1+\theta)^2 + 4\theta)^{1/2} - (1+\theta)\right)\right)$$

$$n_0 \geq \lambda\left(\left(\frac{p_l(0)}{\ln\left(\frac{1}{\delta}\right)}\right)\left(((1+\theta)^2 + 4\theta)^{1/2} - (1+\theta)\right)\right)^{-1}$$

REMARK 5.2.1 *The accuracy of the process is controlled by the parameter $\tilde{\lambda}$, while n and λ can be varied to control the speed of convergence for any given accuracy. Since both the speed and the accuracy can be controlled independently, this algorithm is more flexible when compared with the sequential L_{R-I} algorithm.* ∎

REMARK 5.2.2 *In situations where there are multiple optimal actions and there is at least one suboptimal action, similar analysis holds if p_l is interpreted as the sum of action probabilities corresponding to all the optimal actions. It can be shown that the combined probability of choosing one of the optimal actions at a time, can be made as close to unity as desired with a high probability, by choosing a small enough learning parameter. Such a result intuitively makes sense, since, if there is more than one optimal action, best performance is obtained when any of them is attempted; it is sufficient that the sum of the probabilities of attempting them goes to unity. With p_l interpreted this way, exactly the same analysis holds as earlier for demonstrating ϵ-optimality.* ∎

5.2.3 Speed of Convergence and Module Size

The basic attraction of the parallel algorithm (5.2.3) is that it gives a n-fold increase in speed of convergence, while retaining the same accuracy. This can be seen by deriving the approximating ODE for this algorithm using the method outlined in Appendix A. From (5.2.9),

$$\Delta p_i = n\tilde{\lambda}\sum_{s=1}^{r} p_i p_s (d_i - d_s). \tag{5.2.24}$$

Since $\tilde{\lambda}$ is the learning parameter in (5.2.3), the approximating ODE of the algorithm is given by (see Section A.3.1)

$$\frac{dz_i}{dt} = n\sum_{s=1}^{r} z_i z_s (d_i - d_s); \quad z_i(0) = p_i(0) \quad \forall\, i \in \{1, 2, \ldots, r\} \tag{5.2.25}$$

where $\mathbf{z}(t)$ is the variable in the associated ODE corresponding to $\mathbf{p}(k)$ and z_i is its ith component. It may be noticed that the RHS of (5.2.25) is the RHS of the ODE for the single LA (as given by (A.3.28) in Appendix A) multiplied by the factor n. Again, this is basically a consequence of the summation in (5.2.7). As explained in Appendix A, the solution of the ODE approximates the behavior of the algorithm to any desired accuracy for sufficiently small values of $\tilde\lambda$. Since the solution of ODE (5.2.25) is n times faster than that of the ODE (A.3.28) for a single LA, it follows that the parallel algorithm (5.2.3) is n times faster than the L_{R-I} for a single LA. As the only stable equilibrium point of (5.2.25) is the unit vector $\mathbf{e}_l$, it also follows from [Lak81, WN86] that

$$E[\mathbf{p}(k) - \mathbf{z}(\tilde\lambda k)] = O(\sqrt{\tilde\lambda}) \quad \text{and}$$
$$E[(\mathbf{p}(k) - \mathbf{z}(\tilde\lambda k))^2] = O(\tilde\lambda). \qquad (5.2.26)$$

The above expressions help us to analyze the speed of convergence of the algorithm to the desired point $\mathbf{e}_l$. The quantity $\tilde\lambda = \lambda/n$ can be viewed as an accuracy parameter, relating the averaged process to the associated ODE. However, the speed of convergence of the ODE is completely characterized by $\lambda = n\tilde\lambda$ for a given $\mathbf{p}(0)$. If $\tilde\lambda$ is fixed, algorithm (5.2.3), for a module of size n, tracks an ODE which is n times faster, but retains the same accuracy. This could be regarded as a n-fold compression of the time axis. On the other hand, if λ is fixed and n is varied, we have $\tilde\lambda_1 < \tilde\lambda_2$ whenever $n_1 > n_2$, and the algorithm and the ODE trajectory match better in the former case.

REMARK 5.2.3 *The n-fold increase in speed of convergence for the parallel algorithm (5.2.3) represents only an ideal situation. In practice, implementation of the algorithm involves overheads and these reduce the benefit to some extent [TA98].* ∎

5.2.4 Simulation Studies

A five-action problem (i.e., $r = 5$) with d_i, $i = 1, \ldots, 5$, given by $0.9, 0.81$, 0.7, 0.6 and 0.5, is considered for simulation studies. The results are shown in Table 5.2.1. In the table, $k_{\text{avg}}(a)$ is the average value of k (averaged over 100 runs) at which the probability of the optimal action, $p_l(k)$, exceeded a. To ensure sufficiency conditions, the same value of $\tilde\lambda$ was used for all module sizes in the results shown. The table indicates speedups of roughly the order of module size, thus demonstrating the efficacy of the approach. For this problem, $\theta = 0.09$. With $\delta = 0.01$, the sufficiency condition (5.2.23) yields $\tilde\lambda = 0.0067$, which is in good agreement with the value $\tilde\lambda = 0.01$ employed in simulations. This shows that the sufficiency bounds are quite reasonable.

5.3 Parallel Operation of CALA

The continuous action learning automata (CALA) can also be parallelized by forming a module of n such LA. Since the procedure is similar to that of FALA, only the significant points will be given. We follow the same notation as in Section 1.6. A module of CALA operates as follows.

n	λ	k_{avg} (0.9)	k_{avg} (0.95)	k_{avg} (0.99)
1	0.01	2732	3727	5811
2	0.02	1388	1831	2871
4	0.04	679	942	1412
8	0.08	341	473	745
16	0.16	189	261	365
32	0.32	87	119	176
64	0.64	43	56	82

Table 5.3.1. Performance of the algorithm for various module sizes on the 5 action problem.

Step 1: The mean and variance of the Gaussian distribution for the module are initialized as $\mu(0)$ and $\sigma(0)$ at time step $k = 0$.

Step 2: At the kth step, n random numbers $x^\tau(k)$ ($\tau = 1, \ldots n$) are generated from the Gaussian distribution, $N(\mu(k), \phi(\sigma(k)))$, which is the action probability distribution at k.

Step 3: The reinforcements $\beta_{x^\tau(k)}(\tau = 1, \ldots n)$ and $\beta_{\mu(k)}$ are obtained from the environment.

Step 4: The mean $\mu(k)$ and standard deviation $\sigma(k)$ are updated as below.

$$\mu(k+1) = \mu(k) + \tilde{\lambda} \sum_{\tau=1}^{n} F_1(\mu(k), \sigma(k), x^\tau(k), \beta_{x^\tau(k)}, \beta_{\mu(k)}) \quad (5.3.27)$$

$$\sigma(k+1) = \sigma(k) + \tilde{\lambda} \sum_{\tau=1}^{n} F_2(\mu(k), \sigma(k), x^\tau(k), \beta_{x^\tau(k)}, \beta_{\mu(k)})$$
$$- \tilde{\lambda} K [\sigma(k) - \sigma_L] \quad (5.3.28)$$

where, (as in Section 1.6)

$$F_1(\mu, \sigma, x, \beta, \beta') = \left(\frac{\beta - \beta'}{\phi(\sigma)}\right)\left(\frac{x - \mu}{\phi(\sigma)}\right) \quad (5.3.29)$$

$$F_2(\mu, \sigma, x, \beta, \beta') = \left(\frac{\beta - \beta'}{\phi(\sigma)}\right)\left[\left(\frac{x - \mu}{\phi(\sigma)}\right)^2 - 1\right] \quad (5.3.30)$$

$$\phi(\sigma) = \begin{cases} \sigma_L & \text{for } \sigma \leq \sigma_L \\ \sigma & \text{for } \sigma > \sigma_L \end{cases}$$

Here, $\tilde{\lambda} > 0$ is the step size for learning, K is a large positive constant and σ_L is a small positive constant as required in the CALA algorithm (see Section 1.6).

As in the case of FALA, the strength of parallelization is indicated by the summations in (5.3.27), (5.3.28). It follows from an application of the ODE method of analysis that an n-fold increase in speed is obtained here also. A formal statement of this fact is given in the following theorem.

THEOREM 5.6 *For the algorithm given by (5.3.27), (5.3.28), the approximating ODE is given by*

$$\dot{\mu} = n\frac{\partial J}{\partial \mu}(\mu, \phi(\sigma)) \qquad (5.3.31)$$

$$\dot{\sigma} = n\frac{\partial J}{\partial \sigma}(\mu, \phi(\sigma)) - K[\sigma - \sigma_L] \qquad (5.3.32)$$

where J is the same function as in the case of a single CALA (see equation (1.6.29) in Section 1.6). ∎

Proof: The proof is a straightforward extension of the derivation of the approximating ODE for a single CALA which is explained in Section A.3.2.

5.4 Parallel Pursuit Algorithm

In Section 1.4.3, we considered the pursuit algorithm as an example of estimator algorithms which use estimates of reward probabilities for accelerating the learning process. Here we will consider parallelization of the pursuit algorithm for a further speedup of the process. In fact, a module of LA is well suited for this purpose as a number of simultaneous action choices at each instant naturally help in obtaining better estimates of the reward probabilities.

Since we are considering an FALA here, the notation is same as in Section 5.2. Recall that d_i is the reward probability of ith action and let $\mathbf{d} = [d_1, d_2, \ldots d_r]^T$. As earlier, let there be n automata in the module. Let

$$\hat{\mathbf{d}}(k) = [\hat{d}_1(k), \hat{d}_2(k), \ldots \hat{d}_r(k)]^T \qquad (5.4.33)$$

be the vector of estimates of reward probabilities, i.e., $\hat{d}_i(k)$ is the estimate of d_i at the instant k. The internal state of the module is denoted as

$$Q(k) = (\mathbf{p}(k), \hat{\mathbf{d}}(k)) \qquad (5.4.34)$$

At every instant, the fuser computes the following quantities.

- Total reinforcement for action α_i at k:

$$q_i(k) = \sum_{\tau=1}^{n} \beta^\tau(k) I\{\alpha^\tau(k) = \alpha_i\} \qquad (5.4.35)$$

- Number of times α_i is selected at k:

$$n_i(k) = \sum_{\tau=1}^{n} I\{\alpha^\tau(k) = \alpha_i\} \qquad (5.4.36)$$

- Total reinforcement for action α_i till k:

$$Z_i(k) = \sum_{s=0}^{k} q_i(s) \qquad (5.4.37)$$

- Number of times α_i is selected by the τth member of the module till k:

$$\eta_i^\tau(k) = \sum_{s=0}^{k} I\{\alpha^\tau(s) = \alpha_i\} \qquad (5.4.38)$$

- Number of times α_i is selected till k:

$$\eta_i(k) = \sum_{s=0}^{k} n_i(s) = \sum_{\tau=1}^{n} \eta_i^\tau(k) \qquad (5.4.39)$$

Of the above, $q_i(k)$ and $n_i(k)$ are computed using actions and reinforcements at instant k. For computing $Z_i(k)$, $\eta_i^\tau(k)$ and $\eta_i(k)$, we need not explicitly store past actions and reinforcements. We can update their values at each instant using $q_i(k)$ and $n_i(k)$.

The estimates of reward probabilities are then computed as follows.

$$\hat{d}_i(k) = \frac{Z_i(k)}{\eta_i(k)} \quad \text{(assuming } \eta_i(k) \neq 0\text{)} \qquad (5.4.40)$$

where

$$Z_i(k) = Z_i(k-1) + q_i(k) \qquad (5.4.41)$$
$$\eta_i(k) = \eta_i(k-1) + n_i(k) \qquad (5.4.42)$$

As in the pursuit algorithm for a single FALA, at each instant we update $\hat{d}_i(k)$ and $\mathbf{p}(k)$. We update the estimates using (5.4.40)–(5.4.42). For updating $\mathbf{p}(k)$, as earlier, we need to identify the action with highest estimated reward probability. Hence, let

$$\hat{d}_{M(k)}(k) = \max_{j}\{\hat{d}_j(k)\} \qquad (5.4.43)$$

The algorithm updates $\mathbf{p}(k)$ as follows.

$$\mathbf{p}(k+1) = \mathbf{p}(k) + \lambda(\mathbf{e}_{M(k)} - \mathbf{p}(k)) \qquad (5.4.44)$$

where $0 < \lambda < 1$ is the learning parameter and $\mathbf{e}_{M(k)}$ is the unit vector with unity in the $M(k)$ position and zero in the rest. As in the nonmodule case, all the actions are initially chosen a few times to provide starting estimates of $\hat{d}(k)$.

The parallel pursuit algorithm is ϵ-optimal in the same way as the pursuit algorithm considered in Section 1.4.3. However, an improvement in speed is obtained as there are $n_i(k)$ samples for estimating d_i at the instant k. A formal statement of the convergence property is as follows.

THEOREM 5.7 *Consider a module of n LA using the parallel pursuit algorithm. Then $\forall\, \epsilon, \delta \in (0, 1]$ there exist $K^* = K^*(\delta, \epsilon)$, $\lambda^* = \lambda^*(\delta, \epsilon)$ such that*

$$\text{Prob}\,\{p_l(k) > 1 - \epsilon\} > 1 - \delta \qquad (5.4.45)$$

for $\forall\, k > K^*$ and $\forall\, \lambda$ such that $0 < \lambda < \lambda^*$. ∎

The proof of this theorem follows that of Theorem 1.1 and is outlined in Appendix B.

REMARK 5.4.1 *An interesting feature of the parallel pursuit algorithm is that by increasing the size of the module n sufficiently, one can converge to the optimal action in one step! At $k = 0$, we have $\eta_i(0)$ samples to obtain the estimate $\hat{d}_i(0)$. If n is sufficiently large, then, η_i would be large for all i and hence the estimates would be close to their true values. Thus, with a large probability, $M(0)$ would be same as l, the index of the optimal action. Now, if λ in (5.4.44) is 1, we get $\mathbf{p}(1) = \mathbf{e}_{M(0)}$ and this implies convergence in one step.*

In the above argument, the proof that the algorithm converges to optimal action in one step is essentially a consequence of law of large numbers. (The fact that if $\eta_i(0)$ is large then $\hat{d}_i(0)$ would be close to d_i is a consequence of law of large numbers). However, what is interesting about the algorithm is that one can obtain the proof also by interchanging n and k in the proof of convergence of parallel pursuit algorithm given in Appendix B. (The details are in [Arv96]). Thus, the parallel pursuit algorithm clearly brings out the essential feature of the parallelization framework, namely, that we trade 'space' for 'time'. By having more than one LA (as in a module), we get better samples to estimate reward probabilities in fewer iterations. While the one step convergence is an extreme case (where the updated action probabilities are not used to decide on the sampling strategy for the next instant), this boundary case is helpful in getting a feel for our parallelization procedure. ∎

n	λ	$k_{\text{avg}}(0.9)$	$k_{\text{avg}}(0.95)$	$k_{\text{avg}}(0.99)$
1	0.05	56	69	105
2	0.1	27	34	50
4	0.19	16	24	37
8	0.6	4	6	7
16	0.95	2	3	4
32	1.0	1	1	1
64	1.0	1	1	1

Table 5.4.2. Performance of the Parallel Pursuit Algorithm

REMARK 5.4.2 *As in the case of a single FALA, here also we need not restrict the reinforcement, $\beta(k)$, to take values in the interval $[0, 1]$. This is an advantage of pursuit algorithm as compared to, e.g., the L_{R-I}. This avoids vexing problems of scaling in many applications. Since updating of $\mathbf{p}(k)$ does not directly depend on it, $\beta(k)$ can take values in any bounded interval on the real line for using the algorithm. The only requirement is that higher values of $E[\beta(k)]$ correspond to better actions.*

5.4.1 Simulation Studies

The 5-action problem considered in Section 5.2.4 with reward probabilities 0.9, 0.81, 0.70, 0.6 and 0.5 was used to test the parallel pursuit algorithm. The results are shown in Table 5.4.2. Initial values of all action probabilities were 0.2. The estimates $\hat{d}_i(k)$ were maintained at zero till the corresponding actions were attempted at least once. Table 5.4.2 lists the maximum value of λ for which no wrong convergence resulted in any of the 100 runs. $k_{\text{avg}}(a)$ is the average value of k in 100 runs after which $p_l(k)$ remained beyond a. The results indicate speedups roughly of the order of the module size. The entries corresponding to $\lambda = 1$ exhibit one step convergence.

5.5 General Procedure

In this section, a general parallelization paradigm applicable to a large class of LA algorithms is proposed. The design of the proposed parallelization procedure closely follows the developments of the previous section.

The set up is as follows. There are a number of learning agents forming a module. All agents in the module have a common state which includes a probability distribution over the set of actions available to the agents. All agents have the same set of actions. The agents are operating in a stochastic environment which presents them with a common situation at each instant. Each learning agent responds to the situation by selecting an action randomly ac-

General Procedure

cording to the common probability distribution and the environment responds with a stochastic reinforcement for each such action.

The learning agents belonging to a module *cooperate* in some sense, as they have a common internal state representation. A shared memory system housing the common internal state is particularly suitable for such a set up. Actions selected by each learning agent depend on the common internal state, but the selected actions, conditioned on the internal state, are independent of one another. Similarly, the reinforcement to each agent is generated independently of that for other agents. The distribution of this stochastic reinforcement depends only on the action selected.

First consider a single learning agent which corresponds to the nonparallel situation as in earlier chapters. Let each learning agent use the following algorithm for updating its internal state $X(k)$. (In the automata models, the internal state is usually the action probability vector).

$$X(k+1) = X(k) + \lambda G(X(k), Y(k)), \quad X(0) = x_0. \quad (5.5.46)$$

Here $Y(k)$ denotes the observation vector comprising the chosen action and the reinforcement which would also be referred to as payoff. To avoid needless confusion, Y will be explicitly denoted by the tuple (α, β) in the sequel.

Using the usual procedure outlined in Appendix A (and under the assumptions given in Section A.2.1), the associated ODE of the algorithm (5.5.46) is given by

$$\frac{dx}{dt} = g(x); \quad x(0) = x_0. \quad (5.5.47)$$

where

$$g(x) = E[G(X, Y) \mid X = x]. \quad (5.5.48)$$

The algorithm *tracks* the ODE for as large a time as required, provided the learning parameter is small. (See Theorem A.1 in Appendix A).

To derive the parallel version of algorithm (5.5.46) in the framework just described, it is useful to consider the parallel version of L_{R-I} given by (5.2.3) which can be rewritten as

$$p_i(k+1) = p_i(k) + \tilde{\lambda} \sum_{\tau=1}^{n} \beta^\tau(k)(I\{\alpha^\tau(k) = \alpha_i\} - p_i(k)). \quad (5.5.49)$$

The parallel version of the L_{R-I} algorithm is thus seen to have an update vector which is just the sum of the update vectors of n LA employing the L_{R-I} algorithm individually. Using the same methodology, the parallel version of (5.5.46) is obtained as

$$X(k+1) = X(k) + \tilde{\lambda} \sum_{\tau=1}^{n} G(X(k), (\alpha^\tau(k), \beta^\tau(k))); \quad X(0) = x_0$$

$$(5.5.50)$$

where $\tilde{\lambda} = \frac{\lambda}{n}$.

Following the same method as earlier, the associated ODE of (5.5.50) is

$$\frac{dx}{dt} = ng(x); \quad x(0) = x_0. \tag{5.5.51}$$

For a given n, ODE's (5.5.47) and (5.5.51) have the same set of stationary points with identical properties. Specifically, solutions to both ODE's have the same asymptotic behavior. Further, (5.5.51) is nothing but ODE (5.5.47) with the time axis compressed n-fold. If $\tilde{\lambda}$ were fixed at the same value and n were such that λ is within permissible limits, if any, then for a given T, ODE (5.5.51) which is faster by a factor of n in comparison with ODE (5.5.47), is tracked for a time T within desired levels of accuracy by algorithm (5.5.50). As regards asymptotic accuracies, the same arguments hold for both the algorithms. In particular, the proposed procedure does not alter the asymptotic accuracy properties of the sequential algorithm while parallelizing it.

REMARK 5.5.1 *The comments above would lead one to believe that the speed of convergence could be improved arbitrarily by increasing the value of n as desired. This is in general not possible, as in most cases, the increase in n for a fixed $\tilde{\lambda}$ is limited by a ceiling on λ. A reason for this could be the constraints on the state space; for example, most of the LA algorithms maintain a state vector whose elements are probabilities and it is necessary that $\lambda \in (0, 1]$. It can be seen that the behavior of the scheme when λ is fixed and n, and hence $\tilde{\lambda}$, are variable, is characterized by the strong law of large numbers.* ∎

5.6 Parallel Operation of Games of FALA

The idea of parallel operation can be extended to games of FALA considered in Chapter 2. It may be recalled that a game is played by N players each of which is a LA. In a play of the game, each player chooses an action and the environment puts out a reinforcement signal based on the actions chosen by all the players. The players update their states and choose a new action set at the next instant. The game is a sequence of such plays.

Parallelization of the game can be achieved by having a module of FALA as a player in the place of a single FALA. There will now be N modules of FALA in the game, each module consisting of n Learning Automata. For any τ, $\tau = 1, 2, \ldots, n$, all the τth LAs in each module together are involved in a game of N FALA as considered in Chapter 2. Thus there will be n games played simultaneously.

The ith player (or module) ($i = 1, 2, \cdots, N$) consists of n identical FALA with same action set and action probability vector. Let $\tilde{A}_i^\tau$ denote the τth automata in the ith module. Thus, $\tilde{A}_i^\tau$ ($i = 1, 2, \cdots, N$ and $\tau = 1, 2, \cdots, n$) is simultaneously a member of the ith player (module) and τ^{th} game. The set up

Parallel Operation of Games of FALA 193

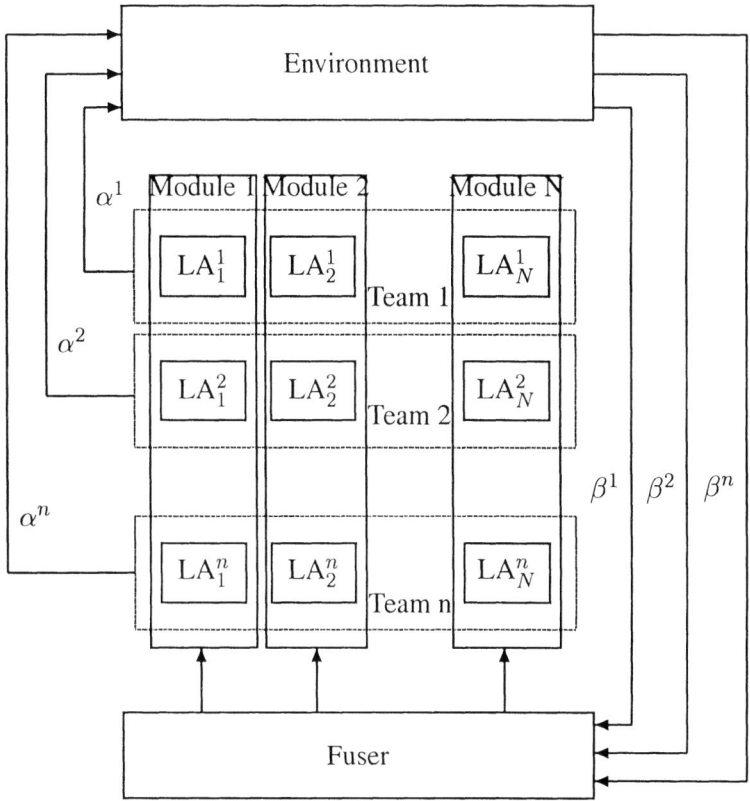

Figure 5.2. A Team of Modules of Learning Automata.

is shown in Fig. 5.2. The environment responds individually to the collection of actions selected in each game. Since each module consists of identical LA, one could regard each play of all the modules as n simultaneous plays of a single game of LA. This is essentially the parallelization achieved here.

We follow the same notation as in Chapter 2. Player i of any game has r_i actions or pure strategies. The action set of ith player module (common to all LA in the module) is

$$A_i = \{\alpha_{i1}, \alpha_{i2}, \cdots, \alpha_{ir_i}\} \quad (i = 1, 2, \cdots, N).$$

The action selected at k by $\tilde{A}_i^\tau$ would be denoted by $\alpha_i^\tau(k)$. The action probability distribution of ith module (common to all LA in the module) is given by

$$\mathbf{p}_i(k) = (p_{i1}(k), p_{i2}(k), \cdots, p_{ir_i}(k))^T \quad (i = 1, 2, \cdots, N)$$

where $p_{ij}(k) \geq 0, \sum_{j=1}^{r_i} p_{ij}(k) = 1, \forall i, j, k$. Here, $p_{ij}(k) = \text{Prob}[\alpha_i^\tau(k) = \alpha_{ij}], \forall \tau$.

The state of the system given by action probability distributions of all the players is given by

$$P(k) = (\mathbf{p}_1(k), \mathbf{p}_2(k), \cdots, \mathbf{p}_N(k)).$$

Each LA in a game receives a payoff for the tuple of actions chosen. The payoff vector

$$\boldsymbol{\beta}^\tau(k) = (\beta_1^\tau(k), \beta_2^\tau(k), \cdots, \beta_N^\tau(k))^T$$

denotes the reinforcements given to the LA in game τ in response to the selection of action tuple

$$\boldsymbol{\alpha}^\tau(k) = (\alpha_1^\tau(k), \alpha_2^\tau(k), \cdots, \alpha_N^\tau(k))^T \quad (\tau = 1, \cdots, n).$$

It is assumed that the payoff to each LA, although random, belongs to the interval $[0, 1]$. Let $\boldsymbol{\beta}(k)$ and $\boldsymbol{\alpha}(k)$ denote the composite column vectors of all the payoffs and actions, respectively. Let

$$q_{ij}(k) = \sum_{\tau=1}^{n} \beta_i^\tau(k) \, I\{\alpha_i^\tau(k) = \alpha_{ij}\} \tag{5.6.52}$$

$$q_i(k) = \sum_{j=1}^{r_i} q_{ij}(k) = \sum_{\tau=1}^{n} \beta_i^\tau(k) \tag{5.6.53}$$

As seen from the above, $q_{ij}(k)$ is the sum of payoffs to the ith players of all those games where the ith player chose the jth action. $q_i(k)$ is the net payoff to the ith players of all the teams.

Now the action probability vectors of all modules are updated by the learning algorithm given below.

$$p_{ij}(k+1) = p_{ij}(k) + \tilde{\lambda}(q_{ij}(k) - p_{ij}(k)q_i(k)) \quad j = 1, 2, \cdots, r_i; \ i = 1, 2, \cdots, N. \tag{5.6.54}$$

Comparing (5.6.54) with (5.2.3) it may observed that since the payoffs to each player could be different in a general situation, the term q_i appears in (5.6.54) instead of q in (5.2.3). As earlier, $\tilde{\lambda} = \frac{\lambda}{n} \in (0, 1/n)$ is the learning parameter.

5.6.1 Analysis

For deriving the approximating ODE of the algorithm, (5.6.54) can be written as

$$P(k+1) = P(k) + \tilde{\lambda} \sum_{\tau=1}^{n} G(P(k), \boldsymbol{\alpha}^\tau(k), \boldsymbol{\beta}^\tau(k)) \tag{5.6.55}$$

where the (i,j)th component of $G(P(k), \boldsymbol{\alpha}^\tau(k), \boldsymbol{\beta}^\tau(k))$ is given by,

$$G_{ij}(P(k), \boldsymbol{\alpha}^\tau(k), \boldsymbol{\beta}^\tau(k)) = \beta_i^\tau(k)[I\{\alpha_i^\tau(k) = \alpha_{ij}\} - p_{ij}(k)]. \quad (5.6.56)$$

Define

$$\mathcal{G}(P) = E[G(P(k), \boldsymbol{\alpha}^\tau(k), \boldsymbol{\beta}^\tau(k)) \mid P(k) = P] \quad (5.6.57)$$

for any $\tau \in \{1, 2, \cdots, n\}$.

Note that $\mathcal{G}(P)$ for given P is not dependent on any particular game τ as the selection of actions made by each game is independent of the others. Thus $\mathcal{G}(P)$ is common to all the n games. Now let us look at the nature of $\mathcal{G}(P)$.

Denote the components of $\mathcal{G}(P)$ by $\mathcal{G}_{ij}(P)$ where $1 \leq j \leq r_i$ and $1 \leq i \leq N$. Then, from (5.6.54), as in (2.3.21)

$$\mathcal{G}_{ij}(P) = p_{ij} \sum_{s=1}^{r_i} p_{is}(h_{ij}(P) - h_{is}(P)) \quad (5.6.58)$$

where

$$h_{ij}(P) = E[\beta_i^\tau(k) | \alpha_i^\tau(k) = \alpha_{ij}, P(k) = P] \quad (5.6.59)$$

for any τ. Also,

$$h_{ij}(P) = \sum_{s_1, s_2, \cdots, s_{i-1}, s_{i+1}, \cdots, s_N} d^i(s_1, s_2, \cdots, s_{i-1}, j, s_{i+1}, \cdots, s_N) \prod_{l \neq i} p_{ls_l} \quad (5.6.60)$$

where

$$d^i(s_1, s_2, \cdots, s_N) = E[\beta_i^\tau(k) | \alpha_l^\tau(k) = \alpha_{ls_l}, 1 \leq l \leq N] \quad \text{for any } \tau. \quad (5.6.61)$$

The following theorem characterizes the behavior of the algorithm.

THEOREM 5.8 *The approximating ODE for the game of module of automata is given by*

$$\frac{d\mathbf{z}}{dt} = n\mathcal{G}(\mathbf{z}); \quad \mathbf{z}(0) = P(0) \quad (5.6.62)$$

Proof: The proof easily follows from Theorem A.1 of Appendix A. The speedup factor n appears in the RHS similar to the case of single module as $\tilde{\lambda} = \frac{\lambda}{n}$ is the learning parameter. ∎

REMARK 5.6.1 *From the above theorem one can conclude results as in Theorem 2.5 about maximal points and equilibrium points. All such results hold with the additional facility of n-fold increase in speed.* ∎

5.6.2 Common Payoff Game

As seen in Chapter 2, the common payoff game is characterized by identical reinforcement given to all the players and we refer to the set of players as a team.

In the game of N modules of LA, we have n teams and N players, each team playing a common payoff game. One could also regard a play of N modules as n simultaneous plays of the same common payoff game.

Since the payoff is identical for each player, algorithm (5.6.54) takes the form

$$p_{ij}(k+1) = p_{ij}(k) + \tilde{\lambda}(q_{ij}(k) - p_{ij}(k)q(k)). \tag{5.6.63}$$

It may be noted that $q(k)$ is the total payoff received by any player (module) at k and hence no subscript is used unlike in (5.6.54).

The approximating ODE for the algorithm (5.6.63) is again of the form

$$\frac{d\mathbf{z}}{dt} = n\mathcal{G}(\mathbf{z}); \quad \mathbf{z}(0) = P(0) \tag{5.6.64}$$

where,

$$\mathcal{G}_{ij}(P) = p_{ij} \sum_{s=1}^{r_i} p_{is}(h_{ij}(P) - h_{is}(P)) \tag{5.6.65}$$

as in (5.6.58) for the general game. The expressions get slightly simplified because $\beta_i(k)$ being the common payoff is the same for all i.

REMARK 5.6.2 *The ODE (5.6.64) is a particular case of the ODE (5.6.62), and hence the remarks of the speed of convergence made earlier (see Remark 5.6.1) hold good here also. There is a n-fold increase in speed in comparison with the game of single LA.* ∎

REMARK 5.6.3 *Since the ODE (5.6.64) is the same as that for a common payoff game of single LA players, except for the speed factor n, the convergence properties shown in Chapter 2 carry over to the present situation. In particular, the following results are evident.*

1 *Unimodal case: Suppose there is a unique optimal pure strategy combination and $\mathbf{z}_0$ is the corresponding action probability unit vector. Then algorithm (5.6.63) is ϵ-optimal. That is, given $P(0)$ with all nonzero entries, for any $\delta \in (0,1), \exists \tilde{\lambda}_0 \in (0,1)$ such that $\forall \tilde{\lambda} \in (0, \tilde{\lambda}_0)$*

$$Pr\{\lim_{k \to \infty} P(k) = \mathbf{z}_0\} > 1 - \delta. \tag{5.6.66}$$

2 *Multimodal case: As shown in Theorems 2.1 and 2.2, the algorithm converges to one of the maximal points. Since we are considering games of*

n	λ	$k_{\text{avg}}(0.95)$
1	0.00625	3995
2	0.0125	2402
4	0.025	1001
8	0.05	520
16	0.1	263
32	0.2	132
64	0.4	65
128	0.8	33

Table 5.6.3. Results of 2-player common payoff game with modules

FALA, these would also be pure maximal points or modal points. Furthermore, they are also Nash equilibria in pure strategies. The convergence is in the same sense as in (5.6.66). ∎

Simulation Results

Simulation results are presented for a two-player, two-action multimodal common payoff game. The game matrix is given by

$$\begin{bmatrix} 0.9 & 0.6 \\ 0.7 & 0.8 \end{bmatrix}.$$

Entries 0.9 and 0.8 are the two modes of the matrix and convergence to either mode is observed to occur. Convergence is assumed when the action probabilities corresponding to either of the modes exceed 0.95. Algorithm (5.6.63) is used for updating the action probabilities of the two modules. For each n, 50 runs of simulation were conducted and $\tilde{\lambda}$ was chosen such that no wrong convergences resulted in any of these runs. $\tilde{\lambda}$ was set to 0.00625 and $\lambda = \tilde{\lambda} n$ in each case. All action probabilities started at 0.5. The results are displayed in Table 5.6.3 which shows speedups of the order of the module size n.

5.7 Parallel Operation of Networks of FALA

The feedforward networks of FALA considered in Chapter 3 could also be operated in parallel to improve their speed. In fact, parallel operation is very much relevant here as we are using a number of LA in tandem and the speed of operation drops with increasing numbers. In this section, we would be using the same terminology as in Chapter 3.

The simple idea is to replace a single LA in the feedforward network by a module of n LA. If there were originally N learning automata in the single network, now there will be N modules each containing n LA. These LA now form n feedforward networks operating in parallel. The ith LA in each such network together form the ith module and share a common action probability vector $\mathbf{p}_i$. The state of the entire network is $\mathbf{P}(k)$ which contains all the individual action probability vectors and is given by

$$\mathbf{P}(k) = (\mathbf{p}_1^T(k), \mathbf{p}_2^T(k), \cdots, \mathbf{p}_N^T(k))^T.$$

Consider a network of modules of FALA. Let us assume for simplicity that each FALA has r actions. Then,

$$\mathbf{p}_i(k) = (p_{i1}(k), p_{i2}(k), \cdots, p_{ir}(k))^T$$

where $p_{ij}(k)$ is the probability of ith LA in each network selecting action j at instant k. More generally, the number of actions r could vary with i.

Each of the n parallel networks interacts with the environment based on the actions chosen by its LA and obtains an SRS common to all its LA. Let $\beta^\tau(k)$ be the SRS obtained by the τth network at k. Then the objective of the parallel network is to

$$\text{maximize } f(\mathbf{P}) = \sum_{\tau=1}^{n} E[\beta^\tau | \mathbf{P}] \qquad (5.7.67)$$

subject to $\sum_{j=1}^{r} p_{ij} = 1, \forall\, i \in \{1, \cdots, N\}$.

Since the n networks share the same action probability vectors $\mathbf{p}_i$ ($i = 1, \cdots, N$),

$$E[\beta^1 | \mathbf{P}] = E[\beta^2 | \mathbf{P}] = \cdots = E[\beta^n | \mathbf{P}] \qquad (5.7.68)$$

and hence

$$f(\mathbf{P}) = nE[\beta^\tau | \mathbf{P}] \quad \text{for any } \tau \in \{1, 2, \cdots, n\} \qquad (5.7.69)$$

even though $\beta^\tau(k)$ can vary with τ. Thus (5.7.67) is equivalent to,

$$\text{maximize } f(\mathbf{P}) = E[\beta^\tau | \mathbf{P}] \quad \text{for any } \tau \in \{1, 2, \cdots, n\}. \qquad (5.7.70)$$

At the instant k, let

- $\alpha_i^\tau(k)$ = Action chosen by i^{th} LA of the τth network
- $\beta^\tau(k)$ = SRS received by the τth network
- $q_{ij}(k) = \sum_{\tau=1}^{n} \beta^\tau(k) I\{\alpha_i^\tau(k) = \alpha_{ij}\}$ = Total reinforcement for the jth action of ith module.
- $q(k) = \sum_{\tau=1}^{n} \beta^\tau(k)$ = total reinforcement to each module.

Let each module operate according to the algorithm given by (5.6.63). That is, for the ith LA in each network,

$$p_{ij}(k+1) = p_{ij}(k) + \tilde{\lambda}(q_{ij}(k) - p_{ij}(k)q(k)) \quad j = 1, 2, \cdots, r;\; i = 1, 2, \cdots, N. \qquad (5.7.71)$$

Parallel Operation of Networks of FALA 199

where $\tilde{\lambda} = \frac{\lambda}{n}$ is the normalized learning parameter as before.

Using the definitions of $q_{ij}(k)$ and $q(k)$, algorithm (5.7.71) can be rewritten as,

$$p_{ij}(k+1) = p_{ij}(k) + \tilde{\lambda} \sum_{\tau=1}^{n} \beta^\tau(k) \left(I\{\alpha_i^\tau(k) = \alpha_{ij}\} - p_{ij}(k) \right). \quad (5.7.72)$$

This alternate form is useful for the analysis.

5.7.1 Analysis

It follows from the analysis in Chapter 3 and in Section 5.2.1 that

$$\begin{aligned} \Delta p_{ij} &= \lambda p_{ij} \sum_s p_{is} \left(\frac{\partial f}{\partial p_{ij}} - \frac{\partial f}{\partial p_{is}} \right) \\ &= n\tilde{\lambda} p_{ij} \sum_s p_{is} \left(\frac{\partial f}{\partial p_{ij}} - \frac{\partial f}{\partial p_{is}} \right) \end{aligned}$$

and hence,

$$\Delta p_{ij} = n\tilde{\lambda} s_{ij}(\mathbf{P}) \quad (5.7.73)$$

where $s_{ij}(\mathbf{P})$ is defined as in (3.5.20). Thus the ODE approximation turns out to be

$$\frac{dz_{ij}}{dt} = ns_{ij}(\mathbf{z}); \quad \mathbf{z}(0) = \mathbf{P}(0) \quad (5.7.74)$$

which has the n-factor in comparison with (3.5.21). Thus all the remarks made in connection with common payoff games carry over to the network situation also.

5.7.2 Modules of Parameterized Learning Automata (PLA)

The FALA in the network could be replaced by Parameterized Learning Automata (PLA) to form a network of modules of PLA. As seen earlier in Section 3.7, PLA are primarily used for attaining global convergence and now we wish to improve the speed as well by using modules of PLA.

For brevity, we will give only the main points of departure from the delineation in Section 3.7. We consider a network of N modules of PLA, each module containing n PLA. These form n parallel networks of PLA. The action probability generation takes place as in (3.7.40) and the maximization problem is now (following (3.7.42))

$$\text{maximize } f(\mathbf{U}) = E[\beta^\tau | \mathbf{U}] \quad (5.7.75)$$

for any $\tau \in \{1, 2, \cdots, n\}$, subject to $|u_{ij}| \leq L$ for $i \in \{1, \cdots, N\}$, $j \in \{1, 2, \cdots, r_i\}$.

The learning algorithm for the ith activated PLA is

$$u_{ij}(k+1) = u_{ij}(k) + \tilde{\lambda}(q_{ij}(k) - p_{ij}(k)q(k)) + \tilde{\lambda}\frac{dh(u_{ij}(k))}{du_{ij}(k)} + \sqrt{\tilde{\lambda}}\zeta_{ij}(k) \tag{5.7.76}$$

where the notation is the same as in Section 3.7.2 and $q_{ij}(k), q(k)$ are as defined earlier in Section 5.7.

The approximating stochastic differential equation has the form,

$$d\mathbf{z} = n\nabla V(\mathbf{z})dt + \sqrt{n}\,\sigma d\mathcal{W}; \quad \mathbf{z}(0) = \mathbf{U}(0) \tag{5.7.77}$$

where,

$$V(\mathbf{z}) = E[\beta^\tau|\mathbf{z}] + \sum_{(i,j)} h(u_{ij}) \tag{5.7.78}$$

and $\mathcal{W}$ is the standard Brownian motion of the appropriate dimension.

It may be observed from (5.7.77) that the first term on the RHS gives the familiar n-fold speedup, but the second term has $\sqrt{n}$ as the speedup factor. Hence even in the case of no overheads, one cannot expect a linear speedup (with module size) in PLA networks.

To see the effect of increasing n, a simulation of a common payoff game is given as an example. Simulation results pertain to algorithm (5.7.76) on a bimodal two player, two actions per player game with the following payoff matrix

$$\begin{bmatrix} 0.9 & 0.7 \\ 0.4 & 0.8 \end{bmatrix}.$$

It is easy to see that options corresponding to 0.9 and 0.8 entries of the matrix are locally optimal; options corresponding to the former being globally optimal. Each player has a scalar internal state, since each of them has only one independent action probability. Designating the internal states as u and v, and i^{th} player's j^{th} action as α_{ij}, the probability generating functions are chosen as

$$\text{Prob}\{\alpha_1^\tau(k) = \alpha_{11}\} = 1 - \text{Prob}\{\alpha_1^\tau(k) = \alpha_{12}\} = \frac{1}{1+e^{-u(k)}}$$

$$\text{Prob}\{\alpha_2^\tau(k) = \alpha_{21}\} = 1 - \text{Prob}\{\alpha_2^\tau(k) = \alpha_{22}\} = \frac{1}{1+e^{-v(k)}}$$

where α_i^τ denotes the action selected by the τth member of player (module) i.

To demonstrate the efficacy of PLA, starting points biased towards convergence to the local but not global optimum are considered. As in the case of the example discussed in Section 4.4.2, the algorithm makes use of a large value of standard deviation, σ, in the beginning, progressively reduces it to a small nonzero value, and maintains this value constant later on. It may be noted that

Parallel Operation of Networks of FALA 201

n	λ	a	$k_{\text{avg}}(0.95)$
4	0.08	0.99	1001
8	0.16	0.95	245
16	0.32	0.8	190
32	0.64	0.5	55

Table 5.7.4. Results with network of modules of PLA on a multimodal game

since the standard deviation of the random walk term remains constant after some finite number of iterations, all the analytical results continue to hold in this case. For each value of n, 20 simulation runs were conducted and $\tilde{\lambda}$ was chosen such that no wrong convergence resulted in any of these runs. The value of σ was initially set to 10 and reduced as

$$\sigma(k+1) = a\sigma(k)$$

where $a \in (0,1)$. σ is maintained constant at 0.01 once $\sigma(k)$ falls below 0.01. The results are given in Table 5.7.4 for the minimum value of a for which correct convergence resulted in every run. The other parameters of the algorithm are $K = 1, L = 3, J = 2$. From the table it can be seen that the improvement in speed is not linear in n. This could be attributed to the effect of the random term which has an effective standard deviation of $\sqrt{n}\sigma$ for a given n, and the different reduction parameters a employed in each case.

5.7.3 Modules of Generalized Learning Automata (GLA)

In Chapter 3 we have considered networks of GLA. In that model also we can employ parallelization. We can consider a network of N modules of Generalized Learning Automata (GLA), each module containing n GLA. The procedure would be similar to that in Section 5.7.2 where each PLA is replaced by a GLA. Much of the description in Section 3.8 applies, but for the essential changes given below.

The global algorithm for the ith GLA is of the form

$$u_{ij}(k+1) = u_{ij}(k) + \tilde{\lambda}\sum_{\tau=1}^{n}\beta^{\tau}(k)\frac{\partial \ln g_i(\mathbf{X}_i^{\tau}(k), y_i^{\tau}(k), \mathbf{u}_i(k))}{\partial u_{ij}}$$
$$+ \tilde{\lambda}h'(u_{ij}(k)) + \sqrt{\tilde{\lambda}}\zeta_{ij}(k)$$

The notation follows that in (3.8.75). The approximating SDE is of the same form as the SDE for PLA, (5.7.77) and hence the same comments on the improvement of speed of convergence are valid here also.

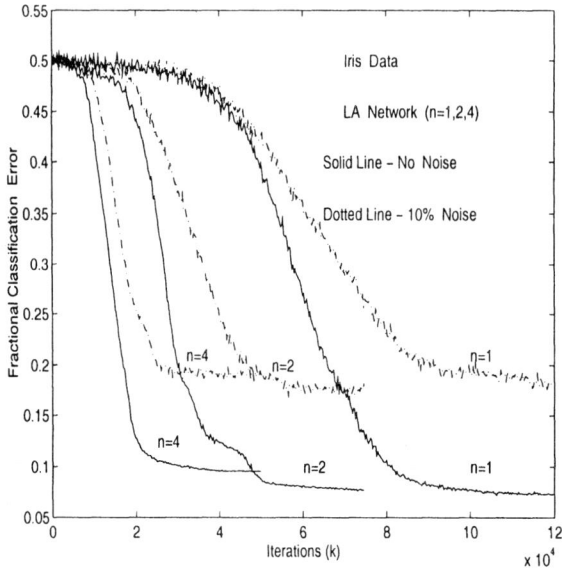

Figure 5.3. Learning curves for classification of Iris data using modules of LA in a 3-layer network.

5.7.4 Pattern Classification Example

As an illustration of the improvement in speed of convergence brought about by parallelization, we return to the iris data problem in pattern classification considered in Chapter 4 (see Section 4.4.1). The same three-layer network is employed for pattern classification, but each LA is replaced by a module. The results obtained in simulation studies are shown in Fig. 5.3 for module sizes of 1, 2, and 4.

The figure shows the fractional error in classification averaged over ten runs as a function of the number of iterations. The algorithm was run in two phases; learning phase and the error computation phase. During the learning phase no error computation is performed. A sample pattern is chosen at random, and its classification is altered with probability 0.1 in the noisy case. Otherwise the given classification itself is maintained. The network of modules classifies the pattern using its internal action probability vectors and based on the payoffs obtained by each module, the probability vectors are updated using (5.7.71). The error computation phase was performed once every 250 iterations. The probability vectors were not updated during this phase. A set of 100 test patterns was presented sequentially to the team of modules. This was repeated over ten cycles and the fraction of patterns classified wrongly was computed. Repeated presentations are meant to average out the stochastic classification effects of the network to yield realistic error estimates. As in the learning phase, classification of the input pattern is altered with probability 0.1 in the noisy case.

The figures indicate faster speed of convergence for larger module sizes, with learning parameters chosen such that the limiting values of the error are approximately same for all the module sizes in the noise–free and noisy cases respectively. The chosen values for the limiting errors in the noise free and noisy cases were 0.1 and 0.2, respectively. The values of $(\tilde{\lambda}_1, \tilde{\lambda}_2)$, the learning parameters used in the first and second layer units are respectively, (0.005, 0.002).

5.8 Discussion

In Chapters 2 and 3 we saw how a number of LA can be configured to form structures such as teams and networks for solving complex learning problems. Chapter 4 illustrated how one can make use of such structures of LA in an application. In general, use of a large number of LA in a system slows down the learning process. One method of improving rate of learning is explained in the present chapter. Interestingly, this method advocates the use of even larger number of LA, though operating in parallel now. Such modules of LA are seen to help improve speed of learning in terms of reduced number of probability updatings.

The parallel operation involves replacing a single LA by a module of LA. The action probability distribution is common to all LA in the module and is updated once per iteration though each LA in the module independently selects an action and elicits reinforcement from the environment, and this constitutes the parallelism in the operation. The learning algorithm for the module is so designed that the speed is improved while the required accuracy is maintained. The technique is feasible only in those situations where several actions can be simultaneously tried and reinforcements elicited. Early applications of LA were in areas such as, e.g., routing in telephone networks where only one action can be tried at a time. In such applications parallel operation described in this chapter is not relevant. However, the modules of LA described here are useful in many other applications. For example, it is well suited in pattern recognition where actions are values of parameters of the classifier and several parameter values can be tested using the next available training sample.

In situations where it is feasible, the technique is applicable to all the learning algorithms and configurations of LA considered earlier in the book. For algorithms such as L_{R-I}, the improvement in speed is nearly linear with the number of LA in the module. The limitation is that the learning parameter λ cannot exceed 1 and the accuracy parameter, $\tilde{\lambda} = \frac{\lambda}{n}$ should not exceed the value needed to maintain the desired accuracy.

The parallelization paradigm extends effortlessly to different LA such as CALA, PLA and GLA. It also extends to games and networks of LA. In all cases it results in improvement in speed while maintaining the desired accuracy. In the case of global algorithms the improvement in speed is not linear in terms of the number of LA in the module because of the presence of the $\sqrt{\lambda}$ in the perturbation term.

In the parallelization schemes considered so far, use of modules containing n learning automata has resulted in approximately n-fold increase in speed of convergence. However, a further increase in speed is possible at the expense of more computation, as explained below.

In the game of N modules, n teams were formed where the τth team consisted of the τth automaton in each module. However, in general, the τth LA in one module could be grouped with any other LA in other modules to form a team. If we permit formation of teams with any one LA from each module without any restriction, we could form a total of n^N teams resulting in an increase in speed by a factor of nearly n^N. This would, however, be accompanied by a corresponding increase in computation to keep track of the performance of all these teams. Having n^N teams is a limiting possibility. One could form lesser number of teams consistent with computational limitations, but obtain increase in speeds with factors higher than n. Such strategies can be employed with networks of LA also.

There are some implementation issues connected with parallelization which have not been discussed in the chapter. There are inevitable computational overheads connected with implementation of any parallel algorithm. The claims of improved speed of convergence made earlier have to be tempered slightly to take into account such factors.

5.9 Supplementary Remarks

The parallelization approach described here is summarized in [TA97] and the material in this chapter closely follows the thesis [Arv96].

This approach is of much wider applicability and can be used with delayed associative reinforcement learning also. For instance, the Temporal Difference learning algorithm (popularly known as $TD(\lambda)$ algorithm) [Sut88] can be parallelized using this approach [Arv96]. Stochastic approximation algorithms also fall within the ambit of this approach. The parallelization technique can be studied in the light of multi-teacher environments [AP99] also. However, it corresponds to the case of teachers with identical characteristics. LA networks for learning probability distribution of output of a system conditioned on the input, using input–output samples only, have been analyzed in [TA00, Arv96]. Such networks form alternatives to the classical Boltzmann machines [AHS85].

Chapter 6

SOME RECENT APPLICATIONS

6.1 Introduction

In this book, we have discussed learning algorithms for different types of multiautomata systems. As we have mentioned at many places, learning automata would be useful in many applications involving decision making under uncertainty or learning parameter values based only on noisy measurements of objective function values etc. In Chapter 4 we have discussed at length one generic application area, namely, pattern classification. We have illustrated in that chapter, how different types of multiautomata methods can be employed for learning optimal pattern classifiers to suit different problem situations. In this chapter, we briefly indicate some other applications where learning automata are successfully employed.

Some of the early applications of learning automata are well documented in [NT89]. Hence, we concentrate here on some of the more recent applications. We first discuss a couple of applications in some detail. Following this we indicate many other areas in which automata are found useful and conclude the chapter.

The first application we discuss in some detail is for parameter optimization in computer vision systems. In a large computer vision system, there would be many modules or algorithms and in each of them, some parameter values are to be fixed. The overall performance of the system depends on these values and often the parameters of different modules interact in different ways. Hence, a systematic procedure for finding optimal parameters in computer vision applications would be useful [SC00]. At the beginning of Chapter 2, we have motivated the game of automata model through a parameter optimization problem. In the next section we discuss an example where a team of FALA is used to find good parameter values in a computer vision application. The second application that we discuss is in the area of communication networks. This illustrates another facet of learning automata, namely, their ability to act as adaptive decision makers that learn on the basis of a stochastic feedback conveying the goodness or otherwise of the decisions made. After discussing these two applications, we briefly point out other generic application areas such as learning controllers, stochastic techniques for constraint satisfaction or combinatorial optimization etc.

6.2 Supervised Learning of Perceptual Organization in Computer Vision

The first application that we discuss is for a computer vision problem. Here learning automata are used for parameter optimization. It is a problem in which we need to fix values of many parameters of a computer vision algorithm. The parameters are not independent of each other. The goodness or otherwise of any specific set of values for the parameters can be roughly judged only through simulating the algorithm on some images. This example illustrates the kind of complicated parameter optimization problems in which learning automata algorithms prove useful.

In a typical computer vision application one is interested in analyzing image(s) to recognize the objects in the scene. The first step is to extract some simple features such as edges, corners etc. The process of such feature extraction comes under the so called *low-level vision*. Given an appropriate feature based representation of the image, recognition of objects through some process of matching features in the image to those of objects in the database, comes under *high-level vision*. An important problem at an intermediate level is that of selecting subsets of features so that each subset belongs to an object. This is referred to in the computer vision community as perceptual organization, feature grouping, object hypothesis generation or segmentation. Such perceptual organization or feature grouping leads to separating objects from background (in the image) and hence is very useful in the subsequent object recognition. As pointed out in [Gri91], the computational complexity of object recognition is considerably reduced by such feature grouping, even if the grouping process makes a few errors.

In this section, we describe a method [SS00] for such perceptual organization which employs a common payoff game of FALA using pursuit algorithm. Since our interest is primarily in illustrating some applications of learning automata, we would not be discussing the image processing issues in any great detail. All material in this section is based on [SS00].

The problem of feature grouping is that of deciding whether or not some subset of low-level feature primitives (such as short edge segments) which are in some geometric relation to each other, can be considered to be part of the same object. One can use object models to decide on the rules needed to effect such grouping. Since the grouping algorithm has to separate the object of interest from background as well as other objects in the scene in a variety of contexts, it is, in general, difficult to design the grouping algorithm based on generic models of isolated objects. An alternate approach (which is what is discussed here) is to use a training set of images of objects in context, but with the object of interest manually outlined. Using such a training set, the importance of various relationships among the features as relevant to the grouping process can be learnt. Such an approach does not require explicit and detailed object models. In this framework, the influence of object models on the final grouping process is implicit through the training set and the influence is mainly statistical in nature.

The overall method is as follows. The input to the grouping algorithm consists of low-level image features such as constant curvature edge segments (arcs and straight lines). The output of the algorithm is salient groups of features. The algorithm consists of two main parts: scene structure graph construction and graph partitioning. First, for every pair of low-level features detected in the image, some measurements such as distances and angles between lines are obtained. Using these, through a probabilistic Bayesian network, probabilities of various relationships such as 'perpendicular', 'parallel' etc. between every pair of features are computed. Then a graph is constructed to represent all the feature tokens in the image. The nodes of the graph are the features and the links between nodes represent relationships. (To avoid confusion with the edges in the image which are the low-level features on which feature grouping algorithm operates, we call the edges in this graph as links). All links (i.e., edges in the graph) have weights attached to them. The weight of a link in the graph between any pair of nodes depends on the probabilities of various relationships between the features corresponding to the two nodes. This graph is referred to as the scene structure graph. In the next step, this graph is partitioned so that nodes connected by links with high weight are in the same partition. The individual partitions obtained are the feature groups output by the algorithm. Both the graph construction and graph partitioning involves choosing values for some parameters. The different parameters may not be independent of each other and we need to choose a good set of parameter values for the grouping algorithm to perform well. The needed parameter values are learnt through a common payoff game of FALA by making use of the training set of images. Below, we first briefly explain the main steps of the algorithm and then discuss how the automata game is used in this application.

The first step in the construction of the scene structure graph is to compute a number of distances between every pair of constant curvature edge segments (straight lines and arcs) which are the low-level features input to the algorithm. These distances include maximum and minimum distances from end points of the smaller line to the larger line for every pair of straight lines, distances between centers of a pair of arcs, the extent of overlap between pairs of lines or arcs etc. To make such distances scale independent, they are expressed as the ratio of the distance to the length of the larger segment in the pair. The angles between every pair of straight lines are also computed. Using all such distances, each pair of edge segments detected in the image is characterized by six geometric attributes. These include, maximum and minimum distances, extent of overlap, angles, distance between center points etc. In addition, for every pair of these low-level features detected in the image, two photometric attributes are also calculated based on the image intensity function around these edge segments. Let us call these attributes, $Z_1, \ldots, Z_8$. (As mentioned earlier, we skip the details of how these are calculated because it is not relevant to our purposes here). Using these attributes, for every pair of edge segments, the probabilities of various relationships between them are calculated. The relations are as follows. Every pair of straight lines can be classified as parallel (*Parl*), T-junction (*Tjun*) or L-junction (*Ljun*). A pair of straight lines or arcs can be classified as continuous (*Con*) which means that the straight lines are continuations of each other or that the arcs are co-circular. A pair of arcs can also be parallel and such a relation is called Ribbon (*Rib*). Two other relations between every pair of edge segments characterize the similarity of image region around them (*Reg*) and how proximal are the two edge segments (*Prx*). The final set of relations thus is: *Parl, Tjun, Ljun, Con, Rib, Reg, Prx* and *Non*. The last one, *Non*, denotes that the pair have no relationship. All these relations are considered as binary random variables. The probabilities of these relations, for every pair of edge segments, are computed in a Bayesian framework using the set of attributes for the pair. For this, we need to model the class conditional densities (that is, probability distributions of $Z_1, Z_2, \ldots$, conditioned on the relations) and the prior probabilities of these relations. Since there are seven relations (each being a binary random variable), we need seven numbers to specify all priors. Since *Parl* and *Rib* are both about parallelism, the same prior is used for them. Thus, there are six parameters to specify the priors. For the class conditional densities, it is to be noted that not all attributes are relevant for all relations. For example, the maximum distance between a pair of lines affects only *Parl, Con* and *Non*. Also, the relations are not all independent. For example, if *Parl*=1 then we must have *Non*=0 and so on. The conditional distributions that are relevant are specified through a Bayesian network. Then the conditional distributions are modeled. For example, if two lines should be considered related by Parallelism, the distance between them should be neither very small nor very large. Thus, the distribution of the attribute of maximum distance between a pair of lines, conditioned on the pair being parallel (that is conditioned on *Parl*=1, *Con*=0, and *Non*=0) can be modeled by a symmetric triangular distribution over $[0, b]$, where b is a parameter. Using such arguments, all conditional distributions are modeled and these have a total of seven parameters. Once the conditional distributions and priors are available, we can calculate the probabilities of various relations between every pair of edge segments. Let Prob($Parl_{ij}$) denote the probability of the relation *Parl* between a pair of edge segments i and j and similarly for other relations. These probabilities are now used to calculate the weight of the link between every pair of nodes in the scene structure graph. Let w_{ij} denote the weight of the link between nodes i and j in this graph. (Recall that each node corresponds to an edge segment). The weight w_{ij} is set as zero if Prob(Non_{ij}) is greater than the probabilities Prob($Parl_{ij}$), Prob(Rib_{ij}), Prob($Ljun_{ij}$), Prob($Tjun_{ij}$) and Prob(Con_{ij}). Otherwise, it is set equal to Prob(Prx_{ij}) + Prob(Reg_{ij}) + Prest where Prest = max(Prob($Parl_{ij}$), Prob(Rib_{ij}), Prob($Ljun_{ij}$), Prob($Tjun_{ij}$), Prob(Con_{ij}))

To sum up, the scene structure graph is constructed as follows. First, each pair of edge segments is characterized by computing some eight attributes. Using these values, the probabilities of seven different relationships between every pair of edge segments is calculated. The calculation of these probabilities is done in a Bayesian framework by modeling the conditional distributions of the attributes given the relations and the prior probabilities of these relations. Specification of these probabilities involves 13 parameters: six for the priors and seven for the conditional distributions. Using the probabilities of different relations computed for a pair of edge segments, the weight of the link joining the two corresponding nodes in the scene structure graph is determined.

The next step is to partition this graph. This is done by recursively bisecting the graph. That is, we first split the graph into two parts then each of them is split into two parts and so on till all parts obtained satisfy some termination criterion.

The bisection of a graph can be represented by a vector **v** where the sign of its ith component, v_i, denotes which of the two parts the node i belongs to. (Note that the dimension of **v** is equal to the number of nodes in the graph). Now the graph bisection can be posed as an optimization problem

$$\min \sum_{i,j} w_{ij}(v_i - v_j)^2 \qquad (6.2.1)$$

subject to the constraints

$$\sum_i v_i = 0, \quad \text{and} \quad \sum_i v_i^2 = 1.$$

The above minimization will tend to assign similar values, v_i and v_j, to nodes i and j between which there is a large link weight, w_{ij}. The two constraints prevent the trivial solution where all components of **v** are assigned the same value. We can merge the second constraint with the objective function by recasting the optimization problem as

$$\min \frac{\sum_{i,j} w_{ij}(v_i - v_j)^2}{\sum_i v_i^2} \quad \text{subject to} \quad \sum_i v_i = 0. \quad (6.2.2)$$

With a little algebra, the numerator in the above objective function can be rewritten as

$$\sum_{i,j} w_{ij}(v_i - v_j)^2 = 2\mathbf{v}^T \mathbf{L}_G \mathbf{v}$$

where $\mathbf{L}_G$, known as the Laplacian matrix of the graph, is a matrix of dimension $M \times M$ (where M is the number of nodes in the graph) given by

$$\mathbf{L}_G = \sum_{j, j \neq i} w_{ij} \quad \text{if} \quad i = j$$
$$= -w_{ij} \quad \text{otherwise.}$$

Now our optimization problem becomes

$$\min 2 \frac{\mathbf{v}^T \mathbf{L}_G \mathbf{v}}{\mathbf{v}^T \mathbf{v}} \quad \text{subject to} \quad \mathbf{v}^T \mathbf{1} = 0$$

where $\mathbf{1}$ is a vector all of whose components are unity. Using a corollary to the Courant Fischer Minimax Theorem [SS00], it can be shown that the solution to the above optimization problem is the eigenvector corresponding to the second largest eigenvalue of the Laplacian matrix, $\mathbf{L}_G$. Thus the graph bisection problem is solved by obtaining the second largest eigenvalue and the corresponding eigenvector of the Laplacian matrix.

The partitioning of the scene structure graph is done by recursively bisecting each part. The recursion is stopped on two termination conditions. One is based on a threshold on the number of nodes in the part. The other is based on a threshold on the strength of a partition in terms of total link weights.

In the feature grouping algorithm as described above, we need to choose values for a number of parameters. There are 13 parameters in the algorithm for constructing the scene structure graph and two parameters (the thresholds used for termination condition) in the graph partitioning algorithm. The input to the grouping process consists of constant curvature edge segments obtained from the image. The algorithms for edge detection and for isolating constant curvature edge segments have another six parameters. Thus there are 21 parameter values to be chosen. The advantage of having such a large number of parameters is that the grouping algorithm can be sufficiently flexible so that it can be adopted for many different classes of objects. The disadvantage is the difficulty in choosing appropriate values for all the parameters.

In many computer vision algorithms, the parameter values are chosen by trial and error procedure or through some heuristics. However, when there are a large number of parameters which are not independent of each other, such a manual choice becomes difficult. Also, it is often desirable to have some automatic method of parameter choice. This is particularly important in applications that have a large number of vision modules or algorithms.

In the grouping method we are discussing, the parameter values are learned through a common payoff game of FALA. Since there are 21 parameters, we need a team of 21 automata. Since we know the implications of each parameter, it is possible to choose an appropriate range for each parameter and discretize it to a few levels. (For example, all prior probabilities take values between 0 and 1, and we do not need the prior probabilities to an accuracy of more than the first decimal place). In the specific method here, the range of possible values for each parameter is discretized to ten levels. Thus each of the FALA in the team have ten possible actions. The total number of action tuples for the team is 10^{21} which is a very large search space. However, the automata team is quite effective in this problem. The automata team algorithm keeps updating the action probabilities based on the goodness of the action tuples chosen so far. This essentially means that the automata algorithm guides the search towards the regions in the parameter space which contain good sets of parameter values. What this would mean, for example, is that if we run the team algorithm for, say, 50,000 iterations, the team would have chosen many more 'good' tuples of values of parameters than would be the case if we had tested the same number of random values by uniformly sampling from the set of 10^{21} values. Thus even if we do not run the algorithm till convergence, we can come up with good values by looking at the trace of action combinations chosen by the team. This is one of the main reasons why the automata method can be effective in such large parameter optimization problems. In a problem such as the current one, we expect that there would be many sets of parameter values that are good and we do not necessarily need to find the combination that is 'best'. (This is particularly so because our evaluation of a set of values for the parameters is also not perfect).

The specific algorithm used in [SS00] is an estimator algorithm for a common payoff game of FALA as described in [TS87]. This algorithm is very similar to the pursuit algorithm described in Section 2.5.2. As discussed in that section, all estimator algorithms for the common payoff game maintain an estimate of the reward probability matrix. Since the number of possible tuples of actions is 10^{21}, the size of this matrix would be prohibitively large. However, we do not need to maintain all elements of this estimated reward probability matrix explicitly. Recall from Section 2.5.2 that the estimated reward probability for any action tuple, $\hat{d}_{i_1 i_2 \ldots i_N}$, is obtained as a ratio of $Z_{i_1 i_2 \ldots i_N}$ (which gives the total amount of reinforcement obtained with this action tuple so far) and $\eta_{i_1 i_2 \ldots i_N}$ (which gives the total number of times this action tuple is chosen so far). The elements of both the matrices Z and η are zero till the corresponding action tuple is chosen for the first time. We also note that the maximum number of elements of these matrices that can become nonzero is bounded above by the number of iterations of the automata algorithm which would be much less than 10^{21}. Hence, we can efficiently maintain these quantities by making use of sparse matrix techniques. There is another point to be noted here. In the estimator algorithms described in Section 2.5.2, it is said that initially each of the action tuples is chosen a few times to initialize the reward probability estimates. When the total number of possible action tuples is very large as in the present case, such a strategy makes the algorithm infeasible. Hence, what is done here is to run the algorithm with $\lambda = 0$ for a few hundred iterations to allow for some initial estimates to be formed. After this, in the algorithm described in [SS00], the learning rate λ is held at 0.1. We note in passing here that another strategy that would be equally effective is to use the stochastic estimator algorithms briefly explained towards the end of Section 1.5.2.

The FALA team algorithm for learning the parameters operates as follows. At each iteration, the team chooses a tuple of values for the parameters. Then the grouping algorithm with these parameter values is run on the training set of images. Based on the performance of the grouping algorithm on the training set, a reinforcement is generated for the team as explained below. And then the cycle repeats. As mentioned earlier, we are interested in getting some good parameter values and are not interested in reaching *the* optimal values. Hence the algorithm need not be run till all action probability vectors converge. The algorithm is terminated when the team does not find any new tuples of values that are better than the ones already found for a fixed number (typically 200) of consecutive iterations. After terminating the algorithm, a set of 100 best values found by the team are output. The idea is that these are all likely to be good and since this is a small set, other parameter tuning strategies can be tried, if necessary, starting from these.

The reinforcement feedback to the team for a given choice of action tuple is generated as follows. The grouping algorithm is run with the chosen parameter values on the training images. Suppose the algorithm generated n groups, $G_1, \ldots, G_n$ for an image with n' objects, $O_1, \ldots, O_{n'}$. In the training images, the objects of interest are manually outlined. Hence, the edge segments close to this outline constitute the 'correct' grouping corresponding to that object. Let ν_{ij}^{go} denote the number of edge primitives that are common to group G_i identified by the algorithm and object O_j. Let ν_i^g denote the number of edge primitives in the group G_i and let ν_j^o denote the corresponding number for object O_j. Let n_f denote the number of groups output by the algorithm which do not have any edges in common with any of the objects and n_v denote the number of groups generated by the algorithm that overlap with some object.

Now, some quantities characterizing the performance are calculated as follows.

$$\rho_{time}(G_i, O_j) = \frac{\nu_{ij}^{go}}{\nu_i^g}$$

$$\rho_{qual}(G_i, O_j) = \frac{\nu_{ij}^{go}}{\nu_j^o}$$

It can be argued [SS00] that ρ_{time} gives a good indication of the efficiency in object matching resulting from the grouping process and ρ_{qual} measures the quality of object matching based on the grouping process. Both these quantities would be in the interval $[0, 1]$ and higher values imply better performance. The reinforcement to the automata team is calculated as

$$\beta = \sqrt{\left(\frac{\sum_{ij} \rho_{time}(G_i, O_j)}{n_v}\right) \left(\frac{\sum_{ij} \rho_{qual}(G_i, O_j)}{n_v}\right) \left(1 - \frac{n_f}{n}\right)}.$$

The automata team algorithm was found to be quite effective in learning good parameter values for feature grouping [SS00]. Also, the values learnt with one set of training images performed very well on images containing similar objects.

Figure 6.1 shows results from one of the simulation experiments from [SS00]. In this experiment there are 40 images of aircraft. Of these, 20 were used as training set. Different combinations of 20 images are tried as training set. The figure shows the final grouping results obtained with the learnt parameters. In the figure, the first column shows the images and the second column shows the set of low-level edge segments detected on the image. These edge primitives are what are input to the grouping algorithm. Third and fourth columns show the grouping obtained using the learnt parameter values. Third column corresponds to the grouping obtained when the training set included that particular image. Fourth column shows the results when the training set for the parameters learnt does not include that image. From the grouping of edge features obtained, it is clear that the algorithm is able to isolate all edge primitives that belong to the object of interest. From the point of view of the learning algorithm, what is interesting is that the groupings seen in third and fourth column are almost identical. What this means is that, if the system is trained on some images, then the learnt parameter values are good for analyzing other similar images. Thus the learning algorithm is achieving proper generalization.

6.3 Distributed Control of Broadcast Communication Networks

The second example of automata applications that we consider is in the area of communication networks. One of the early applications of learning

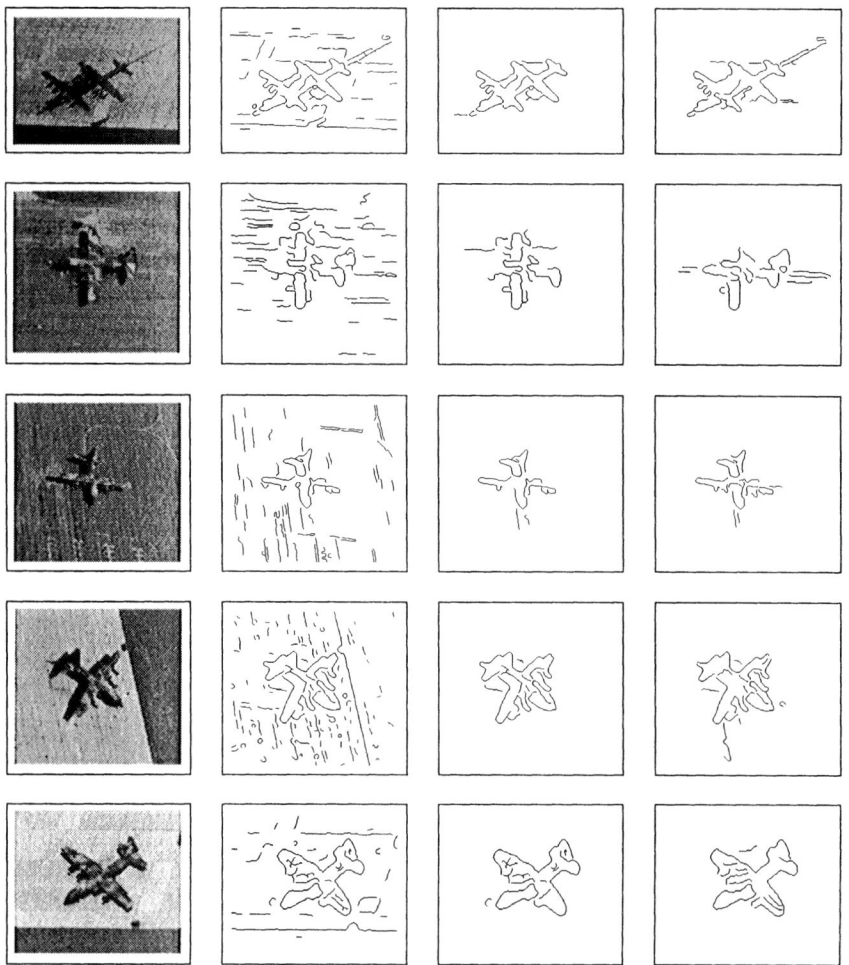

Figure 6.1. An example of results obtained with the learning algorithm. The first column shows some representative images from the set of 40 airplane images. The images in the second column show the edge features that are input to the grouping algorithm. Each image in the third column shows the output group with the best parameter combination obtained by the learning algorithm when the training set *includes* the corresponding image. Each image in the fourth column shows the groups with the best parameter combination learned when the training set *does not include* the corresponding image. (Reproduced with permission from [SS00]).

automata is for routing in circuit switched networks [NT80, NM83]. Here, automata are used for making adaptive decision regarding routing a call. Automata have been used for such adaptive routing decisions in data communication networks also [NN87]. In recent years, automata algorithms have been successfully used for many adaptive decision making problems in wireless and broadcast networks [PM95, PM96, PP00, PP99, OPPL02]. Automata algorithms have also been successfully used in some problems encountered in the design of communication networks [OR02].

In this section we discuss how automata can be used for developing adaptive TDMA (Time Division Multiple Access) protocols. We keep the discussion at a simple qualitative level so that readers unfamiliar with broadcast networks can also appreciate the main ideas. All the material in this section follows [POP02].

Consider a broadcast network, where a number of stations transmit data over a shared medium. Since it is a broadcast channel, all stations can receive what is put on the medium. The main problem is in determining which station gets to access the medium for transmission at any given time. In the well known Time Division Multiple Access (TDMA) protocols, the time axis is divided into a number of slots and, through some mechanism, the stations decide who can access the medium in each time slot. In random access protocols, at the beginning of each time slot, all stations that have data to transmit contend for the access. If more than one station transmits, it results in a collision. The collision can be detected in a distributed manner (since all stations can 'listen' to what is on the medium) and through some arbitration mechanism all this data is scheduled for retransmission. In a fixed assignment protocol, the available bandwidth is shared among the stations by a fixed assignment of only one station at each time slot. For example, each station can be allowed access for one time slot through a round robin scheduling or the assignment can depend on the current demands at various stations as communicated to a controller. In this way, collisions are avoided and hence higher throughput rates can be achieved. However, for achieving good channel utilization through such assignment protocols, the traffic at each station should be stable and known (to the controller). Otherwise, many slots may be wasted by assigning them to stations which do not have anything to transmit at that specific time. This is particularly so if the traffic at the stations is bursty because fixed assignment protocols cannot adapt to sharp changes in the stations' traffic. An alternative would be an adaptive protocol that assigns the channel to only one station at a time; however, this assignment is done in a distributed manner and is done in such a way that the amount of bandwidth assigned to each station is proportional to the expected traffic at the station. Here we describe such an adaptive TDMA protocol using learning automata, which is capable of operating efficiently under bursty traffic conditions.

Let $U = \{u_1, \ldots, u_K\}$ be the set of stations. All stations are connected to a broadcast transmission medium (e.g., a copper wire or an optical coupler). Data are transmitted as packets of equal size. The time axis is slotted with the slot duration being equal to the packet transmission time. Packet transmissions are synchronized with the time slots. Each station has a queue where packets waiting for transmission are stored. When a packet arrives at a station it is placed at the end of the queue. The queue size available is fixed and if the queue is full when a packet arrives, it is discarded. Once a packet is put in the queue, it remains there till it is transmitted. At each slot when the station is granted access to the medium, the station removes the first packet from the queue and transmits it over the medium. If the queue of a station is empty when it is assigned the medium, the station remains idle. Since this is a broadcast network, all stations would know whether or not a packet transmission occurred in any given time slot.

The traffic arriving at each station is assumed to be bursty. That is, the station may have nothing to transmit for some time and then packets arrive in long bursts. Each station may switch between such states randomly and the burst length of packets received would also be random. The algorithm discussed here is particularly effective for such bursty traffic conditions.

The method for deciding on the access to the medium is as follows. We start with a vector $\mathbf{P} = (P(1), \ldots, P(K))$ where $P(i) \in [0, 1], \forall i$. $P(i)$ is viewed as the probability of allowing station u_i to access the medium. We generate a random schedule $(s(1), \ldots, s(K))$. Each $s(i) \in \{0, 1\}$. The schedule is determined by $\text{Prob}[s(i) = 1] = P(i)$. A station u_i is allowed to transmit in this round if we have $s(i) = 1$. That is, after the schedule is generated, each of the stations corresponding to an entry of 1 in the schedule is allowed to access the medium in successive time slots. Every time a station is thus allowed to access the medium, it transmits a packet if it has one in its queue; otherwise it leaves the medium idle during the slot allotted to it. After this, the probabilities $P(i)$ are updated using the algorithm below. Then a new schedule is generated and so on.

The algorithm for updating $\mathbf{P}$ is as follows. Let $u(t)$ denote the station that is granted permission to transmit at time slot t and let $sm(t)$ denote the state of the channel during time slot t. This state can be either 0, meaning that the channel is idle during that slot, or 1 meaning that there was a packet transmission during that slot. At each t, components of $\mathbf{P}$ are updated as follows.

$$P(j) \leftarrow P(j) + \lambda(1 - P(j)) \text{ if } u(t) = u_j \text{ and } sm(t) = 1$$
$$P(j) \leftarrow P(j) - \lambda(P(j) - a) \text{ if } u(t) = u_j \text{ and } sm(t) = 0$$

where $a \in (0, 1)$ is a parameter and λ is the step-size. Note that this algorithm ensures that $P(j)$ is always in the interval $[a, 1]$. For simplicity of notation, we are not explicitly showing any time index in $P(j)$ and hence the above are to be viewed like assignment statements in a programming language. That is, after each t, the value of the appropriate $P(j)$ is replaced by the RHS of the above equation.

Distributed Control of Broadcast Communication Networks 217

We can think of this method as keeping a two action FALA at each station. The actions are 1 and 0. Now, $P(j)$ is the probability for action 1 of the automaton at station u_j and it completely specifies the action probability distribution of this two action automaton. Then, the random schedule generated is like a tuple of action choices by all the automata. The algorithm for updating these action probabilities is such that when the station is granted permission, if it did not have a packet to transmit, then the probability of choosing action 1 for this station is decreased; otherwise it is increased. (We comment on the role of parameter a later on). Thus $sm(t)$ is the reinforcement feedback during the time slot t. When the station is not granted permission, the corresponding $P(j)$ is not updated. This updating algorithm differs from the usual FALA algorithms in this respect.

To implement this method we need a way for all stations to know the schedule so that they know when they can access the channel. We implement this method as follows. We start with $P(i) = 0.5$, $\forall i$. The vector $\mathbf{P}$ is kept at each of the stations. All stations would use the same random number generation algorithm and all stations start with the same seed for the random number generator. All stations generate the random schedule using the probabilities $P(j)$. Since they start with the same seed for random number generation, all of them would generate the same random schedule. Thus all stations know which are the stations that are allowed to transmit in this schedule. Through a fixed ordering of the stations, each station allowed to transmit in this schedule would know when to transmit. For example, suppose in the current schedule only stations u_1, u_3 and u_6 have a 1 and all others have 0. Then, at the next slot u_1 would transmit, at the slot following that u_3 would transmit and in the slot following that u_6 would transmit. Then it is time to generate a new schedule again. After each time slot, all stations know whether or not the channel was idle and hence all of them can update the corresponding $P(j)$. Thus all of them continue to have identical copies of $\mathbf{P}$. Once the current schedule is over, a new schedule is randomly generated by all stations using the current value of $\mathbf{P}$ (and the schedule so generated by all stations would again be identical) and the cycle continues.

When a station u_j is granted permission but has no packets, the value of $P(j)$ decreases under the above learning algorithm. Suppose we make the parameter a have value zero. Then, if u_j has no packets to transmit for a long time then $P(j)$ can become close to zero. This would mean it would not be chosen for transmission in any further schedules. If u_j now has packets, it has no way of communicating this so that it can be included in a schedule. To avoid such a problem, we choose a value greater than zero for a and the algorithm is such that no $P(j)$ can become less than a. This would mean that each station would keep getting a chance to transmit. If the value of a is large, then we may be wasting slots by granting permissions to stations which may not have packets. A small value of a would suffice because once a station is granted permission and if it has packets to transmit, the probability $P(j)$ would increase.

Under this automata-based adaptive protocol, the bandwidth assigned to each station tends to be proportional to the needs of the station. (See [POP02] for details). In [POP02], this automata-based adaptive TDMA protocol is simulated for different network sizes and with bursty traffic with different mean burst lengths. The performance is assessed by analyzing the throughput (in packets/slot) under different values of the load in the network, and the mean delay a packet suffers (in terms of number of slots) at various achieved throughputs. It is found that this protocol achieves higher throughputs and better delays particularly when the mean burst length of the traffic is large. The reader is referred to [POP02] for more details.

An interesting aspect of this approach is that it can be implemented in a totally decentralized fashion without needing either a centralized controller or specific information exchange among the stations. Such methods are useful in other types of networks also. For example, a similar method is explored for wavelength division multiplexed passive star networks [PP99]. In such networks, in addition to deciding which station gets access, one has to also decide on which frequency the station should transmit. All that is needed is to enlarge our vector of probabilities into a matrix. That is, now we have numbers, $P_{ij}, i = 1, \ldots, W, j = 1, \ldots, K$, where K is the number of stations and W is the total number of frequencies available. Now P_{ij} would be the choice probability for station j on frequency i. Once again the access can be controlled in a distributed manner and the probabilities can be adapted to better divide the available bandwidth among stations. The reader is referred to [PP99] for details.

The material in this section illustrates one way of using the basic idea of learning automata, namely, stochastic choice of decisions and updating of choice probabilities using reinforcement feedback, in problems of adaptive decision making in communication networks. Given the importance of the problem of distributed control of resources in a communication network, there would be much scope for using ideas from learning automata theory as the automata algorithms do not need much information about the current state of the system.

6.4 Other Applications

We illustrated above how automata models can be used in optimization and decision making problems. In all such problems, the main attraction of the automata models is that the solution technique is not dependent on the amount of detailed knowledge one has about the system. For example, in the computer vision application, it would be very difficult to model the dependencies among

Other Applications

the parameters or how they affect the quality of the output. Another useful property of automata models is that they learn from reinforcement feedback. That is, all we need is a stochastic scalar feedback signal that indicates the goodness or otherwise of choices made. Thus, for example, no special treatment is needed in the algorithm to tackle noisy situations. We have seen the utility of this in the pattern recognition applications discussed in Chapter 4. Generic problems of optimization where there is a lack of knowledge about the system and where measurements are noise corrupted, come up naturally in many applications involving system identification, control, constraint satisfaction etc. In all such areas, automata techniques should prove useful. In this section we briefly point out some of the reported applications of automata to such problems.

We have seen in Chapter 2 how a common payoff game of CALA can be used for linear system identification. Since the algorithm does not depend on the system model employed, the method would be effective for identification of nonlinear systems as well. One can employ, e.g., a feedforward neural network with a tapped delay line or a memory neuron network [SSU94] as a system model for nonlinear dynamic systems and the CALA algorithm can be used for learning the weights in the network. In cases where the parameters can be discretized, games of FALA could be used for identification. Such online automata-based techniques have been used in adaptive signal processing for learning FIR and IIR filters [TM93, MCN97]. Another interesting way learning automata can be used is in deciding the identification model. For example, FALA have been used for learning the optimal order of a linear regressor model [PNI96].

Another area where automata would be useful is that of learning control. These are problems where appropriate control laws are to be learnt using examples or online experience. An automaton can be used to select a control action (from the allowable set of control actions) at each instant and some suitably normalized value of some performance index can be supplied as the reinforcement. If the control actions can be discretized then an FALA would suffice; otherwise a CALA can be used. For example, an FALA using an estimator algorithm is shown to be useful for controlling a bioprocess [ZZV00]. It is not always necessary to have the actions of automata as the possible values of controller output. The actions of automata could be parameters of a standard controller and the problem could be to learn appropriate parameter settings. For example, Frost et al. [FGHW96] employ an FALA using L_{R-I} algorithm for the problem of learning vehicle suspension control laws. Here the actions of automata are possible feedback gains for a standard LQG (linear quadratic Gaussian) optimal control problem and the automaton learns the optimal gain vectors. To use an FALA, the parameter values have to be discretized. For the same problem, a continuous action automaton (namely, the CARLA discussed in Section 1.6.3) was shown to be effective when we do not want to discretize the parameter values [HFGW97]. Another example where automata models with continuous action sets are used to tune parameters of a controller is in [HB99]. Automata algorithms are also useful in tuning fuzzy controllers. The critical component of a fuzzy control system is the rule base which governs its operation. While a large number of possibilities may exist for forming rule bases from the fuzzy sets defined over the control variables, a team of LA with actions of automata as fuzzy sets, can be used to arrive at the best rule base in an online fashion [Vis95].

The main advantage of learning techniques is that detailed models of the relationship between the performance index and the control actions (or parameters of controllers) are not needed. Both the examples mentioned above are essentially applications of single automaton algorithms. In a large system where multiple controllers are to be deployed, we can adapt all of them together using teams of automata. Like in the computer vision example considered earlier, the advantage is that the interdependence of parameters of different control loops can be handled without any special effort in the design of the learning scheme. For example, [Wu95] discusses the use of a team of FALA for learning coordinated control of a power system. A modern power system consists of many subsystems and there are many controllers for individual components such as generator excitation systems, turbine governors, static voltage compensators, protection relays etc. Building integral models of the system to help design of an integral control strategy is difficult. The paper [Wu95] illustrates how a team of FALA can be used for learning control of synchronous machines in a multi-machine power system. The individual machines are controlled by controllers whose parameters are adapted using the learning automata which are installed locally and which learn through reinforcement based on observed values of an integrated performance index. The optimum parameters of the learning controllers can be obtained after necessary learning. Through some simulation experiments, the advantages of this strategy of learning control and the potential for its application to various aspects of power system control are discussed in [Wu95]. An interesting application for vehicle path control in an automated highway system is described in [UKB99]. The control system here uses one LA to control changing of lanes by the vehicle and another LA to control the speed so as to avoid collisions.

Apart from directly learning the control actions or learning parameters of the controllers, learning automata can be useful in other ways also in a control situation. An interesting example of this can be found in [ARS02] which considers the problem of automatic generation control in multi-area power systems. Like in any learning control application, the objective is to evolve the control law without needing explicit system model. Here, the problem is posed in the framework of Markovian decision processes and a Q-learning strategy is employed to evolve a good control policy through examples (or online experience with a simulation model). For the learning process to be efficient, we need an effective strategy to explore various control options so that we experience all the necessary regions in the state space of the system. In [ARS02], a set of FALA using pursuit algorithm are employed to do the necessary exploration. The strategy is such that as learning proceeds, the exploratory control strategy (represented by the action probability vectors of the FALA) goes to the optimal control strategy.

Learning automata algorithms can also be viewed simply as stochastic search strategies. Even in deterministic problems, sometimes such search strategies can result in efficient suboptimal algorithms. Thus, they can be used to get good approximate solutions to combinatorial optimization problems. For example, FALA are found useful in graph partitioning [OC95]. Another class of problems where automata-based search is effective is that of constraint satisfaction. Teams of FALA are useful in getting efficient solutions to the so called relaxation labeling problem which is a useful framework for many computer vision tasks [TS86b]. Another interesting constraint satisfaction problem where automata have been used is the so called object partitioning problem [OM92]. The problem is that of grouping objects based on similarities as reflected (imperfectly) through a series of user interactions. For example, objects can be records in a database. Based on a series of user queries, we want to organize the records into pages on secondary memory so that groups of records which are often needed together to answer queries are on the same page. Or, the objects can be images in an image database and we may want to group together similar images based on user interactions and relevance feedback. Similar structures would be useful in other problems of information retrieval. Some of the other applications of learning automata include string taxonomy [OC97], learning distributions [AW98] etc.

6.5 Discussion

In this chapter, we have presented a brief review of some of the recent applications of learning automata. We have explained a couple of applications in some detail and have mentioned many others briefly. The learning automaton represents a simple learning machine which can fit in many situations involving learning. The automaton model is very general and hence, the same algorithms can be useful in a wide variety of applications. We have tried to indicate this breadth of application potential in this chapter. While we have tried to include many applications, the discussion in this chapter cannot be viewed as an exhaustive survey of learning automata applications.

Most of the applications described earlier use simple automata structures. Some of the multiautomata algorithms and the associated theory presented in this book (especially the material in Chapter 3) are of recent origin and we expect that these algorithms would be found useful in extending the application domain. Similarly, the parallel operation of automata, discussed in Chapter 5, also constitutes a recent development in the field. Since such a parallel operation can contribute to significant increase in speed, it would also be very useful in applications.

EPILOGUE

In this book we have studied learning automata which are adaptive decision making devices operating in unknown random environments. They learn by updating their action probability distributions in accordance with the reinforcement obtained from the environment. In Chapter 1 we have considered a variety of LA to suit different situations. Chapters 2 and 3 described how collections of LA could be formed into teams and networks to solve more complex learning problems. One example of an application, namely, pattern recognition was considered in detail in Chapter 4. The problem of speed of learning, which becomes crucial with larger assemblage of LA, was handled in Chapter 5 by forming modules which enable parallel operation.

In all the above developments one feature that stands out is the simplicity of the idea of LA which also possesses generality. The applicability of the idea is wide because of the generality of the concept of action set. At the same time, many interesting results concerning the single LA as well as games and networks of LA have been rigorously established. Such mathematical tractability is rather uncommon in other approaches to machine learning and makes the LA approach a promising direction for the development of a general theory of learning.

There is much interest at the present time in multiagent systems, in which several agents act in the same environment to accomplish a task. The complexity here is that each agent is exposed to changes in the environment per se as well as to changes caused by decisions taken by other agents. Studies on the so called ant colonies have recently given rise to interesting algorithms for such multiagent systems [DMC96, VN02]. As we have seen, games and networks of LA handle similar situations of distributed learning and hence would be good candidates for forming novel multiagent systems. It has also recently been suggested that the LA approach could lead to methods of proof for convergence of algorithms found effective in ant systems [VN02]. More investigations are needed in this potentially rich area connecting LA and distributed multiagent systems.

Evolutionary algorithms for optimization which are modeled after Darwinian principles of natural selection and evolution of biological populations, have made much progress in recent times. As evolution in essence is a modification of probability distribution of a population from generation to generation, evolutionary algorithms have a natural affinity to LA algorithms. There have been attempts to cross-fertilize the two areas. One such attempt combines genetic algorithms with the L_{R-I} where the latter is used to update a string of probabilities used to generate binary strings. The optimum binary string is obtained when the probability string itself converges to a binary string. This approach has been found effective in function optimization problems [HGB02] and represents one possible way of combining LA with evolutionary methods. There is a wide spectrum of possibilities here waiting to be explored.

Artificial neural networks have now been established as very useful models for function approximation, pattern recognition and other tasks. A main requirement here is that of learning the weights. While backpropagation and its variants have traditionally been popular for this purpose, they are not particularly effective under noisy conditions. LA algorithms score over them under such situations. A glimpse of this was seen in connection with the Iris data problem in Chapter 4, where the LA network using L_{R-I} performed well even under heavy noise conditions. As has been mentioned in Chapter 2, games of CALA provide an alternative method for learning weights in artificial neural networks. An advantage of the LA methods in this context is that all LA in a team are updated in the same way and thus the learning algorithm is not dependent on the architecture. LA systems used in conjunction with artificial neural networks for performing specific types of tasks, appears to be a promising area of investigation. The use of L_{R-P} algorithm has been investigated in one such situation [SC98], but more comprehensive studies are needed.

In this book we have concentrated on reinforcement learning problems where LA models are appropriate. By combining the basic idea of reinforcement learning with that of Markovian decision processes and dynamic programming, one gets another important class of reinforcement learning techniques [SB98, BT96]. These techniques are particularly relevant in control applications. Consider a network of LA consisting of a single unit where the context vectors represent the measured state of a dynamic system and the actions of the unit represent control signals. Learning here consists of associating appropriate control actions with measured states. However, in such problems, the context vector at $k+1$ would depend on the context vector at k as well as the action output at k. In the models considered here, we have not allowed such dependence among context vectors though we have allowed a lot of complexity in the architecture of the network. As pointed out in Chapter 6, the LA models

Epilogue

have been employed in learning control applications, e.g., in fixing the parameters of a general control law. However, the algorithms discussed here cannot handle dynamics in the environment directly. More work needs to be done for elucidating the relationship between these different classes of algorithms and for combining them effectively.

The theory of learning automata also needs to be developed further. More general interconnections of LA including recurrent networks could perhaps handle a higher order of complexity in learning problems. Suitable algorithms and convergence properties of such networks are open to investigation. In the games and networks considered in this book, the resulting optimization problems are essentially unconstrained problems. (The only constraints involved are related to the algorithm such as keeping probabilities or the internal state variables in suitable ranges). More complicated problems could involve constraints from the problem domain. Some sophisticated games with constraints have been analyzed recently [PN02]. Many such theoretical problems would become important as the applications become more sophisticated.

In Chapter 6 of the book, we have briefly discussed some of the recent applications of LA techniques. It is seen that many of the successful applications of LA are in areas such as Pattern Recognition, Computer Vision, Communication Systems, Adaptive Control, Robotics etc. In most applications, LA are used as tools for decision making and optimization. LA techniques are seen to be closely related to the newly emerging field of Soft Computing where the emphasis is on obtaining approximate rather than optimal solutions in complex applications. The ability of LA to learn with little knowledge of the environment and their ability to handle noise, would make them attractive as a tool for soft computing. A recent and important application area for machine learning techniques is that of Datamining. In a typical datamining problem, one is interested in finding some hidden stochastic dependencies in the data, ideally in a completely model-free manner. LA techniques, with their ability to learn in unknown environments, should be useful in datamining applications and this is one more area worthy of investigation.

Over the last few decades, there has been an impressive amount of progress in the field of machine learning and many techniques are being employed in applications of increasing complexity. The general paradigm of learning to act in a nearly optimal manner by interacting with the system in an online fashion, is a method of wide applicability if suitable learning algorithms are available. The LA models discussed in this book constitute a general theory of such learning techniques. The models are simple and general, can learn under noisy reinforcement feedback and many theoretical results regarding their behavior are available. We hope that LA would contribute to a broad spectrum of developments in the theory and applications of machine learning.

Appendix A
The ODE Approach to Analysis of Learning Algorithms

A.1 Introduction

For most of the learning algorithms considered in this book, the analysis proceeds through two stages: first, we obtain an ordinary differential equation (ODE) that well approximates the behavior of the algorithm (in a sense to be made precise in this appendix); then, we analyze the asymptotic properties of the solutions of the ODE to infer the long-time behavior of the algorithm[1].

In this appendix we show how the approximating ODE can be obtained for a general class of stochastic algorithms. In the process, we state and prove a theorem that is used in some of the chapters of the book for obtaining the approximating ODE for different automata algorithms. Approximating a stochastic algorithm by an ODE to understand the long-time behavior of the algorithm is a well studied method (see, e.g, [Kus84, BMP87, KY97]). In this appendix we would not be discussing this general problem. We focus only on techniques that are sufficient to analyze the automata algorithms considered in this book. The material in this appendix is based on [Pha94].

Consider a stochastic algorithm given by the following update equation:

$$X_{k+1} = X_k + b\, G(X_k, \xi_k) \tag{A.1.1}$$

where $X_k \in \Re^N$, is called the state vector, $\xi_k \in \Re^{N'}$, is called the noise vector and b is the step-size or learning parameter. Here N and N' denote, respectively, the dimensions of X_k and ξ_k.

As a specific example of eqn. (A.1.1), consider the L_{R-I} algorithm for a single automaton considered in Section 1.4.1. Here we update the action probability vector, $\mathbf{p}(k)$, as

$$\mathbf{p}(k+1) = \mathbf{p}(k) + \lambda \beta(k)(\mathbf{e}_i - \mathbf{p}(k)) \tag{A.1.2}$$

---

[1] The only exceptions to this general framework are the estimator algorithms (cf. Section 1.4.3) and the algorithms for converging to global optima (cf. Section 3.7). The proof of convergence for the estimator algorithms is provided in Appendix B. The global algorithms are analyzed by obtaining an approximating stochastic differential equation (SDE) which is explained in Appendix C

where the index i corresponds to the selected action (that is, $\alpha(k) = \alpha_i$) and $\beta(k)$ is the reinforcement received at k. $\mathbf{e}_i$ is the unit vector with i^{th} component unity and all others zero. By choosing $X_k = \mathbf{p}(k)$, $\xi_k = (\xi_k^1, \xi_k^2) = (\alpha(k), \beta(k))$, $b = \lambda$, and $G(X_k, \xi_k) = \xi_k^2(\mathbf{e}_{\xi_k^1} - X_k)$, it is easy to see that the L_{R-I} algorithm given by eqn.(A.1.2) will be the same as the algorithm given by eqn.(A.1.1).

In general, in a FALA algorithm, X_k would consist of the action probabilities of all the automata and ξ_k would consist of all the other quantities (such as actions selected, reinforcements obtained etc.) that are needed for updating the action probabilities. $G(\cdot, \cdot)$ would represent the updating scheme. The step-size parameter b corresponds to the learning parameter (denoted by λ in the text). Similarly, in a CALA algorithm, X_k would consist of the means and standard deviations of the normal distributions that represent the action probability distributions of the automata, and ξ_k would consist of all the other quantities such as actions selected, reinforcements obtained etc. at time step k.

The main idea behind the analysis presented here is the following. We assume that the sequence of random variables $\{(X_k, \xi_k), k \geq 1\}$ is a Markov process. In addition, in automata algorithms we find that ξ_k is completely stochastically determined by X_k. Specifically, conditioned on $\mathbf{p}(k)$, the probability distribution of $(\alpha(k), \beta(k))$ is independent of $(\alpha(k-1), \beta(k-1))$. Hence, we would assume that conditioned on X_k, the random variable ξ_k is stochastically independent of ξ_{k-1}. Then we can define a function, $g : \Re^N \to \Re^N$, by

$$g(x) = E[G(X_k, \xi_k) \mid X_k = x] \quad (A.1.3)$$

We are assuming here that the expectation in the RHS of the above equation is independent of the time step, k, which is true for all our algorithms. Using (A.1.3), we can rewrite (A.1.1) as

$$X_{k+1} = X_k + bg(X_k) + b\theta_k \quad (A.1.4)$$

where $\theta_k = G(X_k, \xi_k) - g(X_k)$ is a zero mean process. If we assume that the step-size, b, is small, then X_k would change very little over a few iterations during which we can expect sufficient 'averaging' of the θ_k terms to take place. Since θ_k is a zero-mean process, we can expect that X_k would be well approximated by solution of the difference equation

$$y_{k+1} = y_k + bg(y_k), \quad y_0 = X_0. \quad (A.1.5)$$

In the next section we show that this intuition is indeed true under some conditions on the various quantities involved.

The reason (A.1.5) is preferable to (A.1.1) is that the former is a good approximation (if $g(\cdot)$ is a 'good' function) to the ODE given by

$$\frac{dz}{dt} = g(z), \quad z(0) = y_0. \quad (A.1.6)$$

It may be noted that (A.1.5) is the so called Euler approximation to the ODE (A.1.6). Specifically, if y_k is a solution of (A.1.5) and $z(t)$ is a solution of (A.1.6), and if $g(\cdot)$ is locally Lipschitz, then, for any $T > 0$, we have

$$\text{Lim}_{b \to 0} \sup_{0 \leq k \leq T/b} \|y_k - z(kb)\| = 0. \quad (A.1.7)$$

What this means is that if we have the solution to (A.1.6) then we can obtain information regarding the behavior of y_k (which is a solution of (A.1.5)) which in turn is a good approximation to X_k.

Derivation of the ODE Approximation

We call (A.1.6) as the approximating ODE for the algorithm given by (A.1.1). Based on the foregoing discussion, for sufficiently small value of b, X_k would well approximate the solution of the ODE given by (A.1.6) in the sense that, for any T, $\epsilon > 0$,

$$\text{Lim}_{b \to 0} \text{Prob}[\sup_{0 \leq k \leq T/b} ||X_k - z(kb)|| > \epsilon] = 0 \qquad (A.1.8)$$

where $z(t)$ is the solution of ODE (A.1.6). To prove this, in view of (A.1.7), it is enough if we prove that

$$\text{Lim}_{b \to 0} \text{Prob}[\sup_{0 \leq k \leq T/b} ||X_k - y_k|| > \epsilon] = 0 \qquad (A.1.9)$$

This is what is done in the next section under some conditions on the algorithm.

As is easy to see, all we need to do for obtaining the approximating ODE is to evaluate the expectation on the RHS of (A.1.3). This is how we derived the ODE approximations for our learning algorithms in the text. In the analysis of the algorithms as presented in the earlier chapters, we used the asymptotic properties of the solutions of ODE to infer the asymptotic properties of the algorithm. Thus, to complete these analyses we need to show that the ODE we used is indeed the approximating ODE and this is what is done in this appendix.

The proofs presented here are largely self contained. However, we would be using some concepts such as martingales and some inequalities associated with them in the proofs. Readers unfamiliar with martingales can refer to standard text books on stochastic processes such [KT75, WH85]. For the sake of completeness, we give below a simplified explanation of these concepts. A stochastic process $\{X_t\}$ is called a `martingale` if for any $s < t$,

$$E[X_t \mid X_1, \ldots, X_s, \ s < t] = X_s, \ almost \ surely.$$

The process is called a `submartingale` (`supermartingale`) if the equality in the above is replaced by $\geq$ (respectively, $\leq$). More generally, for X_t to be a martingale, in the above conditional expectation the conditioning random variables need not have to be the past values of the same process. Let $\{Y_t\}$ be any other process. Let $\mathcal{G}_k = \sigma\{Y_n, \ n \leq k\}$ denote the smallest σ-algebra generated by the indicated random variables. Suppose, for every k, X_k is measurable with respect to the σ-algebra $\mathcal{G}_k$. This essentially means that X_k is 'determined' by the random variables that generate $\mathcal{G}_k$. Thus, expectation of X_k conditioned on these random variables would be X_k itself. (Note that if Y_k is same as X_k then this requirement is trivially satisfied). Now the process X_t is called a martingale (with respect to the sequence of σ-algebras, $\mathcal{G}_t$) if for any $s < t$,

$$E[X_t \mid Y_1, \ldots, Y_s, \ s < t] = X_s, \ almost \ surely.$$

In the proofs, we use the above characterization of martingales. As before, X_t is a submartingale if the equality in the above is replaced by $\geq$ and it is called supermartingale if the equality is replaced by $\leq$.

A.2   Derivation of the ODE Approximation

In the algorithm given by (A.1.1), the evolution of X_k, is dependent on the step-size b. With different values of b, we would be generating different sequences X_k and ξ_k. Since the approximations we derive are in the limit as $b \to 0$, it is necessary to remember this dependence. Thus we rewrite our general algorithm as

$$X^b_{k+1} = X^b_k + bG(X^b_k, \xi^b_k), \ \ X^b_0 = x_0. \qquad (A.2.10)$$

Recall that $X^b_k \in \Re^N$ is the state vector and $\xi^b_k \in \Re^{N'}$ is the noise vector. $G(\cdot, \cdot)$ is a function from $\Re^N \times \Re^{N'}$ to $\Re^N$.

A.2.1 Assumptions

We make the following assumptions on the system.

A1. $\{(X_k^b, \xi_k^b), \ k \geq 1\}$ is a Markov Process.

A2. For any appropriate Borel set, $\mathcal{B}$,

$$\text{Prob}[\xi_k^b \in \mathcal{B} \,|\, X_k^b, \xi_{k-1}^b] = \text{Prob}[\xi_k^b \in \mathcal{B} \,|\, X_k^b].$$

That is, conditioned on X_k^b, ξ_k^b is independent of ξ_{k-1}^b.

A3. Define $g : \Re^N \to \Re^N$ by[2]

$$g(x) = E[G(X_k, \xi_k) \,|\, X_k = x]$$

We assume that $g(\cdot)$ is independent of k and that it is globally Lipschitz.

A4. Define

$$\theta_k^b = G(X_k^b, \xi_k^b) - g(X_k^b)$$

We assume that $E||\theta_k^b||^2 < M < \infty$, for some M and $\forall k$.

Comments on the Assumptions

To begin with we note that the above assumptions are much stronger than those generally needed for deriving the ODE approximation (see, e.g, [Kus84, BMP87]). However, for most of the automata algorithms these assumptions are satisfied. That is the reason we chose to make these assumptions. These assumptions allow us to keep the proofs simple and largely self-contained. Later on we shall see how some of these assumptions can be weakened.

Assumption $A1$ is reasonable and we cannot really weaken the Markovian requirement if we want (A.2.10) to define an easily implementable iterative algorithm. In (A.2.10) we have implicitly assumed that G does not depend on b, k, which is true of all automata algorithms.

Assumption $A2$ is satisfied by all automata algorithms. The same ODE approximation can be derived even if we weaken it to: for each $x \in \Re^N$, if we fix $X_k = x$, $\forall k$, then ξ_k^b is a Markov process with a unique stationary distribution (which can depend on x). However, the proof would be more complicated (see, e.g., [Kus84]).

The restrictive part of assumption $A3$ is that $g(\cdot)$ is globally Lipschitz. In our analysis, we are trying to approximate (A.1.1) with the difference equation given by (A.1.5). For the latter difference equation to be a good approximation to the ODE given by (A.1.6), we anyway need g to be Lipschitz on compact sets. In the FALA algorithms, the state vector, X, would be the vector of all action probabilities. Thus g would have a compact support. Also, the update function G and the noise vector, ξ would be bounded. Thus g would be a bounded function with compact support and thus continuity of g would ensure it is globally Lipschitz. Thus, for FALA algorithms, assumption $A3$ is not restrictive. But in CALA algorithms, ξ and hence G cannot be bounded due to the fact that the action probability distribution is Gaussian. Hence to satisfy this assumption in CALA algorithms we need to put some restrictive assumptions on the unknown reward probability function. We shall see later on how to weaken this assumption.

Assumption $A4$ is also not very restrictive. Note that θ_k^b is a zero mean random variable. Since we want to approximate a stochastic difference equation driven by $G(\cdot, \cdot)$ by a deterministic difference equation driven by $g(\cdot)$, it is natural that we need to ensure that the variance of θ_k^b does not go unbounded. It is easy to see that for FALA algorithms this assumption is easily satisfied because all components of X and ξ are bounded.

---

[2]This function, g, should not be confused with the probability generating function used in GLA and PLA algorithms.

Derivation of the ODE Approximation

A.2.2 Analysis

We are interested in analyzing the algorithm given by (A.2.10) which is reproduced below.

$$X_{k+1}^b = X_k^b + b\, G(X_k^b, \xi_k^b), \quad X_0^b = x_0.$$

The approximating ODE for this algorithm is given by

$$\frac{dz}{dt} = g(z), \quad z(0) = X_0^b = x_0 \qquad (A.2.11)$$

where the function g is given by (A.1.3). We want to show that the solution of this ODE would be a good approximation for X_k^b for sufficiently small b. For this, as discussed in the previous section, we want to show that X_k^b can be approximated by y_k^b given by

$$y_{k+1}^b = y_k^b + b\, g(y_k^b), \quad y_0^b = X_0^b = x_0. \qquad (A.2.12)$$

In this section we prove the following.

THEOREM A.1 *For any initial condition x_0 and for any given finite T, ϵ, $\delta > 0$, there exists $b^* > 0$ such that for all $0 < b \leq b^*$,*

a.
$$Prob[sup_{0 \leq k \leq \frac{T}{b}} \|X_k^b - y_k^b\| \geq \epsilon] \leq \delta, \qquad (A.2.13)$$

b. *and hence,*
$$Prob[sup_{0 \leq k \leq \frac{T}{b}} \|X_k^b - z(kb)\| \geq \epsilon] \leq \delta \qquad (A.2.14)$$

where $z(\cdot)$ is the solution of ODE (A.2.11).

Part **b** of the above theorem easily follows from part **a** in view of (A.1.7). Hence we need to present proof only for part **a**.

Before we present the proof for the above theorem, we briefly explain the approach to the proof and then prove some preliminary lemmas.

Define
$$m_k^b = \|X_k^b - y_k^b\| \qquad (A.2.15)$$

Recall that
$$g(x) = E[G(X_k^b, \xi_k^b) \mid X_k^b = x]. \qquad (A.2.16)$$

Denote
$$\theta_k^b = G(X_k^b, \xi_k^b) - g(X_k^b). \qquad (A.2.17)$$

. Let
$$S_k^b = \sum_{n=0}^{k} \theta_n^b. \qquad (A.2.18)$$

From, (A.2.10) and (A.2.12) we have

$$\begin{aligned}
X_{k+1}^b - y_{k+1}^b &= b \sum_{n=0}^{k} [G(X_n^b, \xi_n^b) - g(y_n^b)] \\
&= b \sum_{n=0}^{k} [G(X_n^b, \xi_n^b) - g(X_n^b) + g(X_n^b) - g(y_n^b)] \\
&= b \sum_{n=0}^{k} \theta_n^b + b \sum_{n=0}^{k} [g(X_n^b) - g(y_n^b)] \\
&= b S_k^b + b \sum_{n=0}^{k} [g(X_n^b) - g(y_n^b)]
\end{aligned}$$

Thus we get

$$m_{k+1}^b = \|X_{k+1}^b - y_{k+1}^b\|$$
$$\leq b\|S_k^b\| + b \sum_{n=0}^{k} \|g(X_n^b) - g(y_n^b)\| \qquad (A.2.19)$$

Now the objective is to bound both terms on the RHS of (A.2.19) so that we can derive an inequality of the form

$$m_{k+1}^b \leq \alpha \sum_{n=0}^{k} m_n^b + C, \quad \alpha > 0, C > 0.$$

As would be seen, this inequality is sufficient to establish our theorem by using the discrete Gronwall lemma (see Lemma A.5). Hence, we would first establish this inequality.

We can bound the second term on the RHS of (A.2.19) using Lipschitz property of g. So, we will first concentrate on bounding S_k^b.

LEMMA A.1 *For every $b > 0$, $\{S_k^b, k \geq 0\}$ is a martingale with respect to the sequence of σ-algebras, $\mathcal{G}_k^b = \sigma\{X_n^b, \xi_n^b, n \leq k\}$.*

Proof: S_k^b is easily seen to be measurable with respect to $\mathcal{G}_k^b$. As a notation, let $\Delta S_k^b = S_{k+1}^b - S_k^b$. We have

$$\begin{aligned}
E[\Delta S_k^b \,|\, X_n^b, \xi_n^b, n \leq k] &= E[\theta_{k+1}^b \,|\, X_n^b, \xi_n^b, n \leq k] \\
&= E[E[G(X_{k+1}^b, \xi_{k+1}^b) - g(X_{k+1}^b) \,|\, X_{k+1}^b, \xi_k^b] \,|\, X_n^b, \xi_n^b, n \leq k] \\
&= E[\{E[G(X_{k+1}^b, \xi_{k+1}^b) \,|\, X_{k+1}^b, \xi_k^b] - g(X_{k+1}^b)\} \,|\, X_n^b, \xi_n^b, n \leq k] \\
&= 0, \quad \text{Assumption } A2 \text{ and (A.2.16).}
\end{aligned}$$

Thus $\{S_k^b\}$ is a martingale.

LEMMA A.2 *For all m, n, $m \neq n$, $E[(\theta_n^b)^T \theta_m^b] = 0$.*

Proof: Without loss of generality, assume $m > n$.

$$\begin{aligned}
E[(\theta_n^b)^T \theta_m^b] &= E[E[(\theta_n^b)^T \theta_m^b \,|\, X_k^b, \xi_{k-1}^b, k = 1, \ldots, m]] \\
&= E[(\theta_n^b)^T E[\theta_m^b \,|\, X_m^b, \xi_{m-1}^b]] \\
&= 0 \quad \text{by assumption } A2, \text{ and (A.2.16) \& (A.2.17).}
\end{aligned}$$

LEMMA A.3 *For all k, b, $E[\|S_k^b\|^2] \leq kK_3$ for some K_3, $0 < K_3 < \infty$.*

Proof:

$$\begin{aligned}
E[\|S_k^b\|^2] &= E[(S_k^b)^T S_k^b] \\
&= E[\sum_{n,\ell} (\theta_n^b)^T \theta_\ell^b] \\
&= \sum_{n=0}^{k} E[\|\theta_n^b\|^2], \quad \text{by Lemma A.2.} \\
&\leq (k+1)M \quad \text{by assumption } A4 \\
&\leq kK_3, \quad \text{where, e.g., } K_3 = 2M.
\end{aligned}$$

Derivation of the ODE Approximation

Now we are ready to bound the first term on the RHS of (A.2.19). Since $\{S_k^b\}$ is a martingale, $\|S_k^b\|^2$ is a submartingale. Hence by using a special inequality applicable for submartingales (see [WH85, Chap.6]), we get the following.

LEMMA A.4 *For any $b > 0$ and $\epsilon > 0$,*

$$Prob[\max_{0 \leq k \leq \frac{T}{b}} b\|S_k^b\| \geq \epsilon] \leq \frac{bTK_3}{\epsilon^2}.$$

Proof: Let n' denote the largest integer smaller than $\frac{T}{b}$.

$$\begin{aligned}
Prob[\max_{0 \leq k \leq \frac{T}{b}} b\|S_k^b\| \geq \epsilon] &= Prob[\max_{0 \leq k \leq \frac{T}{b}} \|S_k^b\|^2 \geq \frac{\epsilon^2}{b^2}] \\
&\leq \frac{b^2}{\epsilon^2} E[\|S_{n'}^b\|^2] \quad \text{by [WH85, Chap.6]} \\
&\leq \frac{b^2}{\epsilon^2} \frac{T}{b} K_3, \quad \text{by Lemma A.3} \\
&= \frac{bTK_3}{\epsilon^2}
\end{aligned}$$

LEMMA A.5 *(Discrete Gronwall Lemma) Let u_k be a sequence of real numbers satisfying the following: $u_k \geq 0 \,\forall k \geq 0$, $u_0 \leq \epsilon$ and there is a γ such that for all n*

$$u_{n+1} \leq \gamma \sum_{k=0}^{n} u_k + \epsilon.$$

Then we have,

$$u_n \leq (1+\gamma)^n \epsilon, \quad \forall n \geq 0.$$

Proof: The proof is by mathematical induction. Let $u_k \leq (1+\gamma)^k \epsilon$, for all $0 \leq k \leq n$. Then

$$\begin{aligned}
u_{n+1} &\leq \gamma \sum_{k=0}^{n} u_k + \epsilon \\
&\leq \gamma \sum_{k=0}^{n} (1+\gamma)^k \epsilon + \epsilon \\
&= \epsilon \left[\gamma \sum_{k=0}^{n} (1+\gamma)^k + 1 \right] \\
&= \epsilon \left[\gamma \frac{(1+\gamma)^{n+1} - 1}{(1+\gamma) - 1} + 1 \right] \\
&= (1+\gamma)^{n+1} \epsilon
\end{aligned}$$

This completes proof of the lemma.
Proof of Theorem A.1

We have from (A.2.15) and (A.2.19)

$$\begin{aligned}
m_{k+1}^b &= \|X_{k+1}^b - y_{k+1}^b\| \\
&\leq b\|S_k^b\| + b \sum_{n=0}^{k} \|g(X_n^b) - g(y_n^b)\|
\end{aligned}$$

$$\leq b\|S_k^b\| + b\sum_{n=0}^{k} K\|X_n^b - y_n^b\| \quad \text{by assumption } A3; K \text{ is the Lipschitz constant}$$

$$= b\|S_k^b\| + bK\sum_{n=0}^{k} m_n^b$$

Now, by Lemma A.4, with probability at least $1 - \frac{bTK_3}{\epsilon_1^2}$, we have

$$m_{k+1}^b \leq \epsilon_1 + bK\sum_{n=0}^{k} m_n^b \tag{A.2.20}$$

By Gronwall lemma (that is, Lemma A.5), (A.2.20) implies that with probability of at least $1 - \frac{bTK_3}{\epsilon_1^2}$,

$$m_{k+1}^b \leq (1 + bK)^{k+1}\epsilon_1 \tag{A.2.21}$$

To prove the theorem we need to show: given any ϵ, δ, T, we can find a b^* such that for all $b \leq b^*$ and $k \leq \frac{T}{b}$,

$$m_k^b \leq \epsilon \quad \text{with prob.} \geq 1 - \delta.$$

Choose $\epsilon_1 = e^{-KT}\epsilon$ and $b^* = \frac{\delta\epsilon_1^2}{K_3 T}$. Then for all $k \leq \frac{T}{b}$, from (A.2.21) we have

$$\begin{aligned} m_k^b &\leq (1 + bK)^k \epsilon_1 \\ &\leq (1 + bK)^{\frac{T}{b}} \epsilon_1 \\ &\leq e^{KT}\epsilon_1 \quad \text{because } (1 + x) \leq e^x \\ &= \epsilon \end{aligned}$$

with probability greater than or equal to

$$1 - \frac{bTK_3}{\epsilon_1^2} \geq 1 - \delta$$

for all $b \leq b^*$. This proves the theorem.

Remark A.1: What theorem A.1 says is that X_k^b, given by (A.2.10) will follow y_k^b given by (A.2.12) as closely as desired and for as long a time as desired by taking the step-size, b, sufficiently small. In turn, y_k^b will closely follow the solution of the ODE (A.1.6). Putting these together, we see that X_k^b given by our algorithm can be made to closely approximate the solution of its approximating ODE. Thus, if the ODE has a locally asymptotically stable equilibrium point, then, we can conclude that (by taking b sufficiently small) X_k^b, for large k, would be close to this equilibrium point if X_0 is within the region of attraction of that point. These are precisely the type of results we presented in the text regarding the behavior of our algorithms.

A.3 Approximating ODEs for Some Automata Algorithms

Obtaining the approximate ODE for any learning algorithm consists of two steps: first verifying that all the assumptions listed in Section A.2.1 are satisfied; and then evaluating the expectation on the RHS of (A.1.3) to determine the RHS of the ODE (A.1.6). In this section we illustrate this for some of the automata algorithms. We shall consider two single automata algorithms, one for FALA and one for CALA. It is easily seen that similar arguments hold for automata team algorithms also.

A.3.1 L_{R-I} Algorithm for a Single Automaton

Consider the L_{R-I} algorithm for a single FALA discussed in Section 1.4.1. Let $\mathbf{p}(k)$ be the action probability distribution at k and let $\alpha(k) = \alpha_i$ be the action selected and $\beta(k)$ be the reinforcement obtained at k. Then the action probabilities are updated by

$$\mathbf{p}(k+1) = \mathbf{p}(k) + \lambda\,\beta(k)\,(\mathbf{e}_i - \mathbf{p}(k)) \qquad (A.3.22)$$

As explained in Section A.1, this is same as our general algorithm.

It is easily seen that this algorithm satisfies $A1$. Since given $\mathbf{p}(k)$, $\alpha(k)$ and $\beta(k)$ are independent of $\alpha(k-1)$ and $\beta(k-1)$, assumption $A2$ is also satisfied.

From (A.3.22), it is easy to see that

$$\begin{aligned}
g(\mathbf{p}) &= E[\beta(k)\,(\mathbf{e}_{\alpha(k)} - \mathbf{p}(k)) \mid \mathbf{p}(k) = \mathbf{p}] \\
&= \sum_{j=1}^{r} E[\beta(k)\,(\mathbf{e}_j - \mathbf{p}(k)) \mid \mathbf{p}(k) = \mathbf{p},\ \alpha(k) = \alpha_j]\,p_j \\
&= \sum_{j=1}^{r} d_j\,p_j\,(\mathbf{e}_j - \mathbf{p}) \qquad (A.3.23)
\end{aligned}$$

where r is the number of actions and d_j is the reward probability for the jth action. If we represent by g_i the ith component of g then we get

$$\begin{aligned}
g_i(\mathbf{p}) &= d_i p_i (1 - p_i) + \sum_{j \neq i} d_j p_j (-p_i) \\
&= d_i p_i \sum_{j \neq i} p_j - p_i \sum_{j \neq i} d_j p_j \\
&= \sum_{j} (d_i - d_j) p_i p_j. \qquad (A.3.24)
\end{aligned}$$

Note that the function $g(\cdot)$ has compact support (it is defined on the r-dimensional simplex) and it is bounded because $-1 \le g_i(\mathbf{p}) \le +1 \ \forall \mathbf{p}$. Since g is also continuous, it is easily seen that g is globally Lipschitz. For completeness, we explicitly verify assumption $A3$ below. Let $d_{ij} = d_i - d_j$. For any probability vectors, $\mathbf{x}, \mathbf{y}$,

$$\begin{aligned}
|g_i(\mathbf{x}) - g_i(\mathbf{y})| &= \left|\sum_{j \neq i} d_{ij}(x_i x_j - y_i y_j)\right| \\
&= \left|\sum_{j \neq i} (-d_{ij})(x_i \Delta x_j + x_j \Delta x_i + \Delta x_i \Delta x_j)\right| \quad \text{where } \Delta x_i = y_i - x_i,\ \forall i \\
&\le 2(\max_j d_{ij}) \sum_{j \neq i} (|\Delta x_i| + |\Delta x_j|) \quad \text{since } x_i, y_i \in [0,1]\ \forall i, \\
&\le 2\bar{d}_i r \sum_{j} |\Delta x_j|, \quad \text{where } \bar{d}_i = \max_j d_{ij}. \qquad (A.3.25)
\end{aligned}$$

Thus we get

$$\begin{aligned}
|g_i(\mathbf{x}) - g_i(\mathbf{y})|^2 &\le 4\bar{d}_i^2 r^2 \left(\sum_j |\Delta x_j|\right)^2 \\
&\le K_i \|\mathbf{x} - \mathbf{y}\|^2, \quad \text{where } K_i = 12\bar{d}_i^2 r^2. \qquad (A.3.26)
\end{aligned}$$

Thus we have

$$\|g(\mathbf{x}) - g(\mathbf{y})\|^2 = \sum_{i=1}^{r} |g_i(\mathbf{x}) - g_i(\mathbf{y})|^2$$
$$\leq (\sum_i K_i) \|\mathbf{x} - \mathbf{y}\|^2 \quad (A.3.27)$$

which proves that $g(\cdot)$ is globally Lipschitz.

Finally, to verify assumption $A4$, we note that for L_{R-I}, we have that $G(\cdot, \cdot) \in [-1, 1]^r$ and hence the variance is bounded as needed.

Since all assumptions are satisfied, we can approximate $\mathbf{p}(k)$ resulting from L_{R-I} algorithm by the solution of the ODE

$$\frac{dx_i}{dt} = \sum_j x_i x_j (d_i - d_j), \quad i = 1, \ldots, r \quad (A.3.28)$$

It is easy to show that when d_i are distinct, this ODE has all unit probability vectors as the only stationary points, and, further, the unit vector $\mathbf{e}_\ell$ is the only locally asymptotically stable one while all other stationary points are unstable. (Here, ℓ is the index of the optimal action).

A.3.2 The CALA Algorithm

In this subsection we take up the analysis of the learning algorithm for a single Continuous Action Set Automaton (CALA) which was described in Section 1.6. Here we are interested only in deriving the approximating ODE. In section 1.6.1 we have analyzed the asymptotic behavior of the algorithm based on this approximating ODE. Once again we note that for deriving the ODE, all we need to do is to verify assumptions $A1$-$A4$.

As described in Section 1.6, the algorithm updates $\mu(k), \sigma(k)$ which represent the mean and standard deviation parameters of the action probability distribution at k as below. (All symbols are as explained in Section 1.6).

$$\mu(k+1) = \mu(k) + \lambda\, F_1(\mu(k), \sigma(k), x(k), \beta_{x(k)}, \beta_{\mu(k)}) \quad (A.3.29)$$
$$\sigma(k+1) = \sigma(k) + \lambda\, F_2(\mu(k), \sigma(k), x(k), \beta_{x(k)}, \beta_{\mu(k)})$$
$$- \lambda\, K[\sigma(k) - \sigma_L], \quad (A.3.30)$$

where

$$F_1(\mu, \sigma, x, \beta_x, \beta_\mu) = \left(\frac{\beta_x - \beta_\mu}{\phi(\sigma)}\right)\left(\frac{x - \mu}{\phi(\sigma)}\right) \quad (A.3.31)$$

$$F_2(\mu, \sigma, x, \beta_x, \beta_\mu) = \left(\frac{\beta_x - \beta_\mu}{\phi(\sigma)}\right)\left[\left(\frac{x - \mu}{\phi(\sigma)}\right)^2 - 1\right] \quad (A.3.32)$$

Recall that $x(k)$ is the action selected at k and $\beta_{x(k)}$ and $\beta_{\mu(k)}$ are the reinforcements for the two actions $x(k)$ and $\mu(k)$ respectively. Let $f(x)$ denote the reward probability function. That is, $f(x) = E[\beta_{x(k)} \mid x(k) = x]$. Define

$$J(\mu, \sigma) = \int f(x)\, dN(\mu, \sigma)$$
$$= \int f(x)\, N(\mu, \sigma) dx \quad (A.3.33)$$

Approximating ODEs for Some Automata Algorithms 237

where $N(\mu, \sigma) = \frac{1}{\sigma\sqrt{2\pi}} e^{-\frac{(x-\mu)^2}{2\sigma^2}}$ is the normal density. (In general, $dN(a,b)$ denotes integration with respect to the normal distribution with mean a and variance b^2). We note that by a simple coordinate transformation, we can rewrite the above as

$$J(\mu, \sigma) = \int f(\sigma x + \mu)) \, dN(0, 1) \quad (A.3.34)$$

This would be used later on in the analysis.

It is easy to see that the above algorithm is in the general form of (A.2.10) by taking $X_n = [\mu_n, \sigma_n]$, $\xi_n = [x_n, \beta_{x_n}, \beta_{\mu_n}]$, $G(X, \xi) = [G_1(X, \xi), G_2(X, \xi)]$, $G_1 = F_1$, and $G_2 = F_2 - K[\sigma - \sigma_L]$. The g function (cf. eqn.(A.1.3)) would also have two components, say g_1 and g_2. Since g is a function of X, we can equivalently denote it as a function of two variables: μ, σ. We can calculate g_1, g_2 as follows.

$$\begin{aligned}
g_1(\mu, \sigma) &= E[F_1(\mu(k), \sigma(k), x(k), \beta_{x(k)}, \beta_{\mu(k)}) \mid \mu(k) = \mu, \sigma(k) = \sigma] \\
&= \int E\left[\frac{\beta_x - \beta_\mu}{\phi(\sigma(k))} \mid x(k) = x, \mu(k) = \mu, \sigma(k) = \sigma\right] \left(\frac{x - \mu}{\phi(\sigma)}\right) dN(\mu, \phi(\sigma)) \\
&= \int \left(\frac{f(x) - f(\mu)}{\phi(\sigma)}\right) \left(\frac{x - \mu}{\phi(\sigma)}\right) dN(\mu, \phi(\sigma)) \\
&= \int \frac{f(x)}{\phi(\sigma)} \left(\frac{x - \mu}{\phi(\sigma)}\right) dN(\mu, \phi(\sigma))
\end{aligned}$$

The last step above follows because $f(\mu)$ is independent of x and hence comes out of the integral and the rest of the integral in that term is zero. By differentiating the RHS of (A.3.33) with respect to μ, we now get

$$g_1(\mu, \sigma) = \frac{\partial J}{\partial \mu}(\mu, \phi(\sigma)) \quad (A.3.35)$$

In a similar fashion, we can show that

$$\begin{aligned}
g_2(\mu, \sigma) &= E[F_2(\mu(k), \sigma(k), x(k), \beta_{x(k)}, \beta_{\mu(k)}) - K(\sigma(k) - \sigma_L) \mid \mu(k) = \mu, \sigma(k) = \sigma] \\
&= \frac{\partial J}{\partial \sigma}(\mu, \phi(\sigma)) - K[\sigma - \sigma_L]. \quad (A.3.36)
\end{aligned}$$

Having calculated the g function, the approximating ODE would be $\dot{Z} = g(Z)$. Since the state here has two components, namely, μ and σ, using (A.3.35) and (A.3.36), the approximating ODE for the CALA algorithm is

$$\begin{aligned}
\frac{d\mu}{dt} &= \frac{\partial J}{\partial \mu}(\mu, \phi(\sigma)) \\
\frac{d\sigma}{dt} &= \frac{\partial J}{\partial \sigma}(\mu, \phi(\sigma)) - K[\sigma - \sigma_L]. \quad (A.3.37)
\end{aligned}$$

This completes the calculation of the approximating ODE for the CALA algorithm. To prove that this is the approximating ODE, all we need to do is to verify assumptions A1 to A4. This is what is done in the rest of this section.

From the structure of the algorithm it is immediately obvious that assumptions $A1$ and $A2$ are satisfied. To satisfy $A3$ and $A4$, we make the following additional assumptions on the unknown reward function, $f(\cdot)$.

B1. $f(\cdot)$ is bounded and is continuously differentiable. Let L denote the bound on f.

B2. The derivative of f, namely, $f'(\cdot)$, is globally Lipschitz, that is, $|f'(x) - f'(y)| \leq K_0 \|x - y\|$.

B3. Define zero mean random variables $\gamma_x = \beta_x - f(x)$. We assume that $\text{Var}(\gamma_x) \leq \sigma_M^2$ for some $\sigma_M < \infty$ and that $E[\gamma_x \gamma_y | x, y] = 0$ for $x \neq y$.

Under the assumptions B1–B3, we verify A3-A4. Thus, we prove that the ODE given by (A.3.37) would approximate the evolution of the CALA algorithm if the unknown reward function satisfies assumptions B1–B3. We mention, in passing, that the result regarding the approximating ODE for the CALA algorithm holds under much weaker assumptions (see [San94]) though such a proof would be involved and beyond the scope of this appendix.

By differentiating (A.3.34) under the integral sign, we get

$$\frac{\partial J}{\partial \mu}(\mu, \sigma) = \int f'(\sigma x + \mu) dN(0,1)$$

$$\frac{\partial J}{\partial \sigma}(\mu, \sigma) = \int f'(\sigma x + \mu) x \, dN(0,1) \quad \text{(A.3.38)}$$

Now we have

$$\begin{aligned}
|g_1(\mu_1, \sigma_1) - g_1(\mu_2, \sigma_2)| &= \left|\frac{\partial J}{\partial \mu}(\mu_1, \sigma_1) - \frac{\partial J}{\partial \mu}(\mu_2, \sigma_2)\right| \\
&\leq \int |f'(\sigma_1 x + \mu_1) - f'(\sigma_2 x + \mu_2)| \, dN(0,1) \\
&\leq K_0 \int |\sigma_1 x + \mu_1 - \sigma_2 x - \mu_2| \, dN(0,1) \quad \text{by B2,} \\
&\leq K_0 |\mu_1 - \mu_2| \int dN(0,1) + K_0 |\sigma_1 - \sigma_2| \int |x| dN(0,1) \\
&\leq K_1 \|(\mu_1, \sigma_1) - (\mu_2, \sigma_2)\|
\end{aligned}$$

where K_1 depends on K_0 and the first absolute moment of standard normal distribution. In a similar fashion we can show that $\left|\frac{\partial J}{\partial \sigma}(\mu_1, \sigma_1) - \frac{\partial J}{\partial \sigma}(\mu_2, \sigma_2)\right| \leq K_2' \|(\mu_1, \sigma_1) - (\mu_2, \sigma_2)\|$ and hence can show that

$$\begin{aligned}
|g_2(\mu_1, \sigma_1) - g_2(\mu_2, \sigma_2)| &\leq K_2' \|(\mu_1, \sigma_1) - (\mu_2, \sigma_2)\| + K|\sigma_1 - \sigma_2| \\
&\leq K_2 \|(\mu_1, \sigma_1) - (\mu_2, \sigma_2)\|
\end{aligned}$$

This proves that g is globally Lipschitz and thus verifies $A3$. We note here that if, instead of assumption $B2$, we had assumed that f' is Lipschitz on compact sets then the above proves that g is Lipschitz on compact sets.

Finally we turn to verifying $A4$. We need to show that $E[(G_i(X, \xi) - g_i(X))^2]$ is bounded for $i = 1, 2$. We show that $E[(G_1 - g_1)^2 | X]$ can be bounded by a constant independent of X and that is enough to show $E[(G_1 - g_1)^2]$ is bounded. Since $E[(G_1 - g_1)^2 | X] = E[(G_1)^2 | X] - (g_1)^2$, we first consider the G_1^2 term.

$$\begin{aligned}
E[(G_1(X, \xi))^2 | X] &= E\left[\left(\frac{\beta_x - \beta_\mu}{\phi(\sigma)} \frac{x - \mu}{\phi(\sigma)}\right)^2 \Big| \mu, \sigma\right] \\
&= E\left[E\left[\left(\frac{\beta_x - \beta_\mu}{\phi(\sigma)}\right)^2 \left(\frac{x - \mu}{\phi(\sigma)}\right)^2 \Big| x, \mu, \sigma\right] \Big| \mu, \sigma\right] \\
&= E\left[\left(\frac{x - \mu}{\phi(\sigma)}\right)^2 E\left[\left(\frac{\beta_x - \beta_\mu}{\phi(\sigma)}\right)^2 \Big| x, \mu, \sigma\right] \Big| \mu, \sigma\right]
\end{aligned}$$

$$\leq E\left[\left(\frac{x-\mu}{\phi(\sigma)}\right)^2 \frac{1}{\phi^2(\sigma)}[(f(x)-f(\mu))^2 + 2(\sigma_M)^2] \mid \mu, \sigma\right]$$

$$= \int \left(\frac{x-\mu}{\phi(\sigma)}\right)^2 \frac{1}{\phi^2(\sigma)}[(f(x)-f(\mu))^2 + 2(\sigma_M)^2] \, dN(\mu, \phi(\sigma))$$

$$\leq \frac{4L^2 + 2\sigma_M^2}{\phi^2(\sigma)} \int y^2 dN(0,1) \quad \text{by } B1$$

This shows that $E[G_1^2(X,\xi)|X] \leq K_4$ for some constant K_4 which is independent of X. Now we have, by the definition of $J(\cdot, \cdot)$,

$$\left|\frac{\partial J}{\partial \mu}(\mu, \sigma)\right| = \left|\frac{\partial}{\partial \mu} \int f(x) \, dN(\mu, \phi(\sigma))\right|$$

$$= \left|\int f(x) \frac{x-\mu}{\phi^2(\sigma)} \, dN(\mu, \phi(\sigma))\right|$$

$$\leq L \int |x| dN(0,1).$$

which shows g_1 is also bounded. Putting all these together we see that $E[(G_1 - g_1)^2] = E[\,E[(G_1 - g_1)^2|X]\,]$ is bounded. We can similarly show that $E[(G_2 - g_2)^2]$ is bounded. Thus assumption $A4$ is satisfied.

Since all assumptions needed for Theorem A.1 are satisfied, we can conclude that (A.3.37) is the approximating ODE for the CALA algorithm. This ODE and characterization of its asymptotic solutions are discussed in Section 1.6.1.

A.3.3 Automata Team Algorithms

Using the same method as in the previous two subsections, we can show that the assumptions needed for Theorem A.1 are satisfied by automata team algorithms also.

For an FALA team, we take the action probabilities of all automata as our state vector, X, and the set of actions selected and reinforcements obtained as the noise vector, ξ. Since the updating algorithms have to keep modifying finite dimensional probability vectors into probability vectors, the function G would be bounded. Further, all the algorithms are such that G is continuous in the components of action probability vectors. This makes g a continuous, bounded function with compact support. Hence we can do the same calculations as earlier to verify assumptions $A1$-$A4$.

The case of CALA teams is also similar. Here X consists of all μ's and σ's while ξ consists of all actions selected and all reinforcements. Once again using assumptions $B1$-$B3$ we can show that assumptions $A1$-$A4$ are satisfied. (In case of an automata team algorithm, the reward function f would be a function of multiple variables and hence, in assumption $B2$, by f' we mean the gradient).

Finally, putting all these together, we can show that this method of approximating algorithms using an ODE works for the hybrid automata team also. Since all these calculations are routine, we do not present the detailed algebra here.

A.4 Relaxing the Assumptions

As mentioned earlier, assumptions $A1$-$A2$ are easily satisfied by all algorithms. $A3$-$A4$ are satisfied by all FALA algorithms; but for CALA algorithms we need to impose some conditions on the unknown reward function as given by $B1$-$B3$.

Assumption $B3$ is not very restrictive. We need to control the variance in the noisy measurements, β_x of the function $f(x)$. Since we assumed f to be bounded (assumption $B1$), assuming this variance to be bounded is not very restrictive. For the CALA algorithm, assuming $f(\cdot)$ to be bounded is also not very restrictive. Normally, one is interested in local maxima only inside a (large) compact set. Thus we can always assume that outside a large enough compact set, the function f monotonically goes to zero. This can be easily achieved by a suitably defined penalty function for the reinforcement. Since we anyway need continuous differentiability of f, boundedness of f follows. Thus the only restrictive assumption is that f' is globally Lipschitz. This was needed to satisfy assumption $A3$. In this section we consider a simple modification to the proof of Theorem A.1 so that $A3$ can be replaced by the following two assumptions:

A31. Define $g : \Re^N \to \Re^N$ by

$$g(x) = E[G(X_k, \xi_k) \mid X_k = x].$$

We assume that $g(\cdot)$ is independent of k and that it is Lipschitz on compact sets. That is, given any compact set $\mathcal{B}$, there is a constant, K_B such that $|g(x) - g(y)| \leq K_B \|x - y\| \; \forall x, y \in \mathcal{B}$.

A32. Given any $\epsilon > 0$ and N_o, there exists a compact set $\mathcal{B}$ such that $\text{Prob}[G(X_k, \xi_k) \in \mathcal{B}] > 1 - \epsilon, \; \forall k \leq N_o$.

We will now replace assumption $B2$ in Section A.3.2 by

$B2'$. The derivative of f, namely, $f'(\cdot)$ is Lipschitz on compact sets. That is, given any compact set, $\mathcal{B}$, there is a constant, K_B such that $|f'(x) - f'(y)| \leq K_B \|x - y\| \; \forall x, y \in \mathcal{B}$.

First we note that for CALA algorithm, assumption $B2'$ implies assumption $A31$. (See the discussion in Section A.3.2 at the end of the proof that g is Lipschitz). It can be shown that assumption $A32$ is also satisfied by the CALA algorithm.

Now we consider how to modify the proof of Theorem A.1 to accommodate these new assumptions in place of old $A3$.

In the proof we started with equation (A.2.19) and wanted to obtain some bounds to satisfy (A.2.20). We obtained a bound, with probability greater than $1 - \epsilon$ for the first term in (A.2.19) through Lemma A.4. This entire chain of reasoning did not use the assumption $A3$ and hence it is still valid if we replaced $A3$ with $A31$-$A32$. We used the Lipschitz property of g only for bounding the second term in (A.2.19). We first note that we need this bound only for $k \leq \frac{T}{b}$. If we take N_o in $A32$ to be the largest integer smaller than or equal to $\frac{T}{b}$, then all X_k, ξ_k in (A.2.19) would be inside a compact set with a probability greater than $1 - \epsilon$ and on this compact set we can use the Lipschitz property of g to bound the second term of (A.2.19) as before. Thus except on an event of probability less than ϵ, we can get the bound on m_k^b as before and hence Theorem A.1 remains valid.

Thus we see that we can relax the assumption that g is globally Lipschitz to the one that it is Lipschitz on compact sets. Within the general proof structure used here we cannot relax this further. This is because, we only showed that the original algorithm closely approximates y_k^b given by (A.2.12). The ODE approximation results from the fact that (A.2.12) defines a close approximation to the ODE (A.1.6). This latter result (given by equation (A.1.7)) needs that g be Lipschitz on compact sets.

Appendix B
Proofs of Convergence for Pursuit Algorithm

B.1 Proof of Theorem 1.1

The proof of theorem 1.1 depends on two lemmas given in the sequel.

LEMMA B.1 *Given any $\delta \in (0,1]$ and a positive integer N, for each action α_i of the LA following the Pursuit algorithm, there exist $K_1(N,\delta)$ and $\lambda_1(N,\delta)$ such that*

$$\text{Prob}\{\alpha_i \text{ is chosen utmost } N \text{ times till instant } k\} < \delta \quad \forall k > K_1(N,\delta) \quad \text{(B.1.1)}$$

$\forall \lambda$ *such that* $0 < \lambda < \lambda_1(N,\delta)$. ∎

Proof: Let $\eta_i(k)$ denote the number of times α_i is chosen up to (and including) instant k. It is required to show

$$\text{Prob.}\{\eta_i(k) \leq N\} < \delta. \quad \text{(B.1.2)}$$

Since the events $\{\eta_i(k) = j_1\}$ and $\{\eta_i(k) = j_2\}$ are mutually exclusive for $j_1 \neq j_2$, (B.1.2) is equivalent to

$$\sum_{s=1}^{N} \text{Prob.}\{\eta_i(k) = s\} \leq \delta \quad \text{(B.1.3)}$$

which follows if, for all s, $1 \leq s \leq N$,

$$\text{Prob}\{\eta_i(k) = s\} < \frac{\delta}{N}. \quad \text{(B.1.4)}$$

Under the Pursuit algorithm, any action probability can decrease at most by $(1-\lambda)$ times in one iteration. Hence,

$$1 - p_i(k) = \text{Prob}\{\alpha(k) \neq \alpha_i\} \leq (1 - (1-\lambda)^k p_i(0)). \quad \text{(B.1.5)}$$

Using (B.1.5), the fact that $p_i(k) \leq 1$, and the binomial distribution,

$$\begin{aligned}\text{Prob}\{\eta_i(k) = s\} &\leq C_s^k(1)^s[1-(1-\lambda)^k p_i(0)]^{k-s} \\ &< k^s[1-(1-\lambda)^k p_i(o)]^{k-s}.\end{aligned} \quad \text{(B.1.6)}$$

It is now sufficient to prove

$$k^s[1-(1-\lambda)^k p_i(0)]^{k-s} < \frac{\delta}{N}, \quad 1 \leq s \leq N. \quad \text{(B.1.7)}$$

holds for all $k > K_1(N,\delta)$ and $\lambda < \lambda_1(N,\delta)$. Fix $k = K > N$ (say).
Let
$$f(K,\lambda) = K^s[1 - (1-\lambda)^K p_i(0)]^{K-s} \qquad (B.1.8)$$
It can be checked that $\frac{\partial f}{\partial \lambda} > 0, \forall K, \lambda$. That is, f is monotonically increasing with λ. Now suppose we choose,
$$\lambda = \lambda_1(K) = 1 - 2^{-1/K}. \qquad (B.1.9)$$
Then $(1-\lambda_1)^K = 1/2$ and hence
$$f(K,\lambda_1) = K^s \left(1 - \frac{p_i(0)}{2}\right)^{K-s} \qquad (B.1.10)$$

Furthermore, using L'Hopital's rule
$$\lim_{K \to \infty} f(K,\lambda_1) = 0. \qquad (B.1.11)$$

It follows that there exists $K_1 = K_1(N,\delta)$ such that
$$f(K_1,\lambda_1) < \frac{\delta}{N}. \qquad (B.1.12)$$

Now suppose we choose λ such that $0 < \lambda < \lambda_1(K_1)$ where $\lambda_1(K_1) = 1 - 2^{-1/K_1}$. Because $\frac{\partial f}{\partial \lambda} > 0$,
$$f(K_1,\lambda) < f(K_1,\lambda_1) < \frac{\delta}{N} \qquad (B.1.13)$$
for all λ such that $0 < \lambda < \lambda_1(K_1)$. Thus (B.1.2) is true for $k = K_1$. Now if $k_1 \geq k_2$, then the event $\{\eta_i(k_1) \leq N\}$ is a subset of the event $\{\eta_i(k_2) \leq N\}$. Hence,
$$\text{Prob}\{\eta_i(k) \leq N\} \leq \text{Prob}\{\eta_i(K_1) \leq N\} < \frac{\delta}{N} \qquad (B.1.14)$$
for all $k > K_1$. ∎

The next lemma considers the closeness of the estimates to the true reward probabilities.

LEMMA B.2 *Given any $\epsilon, \delta \in (0,1]$, for all $i, 1 \leq i \leq r$, there exist K_2 and λ_2 satisfying*
$$\text{Prob}\{|\hat{d}_i(k) - d_i| > \epsilon\} < \delta \qquad (B.1.15)$$
$\forall k > K_2(\epsilon,\delta), \forall \lambda$ *such that* $0 < \lambda < \lambda_2(\epsilon,\delta)$. ∎

Proof: Consider any specific i. By definition, $\hat{d}_i(k)$ is given by
$$\hat{d}_i(k) = \frac{\sum_{s=1}^k I\{\alpha(s) = \alpha_i\}\beta(s)}{\eta_i(k)}. \qquad (B.1.16)$$

Let m_s^i denote the time instant at which α_i is chosen for the sth time. Taking $1 \leq s \leq \eta_i(k)$ and hence
$$\hat{d}_i(k) = \frac{\sum_{s=1}^{\eta_i(k)} \beta(m_s^i)}{\eta_i(k)}. \qquad (B.1.17)$$

For fixed i, the sequence of random variables $\{\beta(m_s^i)\}$ is independent and identically distributed (*iid*) and each member of the sequence is bounded above, say, by M. Applying Hoeffding's inequality [Hoe63] for any N such that $\eta_i(k) = N$,
$$\text{Prob}\left\{\left|\frac{\sum_{s=1}^N \beta(m_s^i)}{N} - d_i\right| > \epsilon\right\} < 2\exp\left(-\frac{2N\epsilon^2}{M^2}\right). \qquad (B.1.18)$$

Proof of Theorem 1.1

Now, (B.1.17), (B.1.18) and Lemma B.1 essentially complete the proof. Formally, let us define events,

$$A = \{|\hat{d}_i(k) - d_i| > \epsilon\}$$
$$B = \{\eta_i(k) > N\}. \quad \text{(B.1.19)}$$

Since probability of any of the above events is less than one, we have

$$P(A) = P(A|B)P(B) + P(A|\bar{B})P(\bar{B})$$
$$< P(A|B) + P(\bar{B}) \quad \text{(B.1.20)}$$

Taking[1]

$$N = \left\lceil \frac{M^2}{2\epsilon^2} \ln \frac{4}{\delta} \right\rceil, \quad \text{(B.1.21)}$$

in (B.1.18), $P(A|B) < \frac{\delta}{2}$ and by lemma B.1,

$$P(\bar{B}) < \frac{\delta}{2} \quad \text{for } k > K_2(\epsilon, \delta) \quad \text{and } \lambda < 1 - 2^{-1/K_2} \quad \text{(B.1.22)}$$

where

$$K_2(\epsilon, \delta) = K_1\left(N, \frac{\delta}{2}\right). \quad \text{(B.1.23)}$$

Hence, from (B.1.20),

$$P(A) < \frac{\delta}{2} + \frac{\delta}{2} = \delta.$$

∎

REMARK B.1.1 *Some conservative estimates have been made for K_1. For instance,*

$$K_1(N, \delta) = \left\lceil \frac{2N}{\ln(1/\sigma)} \ln \left[\frac{N}{\ln(1/\sigma)} \cdot \frac{1}{\sigma} \cdot \left(\frac{N}{\delta}\right)^{1/N} \right] \right\rceil$$

where $\sigma = \frac{2r-1}{2r}$ and r is the number of actions. Consequently,

$$\lambda_1(N, \delta) = 1 - 2^{-\frac{1}{K_1(N,\delta)}}$$
$$K_2(\epsilon, \delta) = K_1\left(\left\lceil \frac{M^2}{2\epsilon^2} \ln \frac{4}{\delta} \right\rceil, \frac{\delta}{2}\right)$$
$$\lambda_2(\epsilon, \delta) = 1 - 2^{-\frac{1}{K_2(N,\delta)}}.$$

For details see [RS96]. ∎

Proof of Theorem 1.1: Now we proceed to prove Theorem 1.1 using lemmas B.1 and B.2. Recall that α_l is the optimal action. Define the events,

$$E_1(k) = \{p_l(k) > 1 - \epsilon\} \quad \text{(B.1.24)}$$
$$E_2(k) = \left\{\sup_{s \geq k} \max_i \left|\hat{d}_i(s) - d_i\right| \leq \frac{\theta}{2}\right\} \quad \text{(B.1.25)}$$

---

[1] $\lceil x \rceil$ denotes the smallest integer greater than or equal to x.

where,
$$\theta = d_l - \max_{i \neq l}\{d_i\} \tag{B.1.26}$$
is the difference between the highest two reward probabilities. The value of θ can be regarded as a measure of the difficulty of the learning problem.

Note that the theorem can be restated as
$$\text{Prob}\{E_1(k)\} \geq 1 - \delta \quad \forall k > K^*.$$

Following lemma B.2 and (B.1.18), set the ϵ of the lemma as $\frac{\theta}{2}$ and
$$K_2 = K_1\left(\left\lceil\frac{2M^2}{\theta^2}\ln\frac{4}{\delta}\right\rceil, \frac{\delta}{2}\right)$$
as in (B.1.21), (B.1.23). Set $\lambda_2 = 1 - 2^{-1/K_2}$ as in (B.1.22). Then it follows that
$$\text{Prob}(E_2(K_2)) \geq 1 - \delta. \tag{B.1.27}$$

Now
$$\begin{aligned}\text{Prob}\{P_l(k+K_2) > 1-\epsilon\} &= \text{Prob}\{E_1(k+K_2)\}\\ &\geq \text{Prob}\{E_1(k+K_2)|E_2(K_2)\}.\\ &\quad \text{Prob}\{E_2(K_2)\}. \end{aligned} \tag{B.1.28}$$

We will now find a K_0 such that
$$\text{Prob}\{E_1(k+K_2)|E_2(K_2)\} = 1 \quad \text{if} \quad k > K_0. \tag{B.1.29}$$

From the algorithm, since $p_l(0) = 1/r$,
$$p_l(K_2) \geq \frac{(1-\lambda)^{K_2}}{r} = a(K_2) \quad \text{(say)}.$$

Given $E_2(K_2)$, we have
$$\hat{d}_l(k) > \hat{d}_j(k), \quad \forall j \neq l, \forall k > K_2.$$

From the algorithm,
$$p_l(k+K_2) \geq a(K_2)(1-\lambda)^k + 1 - (1-\lambda)^k \quad \forall k > 0.$$

Hence
$$p_l(k+K_2) > 1 - (1-\lambda)^k.$$
Set $1 - (1-\lambda)^{K_0} = 1 - \epsilon$, i.e.,
$$(1-\lambda)^{K_0} = \epsilon \quad \text{or} \quad K_0 = \frac{\ln \epsilon^{-1}}{\ln(1-\lambda)^{-1}}. \tag{B.1.30}$$

It follows that $p_l(k+K_2) > 1 - \epsilon$ for all $k > K_0$. Thus (B.1.29) is true for K_0 given by (B.1.30). Hence from (B.1.27) and (B.1.28),
$$\text{Prob}\{p_l(k+K_2) > 1-\epsilon\} = \text{Prob}\{E_2(K_2)\} \geq 1 - \delta$$
or
$$\text{Prob}\{E_1(k)\} \geq 1 - \delta \quad \text{for all} \quad k > K_2 + K_0. \tag{B.1.31}$$

The proof is complete by setting $K^* = K_2 + K_0$.

B.2   Proof of Theorem 5.7

The proof of Theorem 5.7 for the parallel pursuit algorithm follows along the same lines as that of Theorem 1.1. The only difference is that the arguments are applied to each of the LA members in the module. The equivalents of Lemmas B.1 and B.2 follow with K_1 remaining the same and

$$K_2 = K_1\left(\left\lceil \frac{M^2}{2n\epsilon^2} \ln \frac{4}{\delta}\right\rceil, \frac{\delta}{2}\right) \qquad (B.2.32)$$

Appendix C
Weak Convergence and SDE Approximations

C.1 Introduction

A significant undercurrent in all the convergence results in this book is the concept of weak convergence. All convergence theorems stated here can be interpreted in terms of this concept. Understanding or proving our convergence results in this manner would have necessitated more sophisticated mathematical background from the reader. Hence, for all algorithms whose asymptotic behavior is approximated by an ODE, we presented our proofs in a considerably simple manner as detailed in Appendix A. Though the convergence to the solution of ODE has been handled this way, it could also be understood in terms of weak convergence. There are other situations, as in the case of global algorithms, where convergence to a stochastic differential equation (SDE) is involved and at the present time, weak convergence appears to be the only convenient approach. This appendix gives a glimpse of the ideas involved in this approach. The reader should note that a proper understanding of this topic needs a knowledge of measure-theoretic probability and this appendix provides only a rough idea of the approach and pointers to further study needed.

C.2 Weak Convergence

Weak convergence is a generalization of convergence in distribution on the real line. A formal definition is as follows. Let Γ be a metric space and let $\hat{\Gamma}$ denote the σ-algebra generated by the open sets of Γ. Let $\{P_i, i \in I\}$ where I is the set of positive integers and P be probability measures on $(\Gamma, \hat{\Gamma})$.

DEFINITION C.2.1

a. $\{P_i\}$ *is said to weakly converge to* P *if for each bounded and continuous real valued function* $f(\cdot)$ *on* Γ,

$$\int f(x) P_i(dx) \to \int f(x) P(dx) \quad \text{as} \quad i \to \infty. \tag{C.2.1}$$

b. *A sequence of* Γ-*valued random variables* $\{X_i\}$ *is said to weakly converge to* X, *if*

$$E(f(X_i)) \to E(f(X)) \quad \text{as} \quad i \to \infty \tag{C.2.2}$$

for every bounded and continuous real valued function $f(\cdot)$ on Γ.

∎

We denote the above convergence by $\{X_i\} \Rightarrow X$. An important property associated with weak convergence is that of tightness of a set of probability measures defined as follows.

DEFINITION C.2.2 *A set of probability measures $\{P_i, i \in I\}$ on $(\Gamma, \hat{\Gamma})$ is said to be* **tight** *if for every $\delta > 0$ there exists a compact set $K(\delta)$ such that*

$$P_i(K(\delta)) \geq 1 - \delta \quad \text{(C.2.3)}$$

for each $i \in I$.
∎

REMARK C.2.1 *Qualitatively, tightness implies that as much of the probability mass as desired (under **all** the measures in the set) is concentrated in a big enough compact set. In other words, no probability mass 'escapes to infinity'.*
∎

A common example of a set of probability measures which is not tight is the set of normal distributions $\{N(\mu_i, \sigma_i), i \in I\}$ where $\mu_i = i$, and $\sigma_i = \sigma$ (constant). The tightness condition is violated here because $\mu_i \to \infty$ with i.

The algorithms considered in this book can be broadly divided into two categories. The first, where the algorithm can be approximated by an ODE. The second, where the approximation is to an SDE, specifically in the case of global algorithms. We will consider the two cases separately.

One important feature that we have to bear in mind in relation to the algorithms we consider is that the weak convergence is w.r.t a real parameter which goes to zero. The dependence of a sequence $\{X_k\}$ on a parameter λ is represented as $\{X_k^\lambda\}$ and we study the weak convergence of $\{X_k\}$ as $\lambda \to 0$. For comparison with previous definitions one could visualize, for example, $\lambda = \frac{1}{i}$ where $i \in I$. We will consider convergence to SDE first.

C.3 Convergence to SDE

In showing the convergence of global type learning algorithms to stochastic differential equations, we take the weak convergence route, as simple methods of approximation used in Appendix A are not available here.

Consider algorithms of the form

$$X^\lambda(k+1) = X^\lambda(k) + \lambda G(X^\lambda(k), \xi^\lambda(k)) + \sqrt{\lambda}\zeta(\xi^\lambda(k)). \quad \text{(C.3.4)}$$

The following assumptions are made on the quantities involved.

CA.1 $\{X^\lambda(k), \xi^\lambda(k-1), k \geq 0\}$ is a Markov process with a possibly nonhomogeneous transition function

$$P^\lambda(x, \xi, k, A) = \text{Prob}\{(X^\lambda(k+1), \xi(k)) \in A | X^\lambda(k) = x, \xi^\lambda(k-1) = \xi\}$$

CA.2 $\xi^\lambda(\cdot)$ take values in a compact metric space S.

CA.3 For each $T < \infty$ and compact set Q

$$\sup_{\substack{\lambda, k \\ \lambda k \leq T}} E\left[\sup_{x \in Q} \|G(x, \xi^\lambda(k))\|^2\right] < \infty.$$

Convergence to SDE

CA.4 Define the one step transition function on the Borel sets of S by

$$P^\lambda(\xi, 1, \mathcal{B}|x) = \text{Prob}\{\xi^\lambda(k) \in \mathcal{B}|\xi^\lambda(k-1) = \xi, X^\lambda(k) = x\}$$

and suppose it does not depend on k. For each x, there is a transition function $P(\xi, 1, \cdot|x)$ on the Borel sets of S such that for each bounded and continuous $f(\cdot)$, $\int f(\xi')P(\xi, 1, d\xi'|x)$ is continuous in (x, ξ) and $\int f(\xi')P^\lambda(\xi, 1, d\xi'|x)$ converges to it uniformly as $\lambda \to 0$, on each compact (x, ξ) set.

CA.5 There is a unique invariant probability measure $P(\cdot|x)$ corresponding to the transition function $P(\xi, 1, \cdot|x)$ and for each compact Q, the set of measures $\{P(\cdot|x) : x \in Q\}$ is tight.

CA.6 $\int G(x, \xi')P(\xi, 1, d\xi'|x)$ is continuous in (x, ξ) and equals

$$\lim_{\lambda \to 0} \int G(x, \xi')P^\lambda(\xi, 1, d\xi'|x)$$

where the limit is uniform on each compact (x, ξ) set.

REMARK C.3.1 *The assumptions look formidable at first sight, but many of them are trivially satisfied in the context of the algorithms we have in mind. We shall see more details later.* ∎

We shall now invoke a theorem from [Kus84] for establishing weak convergence to a SDE. As a preparation for this, define

$$\begin{aligned}\mathbf{F}(\mathbf{x}) &= E[G(X(k), \xi(k))|X(k) = \mathbf{x}] \\ &= \int G(\mathbf{x}, \xi)P(d\xi|\mathbf{x})\end{aligned} \quad (C.3.5)$$

where $P(\cdot|\mathbf{x})$ is the invariant probability measure defined in CA.5. The following theorem then holds. We consider a continuous time interpolation of $X^\lambda(k)$ as,

$$\tilde{X}^\lambda(t) = X^\lambda(k) \text{ if } t \in (k\lambda, (k+1)\lambda].$$

THEOREM C.1 *[Kus84] Let assumptions CA.1 to CA.6 hold for algorithm (C.3.4). Let the sequence*[1]

$$\left\{\sqrt{\lambda} \sum_{i=0}^{[t/\lambda]} \zeta(\xi^\lambda(i)) = \tilde{\zeta}^\lambda(t)\right\}$$

converge weakly to a Wiener process $\mathcal{W}(\cdot)$ with covariance tA. Let $\tilde{X}^\lambda(0) \Rightarrow \mathbf{x}_0$ and suppose that the Itô equation given by (C.3.6) has a solution on $[0, \infty)$ for each initial condition $\mathbf{x}_0$, which is unique in the sense of distributions. Then $\tilde{X}^\lambda(\cdot) \Rightarrow \mathbf{z}(\cdot)$ where $\mathbf{z}(\cdot)$ satisfies the SDE

$$d\mathbf{z} = \mathbf{F}(\mathbf{z})dt + \sqrt{A}d\mathcal{W} \quad (C.3.6)$$

in which $\mathbf{F}(\mathbf{z})$ is defined in (C.3.5)

---

[1] $[x]$ is the greatest integer in the interval $[0, x]$

C.3.1 Application to Global Algorithms

Consider the application of the above theorem to a global learning algorithm, for example, (3.7.44) which refers to a PLA in a network. As given, the ith PLA is considered and the updating is done as follows

$$u_{ij}(k+1) = u_{ij}(k) + \lambda\beta(k)\frac{\partial \ln g_i(\mathbf{u}_i(k), \alpha_i(k))}{\partial u_{ij}}$$
$$+\lambda h'(u_{ij}(k)) + \sqrt{\lambda}\zeta_{ij}(k). \tag{C.3.7}$$

For other details, refer to Chapter 3. This algorithm can be put in the format of (C.3.4) with the following identifications. Suppressing λ, $X(k) = \mathbf{u}(k)$, the supervector whose ijth component is $u_{ij}(k)$.

$$\xi(k) = (\alpha(k), \beta(k))$$

where $\alpha(k)$ is the set of actions chosen by all the PLA at the instant k. $G(X(k), \xi(k))$ consists of 2nd and 3rd terms on the RHS of (C.3.7) with λ removed.

Now one can check the satisfaction of assumptions CA.1 to CA.6 needed for the validity of Theorem C.1.

For doing this, it is useful to recall the operation of the algorithm (C.3.7). Given $\mathbf{u}_i(k)$, the probability generating function g_i determines the probability of selection of $\alpha_i(k)$. Together with $\beta(k)$, these determine $\mathbf{u}_i(k+1)$ from (C.3.7) probabilistically because of the uncertainty of $\zeta_{ij}(k)$. Thus $(\mathbf{u}_i(k+1), \alpha_i(k), \beta(k))$ or equivalently $(\mathbf{u}_i(k), \alpha_i(k-1), \beta(k-1))$ is a Markov process as needed in CA.1.

We have a finite number of actions $\alpha(k)$ and $\beta(k) \in [0, 1]$. Thus CA.2 is satisfied.

In view of continuity of first partials of g_i w.r.t u_{ij} (as per assumption $A1$ in Section 3.7.3) and the continuity of $h'(\cdot)$, $G(\mathbf{x}, \xi^\lambda(k))$ is continuous. Hence it is bounded over every compact set. Thus CA.3 is satisfied.

One important property to be noted in this context (also stated in Appendix A) is that Prob $(\xi(k) \in \mathcal{B}|X(k), \xi(k-1))$ for any set $\mathcal{B}$, is independent of $\xi(k-1)$. This is because once the internal state $\mathbf{u}(k)$ is given, the probabilities of $\alpha(k)$ and $\beta(k)$ are fixed and are independent of $\alpha(k-1)$ and $\beta(k-1)$.

Now consider CA.4. From the above property of $\xi(k)$,

$$P^\lambda(\xi, 1, \mathcal{B}|\mathbf{x}) = \text{Prob}\{\xi^\lambda(k) \in \mathcal{B}|\ \xi^\lambda(k-1) = \xi, X^\lambda(k) = \mathbf{x}\}$$

does not depend on ξ and k. Also λ enters as a step-size parameter and hence P^λ is independent of λ. Thus $P(\xi, 1, d\xi'|\mathbf{x})$ is continuous w.r.t ξ and also w.r.t $\mathbf{x}$ because of the continuity of the probability generating function g_i w.r.t $\mathbf{u}_i$. The continuity of the integral $\int f(\xi')P(\xi, 1, d\xi'|\mathbf{x})$ follows. The later part of CA.4 follows trivially as $P^\lambda = P$ for all λ.

The invariant probability measure is $P(\cdot|\mathbf{x})$ itself as it is independent of ξ. Since S is compact, the condition on tightness in CA.5 is trivially satisfied.

As seen earlier, $P(\xi, 1, d\xi'|\mathbf{x})$ and $G(\mathbf{x}, \xi')$ are continuous and $P^\lambda(\cdot)$ is independent of λ. Again, CA.6 is trivially satisfied.

Now consider the part of Theorem C.1 concerning convergence to a Wiener process. In the algorithm, $\zeta(\xi^\lambda(k))$ are iid random variables with mean zero and variance σ^2. To show that $\bar{\zeta}^\lambda(t)$ converges to a Wiener process, we use Donsker's theorem [Bil68] given below.

Donsker's Theorem [Bil68, Theorem 16.2] Let $\{X_k\}$ be a sequence of random variables with mean zero and finite positive variance σ^2. Define

$$Y_k(t) = \left(\frac{1}{\sigma\sqrt{k}}\right) \sum_{i=0}^{[kt]} X_i.$$

Convergence to ODE

Then, $\{Y_k\}$ converges weakly to the standard Wiener process $\mathcal{W}(\cdot)$ as $k \to \infty$.

In our case, a small modification is needed. Y^λ is not defined over the natural numbers, but over positive real numbers by

$$Y^\lambda(t) = (\sqrt{\lambda}/\sigma) \sum_{i=0}^{[t/\lambda]} X_i.$$

The proof that Y^λ converges weakly to $\mathcal{W}$ as $\lambda \to 0$ can be obtained directly by replacing k by $1/\lambda$ in the proof of Donsker's theorem [Pha91].

Identifying X_i by $\zeta(\xi^\lambda(i))$, Donsker's theorem ensures that $\bar{\zeta}^\lambda(t)(= \sigma Y^\lambda)$ converges weakly to a Wiener process $\mathcal{W}(\cdot)$ with covariance $t\sigma^2 I$ where I is the unit matrix.

Now all the conditions for the validity of Theorem C.1 are satisfied and the given algorithm (C.3.7) can be approximated by the SDE (corresponding to (C.3.6))

$$d\mathbf{z} = \mathbf{F}(\mathbf{z})dt + \sigma d\mathcal{W}. \tag{C.3.8}$$

As shown in (3.7.62), this can also be written as,

$$d\mathbf{z} = \nabla V(\mathbf{z})dt + \sigma d\mathcal{W}; \quad \mathbf{z}(0) = \mathbf{u}(0) \tag{C.3.9}$$

where V is defined in (3.7.49).

REMARK C.3.2 *The analysis of other global algorithms such as those for GLA (cf. equation (3.8.75)) can be carried out in exactly similar manner. Only the dependence of the probability generating function on the context vector is to be taken into account.* ■

C.4 Convergence to ODE

Consider an algorithm of the form (C.3.4) without the random perturbation term

$$X^\lambda(k+1) = X^\lambda(k) + \lambda G(X^\lambda(k), \xi^\lambda(k)). \tag{C.4.10}$$

For this algorithm, the earlier assumptions CA.1 to CA.6 ensure weak convergence of the continuous–time interpolated process $\tilde{X}^\lambda(t)$ as $\lambda \to 0$, to the solution of an ordinary differential equation given below [Kus84].

$$\frac{d\mathbf{z}}{dt} = \mathbf{F}(\mathbf{z}); \quad \mathbf{z}(0) = X^\lambda(0) \tag{C.4.11}$$

where, as in (C.3.5),

$$\mathbf{F}(\mathbf{z}) = E[G(X^\lambda(k), \xi^\lambda(k))|X^\lambda(k) = \mathbf{z}]. \tag{C.4.12}$$

All the FALA algorithms and the local algorithms for GLA can be handled this way also. Appendix A handled such algorithms in a more transparent manner with almost no knowledge of measure theoretic probability theory. However, it needed somewhat more stringent assumptions.

In Appendix A, it was shown that (Theorem A.1) for any initial condition X_0 and for any given finite $T, \epsilon, \delta > 0$, there exists a $\lambda^* > 0$ such that for all $0 < \lambda < \lambda^*$,

$$\text{Prob}\{\sup_{0 \leq k \leq T/\lambda} \|X^\lambda(k) - \mathbf{z}(k\lambda)\| \geq \epsilon\} \leq \delta. \tag{C.4.13}$$

This inequality represents convergence in probability of the sequence $\{X^\lambda(k) - \mathbf{z}(k\lambda)\}$ to zero under the sup norm. It is well known that this mode of convergence implies weak convergence [Bil91]. There is thus a connection between the two approaches.

References

[AB65]   R. C. Atkinson and G. H. Bower. *An Introduction to Mathematical Learning Theory*. Wiley, New York, 1965.

[AHS85]   D. H. Ackley, G. E. Hinton, and T. J. Sejnowski. A learning algorithm for Boltzmann machines. *Cognitive Science*, 9:147–169, 1985.

[AK76]   H. Aso and M. Kimura. Absolute expediency of learning automata. *Information Sciences*, 17:91–122, 1976.

[AO02]   M. Agache and B. J. Oommen. Generalized pursuit learning schemes: New families of continuous and discretized learning automata. *IEEE Transactions on Systems, Man and Cybernetics: Part B*, 32:738–749, 2002.

[AP99]   A. Ansari and G. P. Papavassilopoulos. A generalized learning algorithm for an automaton operating in a multiteacher environment. *IEEE Transactions on Systems, Man and Cybernetics: Part B*, 29:164–178, 1999.

[APPZ85]   F. Aluffi-Pentini, V. Parisi, and F. Zirilli. Global optimization and stochastic differential equations. *Journal of Optimization Theory and Applications*, 47(1):1–26, September 1985.

[ARS02]   T. P. Imthias Ahamed, P. S. Nagendra Rao, and P. S. Sastry. A reinforcement learning approach to automatic generation control. *International Journal of Electric Power Systems Research*, 63:9–26, 2002.

[Arv96]   M. T. Arvind. *Stochastic Learning Algorithms with Improved Speed Performance*. PhD thesis, Dept. Electrical Engineering, Indian Institute of Science, Bangalore, India, 1996.

[AW98]   N. Abe and M. Warmuth. On the computational complexity of approximating distributions by probabilistic automata. *Machine Learning*, 9:205–260, 1998.

[BA85]   A. G. Barto and P. Anandan. Pattern-recognizing stochastic learning automata. *IEEE Transactions on Systems, Man and Cybernetics*, 15(3):360–374, 1985.

[BA91]   W. Bechtel and A. Abrahamsen. *Connectionism and the Mind*. Basil Blackwell, Cambridge, MA, 1991.

[BB99]	B. Bharath and V. S. Borkar. Stochastic approximation algorithms: Overview and recent trends. *Sadhana*, 24:425–452, 1999.
[BFOS84]	L. Breiman, J.H. Friedman, R.A. Olshen, and C.J. Stone. *Classification and Regression trees*. Belmont, Wadsworth, 1984.
[Bil68]	P. Billingsley. *Convergence of Probability Measures*. Wiley, New York, 1968.
[Bil91]	P. Billingsley. *Probability and Measure*. Wiley, New York, 1991.
[BL99]	E. Billard and S. Lakshmivarahan. Learning in multilevel games with incomplete information: Part I. *IEEE Transactions on Systems, Man and Cybernetics*, 29:329–339, 1999.
[BM55]	R. R. Bush and F. Mosteller. *Stochastic Models for Learning*. Academic Press, New York, 1955.
[BM02]	N. Baba and Y. Mogami. A new learning algorithm for the hierarchical structure learning automata operating in nonstationary S-model environment. *IEEE Transactions on Systems, Man and Cybernetics: Part B*, 32:750–758, 2002.
[BMP87]	A. Benveniste, M. Metivier, and P. Priouret. *Adaptive Algorithms and Stochastic Approximations*. Springer Verlag, New York, 1987.
[BO82]	T. Basar and G. J. Olsder. *Dynamic Noncooperative Game Theory*. Academic Press, New York, 1982.
[BS70]	N.P. Bhatia and G.P. Szego. *Stability Theory of Dynamical Systems*. Springer Verlag, Berlin, 1970.
[BT96]	D. P. Bertsekas and J. N. Tsitsikilis. *Neuro Dynamic programming*. Athena Scientific, Belmont, MA, 1996.
[CS68]	B. Chandrasekaran and D. W. C. Shen. On expediency and convergence in variable-structure automata. *IEEE Trans. Systs. Sci. Cybern.*, 4:52–60, 1968.
[CS69]	B. Chandrasekharan and D. W. C. Shen. Stochastic automata games. *IEEE Transactions on Systems Science and Cybernetics*, 5:145–149, 1969.
[DH73]	R. O. Duda and P. E. Hart. *Pattern Classification and Scene Analysis*. Wiley, New York, 1973.
[DMC96]	M. Dorigo, V. Maniezzo, and A. Colorni. The ant system: Optimization by a colony of cooperating agents. *IEEE Transactions on Systems, Man and Cybernetics: Part B*, 26:29–41, 1996.
[EK02]	A. A. Economides and A. Kehagias. The star automaton: Expediency and optimality properties. *IEEE Transactions on Systems, Man and Cybernetics: Part B*, 32:723–737, 2002.
[FGHW96]	G. P. Frost, T. J. Gordon, M. N. Howell, and Q. H. Wu. Reinforcement learning of active and semi-active vehicle suspension control laws. *Proc. Inst. Mech. Eng. A. Power Process Eng.*, 210:249–257, 1996.

REFERENCES

[FI92] U. M. Fayyad and K. B. Irani. On the handling of continuous valued attributes in decision tree generation. *Machine Learning*, 8:87–102, 1992.

[FM66] K. S. Fu and G. J. McMurtry. A study of stochastic automata as a model for learning and adaptive controllers. *IEEE Transactions on Automatic Control*, 11:379–387, 1966.

[For89] R. Forsyth, editor. *Machine Learning*. Chapman and Hall, London, 1989.

[Fu70] K. S. Fu. Learning control systems – review and outlook. *IEEE Transactions on Automatic Control*, 15:210–221, 1970.

[GQ90] R. Ge and Y. Qin. The globally convexized filled functions for global optimization. *Applied Mathematics and Computation*, 35:131–158, 1990.

[Gra88] S. R. Graubard, editor. *The Artificial Intelligence Debate*. MIT Press, Cambridge, MA, 1988.

[Gri91] W. E. L. Grimson. The combinatorics of heuristic search termination for object recognition in cluttered environments. *IEEE Transactions on Pattern Analysis and Machine Intelligence*, 13:920–935, 1991.

[Hal50] P. Halmos. *Measure Theory*. Van Nostrand Reinhold, New York, 1950.

[HB99] M. N. Howell and M. C. Best. On-line PID tuning for engine idle-speed control using continuous action reinforcement learning automata. *J. Control Eng. Prac.*, 8:147–154, 1999.

[HFGW97] M. N. Howell, G. P. Frost, T. J. Gordon, and Q. H. Wu. Continuous action reinforcement learning applied to vehicle suspension control. *Mechatronics*, 7:263–276, 1997.

[HGB02] M. N. Howell, T. J. Gordon, and F. V. Brandao. Genetic learning automata for function optimization. *IEEE Transactions on Systems, Man and Cybernetics: Part B*, 32:804–815, 2002.

[Hoe63] W. Hoeffding. Probability inequalities for sums of bounded random variables. *J. American Statistical Association*, 58:13–30, 1963.

[HTF01] T. Hastie, R. Tibshirani, and J. Friedman. *The Elements of Statistical Learning: Datamining, Inference and Prediction*. Springer, New York, 2001.

[Jav94] S. Javale. Generalized learning automata algorithms for pattern recognition. ME Project Report, Dept. EE, Indian Institute of Science, Bangalore, 1994.

[KGV83] S. Kirkpatrick, C.D. Gelatt, and M. Vecchi. Optimisation by simulated annealing. *Science*, 220:621–680, 1983.

[KT63] V. Y. Krylov and M. L. Tsetlin. Games between automata. *Automation and Remote Control*, 24:889–899, 1963.

[KT75] S. Karlin and H. M. Taylor. *A First Course in Stochastic Processes*. Academic Press, New York, 1975.

[Kus84] H. J. Kushner. *Approximation and Weak Convergence Methods for Random Processes*. MIT Press, Cambridge, 1984.

[KY97] H. J. Kushner and G. G. Yin. *Stochastic Approximation Algorithms and Applications*. Springer-Verlag, New York, 1997.

[Lak81] S. Lakshmivarahan. *Learning Algorithms: Theory and Applications*. Springer-Verlag, New York, 1981.

[Lew95] F. L. Lewis. *Optimal Control*. Wiley, New York, 1995.

[Lip87] R. P. Lippmann. An introduction to computing with neural nets. *IEEE ASSP Magazine*, pages 4–22, April 1987.

[LN81] S. Lakshmivarahan and K.S. Narendra. Learning algorithms for two-person zero-sum stochastic games with incomplete information. *Mathematics of Operations Research*, 6:379–386, November 1981.

[LN82] S. Lakshmivarahan and K. S. Narendra. Learning algorithms for two-person zero-sum games with incomplete information: A unified approach. *SIAM J. Control and Optimization*, 20:541–552, 1982.

[LO92] J.K. Lanctot and B.J. Oommen. Discretized estimator learning automata. *IEEE Transactions on Systems, Man and Cybernetics*, 22(6):1473–1483, 1992.

[LT73] S. Lakshmivarahan and M. A. L. Thathachar. Absolutely expedient learning algorithms for stochastic automata. *IEEE Transactions on Systems, Man and Cybernetics*, 3:281–286, 1973.

[LT76] S. Lakshmivarahan and M. A. L. Thathachar. Bounds on the probability of convergence of learning automata. *IEEE Transactions on Systems, Man and Cybernetics*, 6:756–763, 1976.

[MA94] P. M. Murphy and D. W. Aha. UCI repository of machine learning databases. Machine-readable data repository, Department of Information and Computer Science, University of California, Irvine, CA, 1994.

[Mag98] M. Magesh. A study of stochastic optimization algorithms for system identification. ME Project Report, Dept. EE, Indian Institute of Science, Bangalore, 1998.

[MCN97] P. Mars, J. R. Chen, and R. Nambiar. *Learning Algorithms: Theory and Applications in Signal Processing, Control and Communications*. CRC Press, London, 1997.

[Mit97] T. M. Mitchell. *Machine Learning*. McGraw-Hill, New York, 1997.

[MKS94] S. K. Murthy, S. Kasif, and S. Salzberg. A system for induction of oblique decision trees. *J. Artificial Intelligence Research*, 2:1–32, 1994.

[MT89] S. Mukhopadhyay and M. A. L. Thathachar. Associative learning of Boolean functions. *IEEE Transactions on Systems, Man and Cybernetics*, 19:1008–1015, 1989.

REFERENCES

[NA89] K. S. Narendra and A. M. Annaswamy. *Stable Adaptive Systems*. Prentice-Hall, Englewood Cliffs, NJ, 1989.

[Nag97] G. D. Nagendra. PAC learning with noisy samples. ME Project Report, Dept. EE, Indian Institute of Science, Bangalore, 1997.

[Nil65] N. J. Nilsson. *Learning Machines*. McGraw-Hill, New York, 1965.

[NL77] K. S. Narendra and S. Lakshmivarahan. Learning automata: A critique. *J. Cybernetics and Information Science*, 1:53–71, 1977.

[NM83] K. S. Narendra and P. Mars. The use of learning algorithms in telephone traffic routing: A methodology. *Automatica*, 19:495–502, 1983.

[NN87] O. V. Nedzelnitsky and K. S. Narendra. Nonstationary models of learning automata routing in data communication networks. *IEEE Transactions on Systems, Man and Cybernetics*, 17:1004–1015, 1987.

[Nor68] M. F. Norman. On linear models with two absorbing barriers. *J. Math. Psychol.*, 5:61–101, 1968.

[Nor72] M. F. Norman. *Markov Processes and Learning Models*. Academic, New York, 1972.

[NP94] K. Nazim and A. S. Poznyak. *Learning Automata: Theory and Applications*. Pergamon, New York, 1994.

[NT74] K. S. Narendra and M. A. L. Thathachar. Learning automata: A survey. *IEEE Transactions on Systems, Man and Cybernetics*, 14:323–334, 1974.

[NT80] K. S. Narendra and M. A. L. Thathachar. On the behaviour of learning automata in a changing environment with applications to telephone traffic routing. *IEEE Transactions on Systems, Man and Cybernetics*, 10:262–269, 1980.

[NT89] K. S. Narendra and M. A. L. Thathachar. *Learning Automata: An Introduction*. Prentice Hall, Englewood Cliffs, 1989.

[NW83] K. S. Narendra and R. M. Wheeler. An N-player sequential stochastic game with identical payoffs. *IEEE Transactions on Systems, Man and Cybernetics*, 13:1154–1158, 1983.

[OA01] B. J. Oommen and M. Agache. Continuous and discretized pursuit learning schemes: Various algorithms and their comparison. *IEEE Transactions on Systems, Man and Cybernetics: Part B*, 31:277–287, 2001.

[OC88] B. J. Oommen and J. R. P. Christensen. Epsilon-optimal discretized reward-penalty learning automata. *IEEE Transactions on Systems, Man and Cybernetics*, 18:451–458, 1988.

[OC95] B. J. Oommen and T. De St. Croix. Graph partitioning using learning automata. *IEEE Transactions on Computers*, 45:195–208, 1995.

[OC97] B. J. Oommen and T. De St. Croix. String taxonomy using learning automata. *IEEE Transactions on Systems, Man and Cybernetics: Part B*, 27:354–365, 1997.

[OL90]   B.J. Oommen and J.K. Lanctot. Discretized pursuit learning automata. *IEEE Transactions on Systems, Man and Cybernetics*, 20:931–938, 1990.

[OM88]   B. J. Oommen and D. C. Y. Ma. Deterministic learning automata solutions to the equipartitioning problem. *IEEE Trans. Computers*, 37:2–14, 1988.

[OM92]   B. J. Oommen and D. C. Y. Ma. Stochastic automata solutions to the object partitioning problem. *Computer Journal*, 35:A105–A120, 1992.

[Oom86]   B. J. Oommen. Absorbing and ergodic discretized two–action learning automata. *IEEE Transactions on Systems, Man and Cybernetics*, 16:282–296, 1986.

[OPPL02]   M. S. Obaidat, G. I. Papadimitriou, A. S. Pompartsis, and H. S. Laskaridis. Learning automata based bus arbitration schemes for for shared-medium ATM switches. *IEEE Transactions on Systems, Man and Cybernetics: Part B*, 32:815–821, 2002.

[OR02]   B. J. Oommen and T. D. Roberts. Discretized learning automata solutions to the capacity assignment problem in prioritized networks. *IEEE Transactions on Systems, Man and Cybernetics: Part B*, 32:821–831, 2002.

[Pap94a]   G. I. Papadimitriou. Hierarchical discretized pursuit nonliner learning automata with rapid convergence and high accuracy. *IEEE Trans. Knowl. Data Eng.*, 6:654–659, 1994.

[Pap94b]   G. I. Papadimitriou. A new approach to the design of reinforcement schemes for learning automata: Stochastic estimator learning algorithms. *IEEE Trans. Knowl. Data Eng.*, 6:649–654, 1994.

[Pha91]   V. V. Phansalkar. *Learning Automata Algorithms for Connectionist systems - local and global convergence*. PhD thesis, Dept. of Electrical Engineering, Indian Institute of Science, 1991.

[Pha94]   V. V. Phansalkar. ODE analysis of learning automata algorithms. Technical report, Department of Electrical Engineering, Indian Institute of Science, Bangalore, India, 1994.

[PM95]   G. I. Papadimitriou and D. G. Maritsas. Self-adaptive random access protocols for WDM passive star networks. *Proceedings of IEE: Computer and Digital Techniques*, 142:306–312, 1995.

[PM96]   G. I. Papadimitriou and D. G. Maritsas. Learning automata based receiver conflict avoidance algorithms. *IEEE/ACM Transactions on Networking*, 4:407–412, 1996.

[PN97]   A. S. Poznyak and K. Nazim. *Learning Automata and Stochastic Optimization*. Springer-Verlag, New York, 1997.

[PN02]   A. S. Poznyak and K. Najim. Learning through reinforcement for N–person repeated constrained games. *IEEE Transactions on Systems, Man and Cybernetics: Part B*, 32:759–771, 2002.

REFERENCES

[PNI96] A. S. Poznyak, K. Najim, and E. Ikone. Adaptive selection of optimal order of linear regression models using learning automata. *International Journal of Systems Science*, 27:151–159, 1996.

[POP02] G. I. Papadimitriou, M. S. Obaidat, and A. S. Pomportsis. On the use of learning automata in the control of broadcast networks: A methodology. *IEEE Transactions on Systems, Man and Cybernetics: Part B*, 32:781–790, 2002.

[PP99] G. I. Papadimitriou and A. S. Pomportsis. Self-adaptive TDMA protocols for WDM star networks: A learning automata based approach. *IEEE Photonics Technology Letters*, 11:1322–1324, 1999.

[PP00] G. I. Papadimitriou and A. S. Pomportsis. Learning-automata-based TDMA protocols for broadcast communication systems. *IEEE Communication Letters*, 4:107–109, 2000.

[Pra03] Ambika Prasad. Learning automata based decision tree induction. ME Project Report, Dept. EE, Indian Institute of Science, Bangalore, 2003.

[PSP03] G. I. Papadimitriou, M. Sklira, and A. S. Pomportsis. A new class of ϵ-optimal learning automata. *IEEE Transactions on Systems, Man and Cybernetics: Part B*, 33, 2003. to appear.

[PST94] V. V. Phansalkar, P. S. Sastry, and M. A. L. Thathachar. Absolutely expedient algorithms for learning Nash equilibria. *Proc. Indian Academy of Sciences: Mathematical Sciences*, 104:279–294, 1994.

[PT95] V. V. Phansalkar and M. A. L. Thathachar. Local and global optimization algorithms for generalized learning automata. *Neural Computation*, 7:950–973, 1995.

[PT96] V. V. Phansalkar and M. A. L. Thathachar. Learning automata in feedforward connectionist systems. *Int. J. Systems Science*, 27(2):145–150, 1996.

[Qui83] J.R. Quinlan. Learning efficient classification procedures and their applications to chess end games. In R.S.Michalski, J.G.Carbonell, and T.M.Mitchell, editors, *Machine Learning: An Artificial Intelligence Approach*, volume 1. Tiogo, Palo Alto, California, 1983.

[Qui86] J.R. Quinlan. Effect of noise on concept learning. In R.S.Michalski, J.G.Carbonell, and T.M.Mitchell, editors, *Machine Learning: An Artificial Intelligence Approach*, volume 2. Morgan Kaufmann, Los Altos, California, 1986.

[Raj91] K. Rajaraman. *Learning Automata Models for Concept Learning*. Bangalore, India, ME Thesis, Dept. of Electrical Engineering, Indian Institute of Science, 1991.

[Raj96] K. Rajaraman. *Robust Distribution Free Learning of Logic Expressions*. PhD thesis, Dept. of Electrical Engineering, Indian Institute of Science, Bangalore, India, 1996.

[Rao84] T. Vishwanatha Rao. Learning solutions to stochastic non-cooperative games. ME Project Report, Dept. EE, Indian Institute of Science, Bangalore, 1984.

[RS96] K. Rajaraman and P. S. Sastry. Finite time analysis of the pursuit algorithm for learning automata. *IEEE Transactions on Systems, Man and Cybernetics*, 26:590–598, 1996.

[RS97] K. Rajaraman and P. S. Sastry. A parallel stochastic algorithm for learning logic expressions under noise. *Jl. Indian Institute of Science*, 37:21–55, 1997.

[RS99] K. Rajaraman and P. S. Sastry. Stochastic optimization over continuous and discrete variables with applications to concept learning under noise. *IEEE Transactions on Systems, Man and Cybernetics: Part A*, 29:542–553, 1999.

[San94] G. Santharam. *Distributed Learning with Connectionist Models for Optimization and Control*. PhD thesis, Dept. of Electrical Engineering, Indian Institute of Science, Bangalore, India, 1994.

[SB98] R. S. Sutton and A. G. Barto. *Reinforcement Learning*. MIT Press, Cambridge, MA, 1998.

[SC98] M. K. Sundareshan and T. A. Condarcure. Recurrent neural–network training by a learning automaton approach for trajectory learning and control system design. *IEEE Transactions on Neural Networks*, 9:354–368, 1998.

[SC00] S. Sarkar and S. Chavali. Modelling parameter space behaviour of vision systems using Bayesian networks. *Computer Vision and Image Understanding*, 79:185–223, 2000.

[SK89] R. Simha and J. F. Kurose. Reletive reward strength algorithms for learning automata. *IEEE Transactions on Systems, Man and Cybernetics*, 19:388–398, 1989.

[SN69] I. J. Shapiro and K. S. Narendra. Use of stochastic automata for parameter self-optimization with multi-modal performance criteria. *IEEE Transactions on Systems Science and Cybernetics*, 5:352–360, 1969.

[Spa92] J. C. Spall. Multivariate stochastic approximation using a simultaneous perturbation gradient approximation. *IEEE Transactions on Automatic Control*, 37:332–341, 1992.

[Spa98] J. C. Spall. An overview of the simultaneous perturbation method for efficient optimization. *John Hopkins APL Technical Digest*, 19:482–492, 1998.

[SPT94] P. S. Sastry, V. V. Phansalkar, and M. A. L. Thathachar. Decentralised learning of Nash equilibria in multi-person stochastic games with incomplete information. *IEEE Transactions on Systems, Man and Cybernetics*, 24:769–777, May 1994.

[SRR93] P. S. Sastry, K. Rajaraman, and S. R. Ranjan. Learning optimal conjunctive concepts through a team of stochastic automata. *IEEE Transactions on Systems, Man and Cybernetics*, 23:1175–1184, 1993.

[SS99] S. Shah and P. S. Sastry. New algorithms for learning and pruning oblique decision trees. *IEEE Transactions on Systems, Man and Cybernetics: Part C*, 29:494–505, 1999.

REFERENCES

[SS00]   S. Sarkar and P. Soundararajan. Supervised learning of large perceptual organization: Graph spectral partitioning and learning automata. *IEEE Transactions on Pattern Analysis and Machine Intelligence*, 22:504–525, 2000.

[SST94] G. Santharam, P. S. Sastry, and M. A. L. Thathachar. Continuous action set learning automata for stochastic optimization. *Journal of the Franklin Institute*, 331B(5):607–628, 1994.

[SSU94] P. S. Sastry, G. Santharam, and K.P. Unnikrishnan. Memory neuron networks for identification and control of dynamical systems. *IEEE Transactions on Neural Networks*, 5:306–319, March 1994.

[ST94]   P. S. Sastry and M. A. L. Thathachar. Analysis of stochastic automata algorithm for relaxation labelling. *IEEE Transactions on Pattern Analysis and Machine Intelligence*, 16:538–543, 1994.

[ST99]   P. S. Sastry and M. A. L. Thathachar. Learning automata algorithms for pattern classification. *Sadhana*, 24:261–292, 1999.

[Sut88] R. S. Sutton. Learning to predict by the method of temporal differences. *Machine Learning*, 2:39–44, 1988.

[SW81]   J. Sklansky and G. N. Wassel. *Pattern Classification and Trainable Machines*. Springer-Verlag, New York, 1981.

[TA97]   M. A. L. Thathachar and M. T. Arvind. Solution of Goore game using models of stochastic learning automata. *J. Indian Institute of Science*, 76:47–61, Jan–Feb 1997.

[TA98]   M. A. L. Thathachar and M. T. Arvind. Parallel algorithms for modules of learning automata. *IEEE Transactions on Systems, Man and Cybernetics: Part B*, 28:24–33, 1998.

[TA00]   M. A. L. Thathachar and M. T. Arvind. Parallel learning automata algorithms for obtaining conditional distributions from input–output samples. In *Proc. Int. Conf. communications, control and signal processing (CCSP2000)*, Bangalore, India, July 2000.

[TM93]   C. K. K. Tang and P. Mars. Games of stochastic learning automata and adaptive signal processing. *IEEE Transactions on Systems, Man and Cybernetics*, 23:851–856, 1993.

[TO79]   M. A. L. Thathachar and B. J. Oommen. Discretized reward-inaction learning automata. *Journal of Cybernetics and Information Sciences*, pages 24–29, 1979.

[TP72]   Ya. Z. Tsypkin and A. S. Poznyak. Finite learning automata. *Engineering Cybernetics*, 10:478–490, 1972.

[TP95a] M. A. L. Thathachar and V. V. Phansalkar. Convergence of teams and hierarchies of learning automata in connectionist systems. *IEEE Transactions on Systems, Man and Cybernetics*, 25(11):1459–1469, 1995.

[TP95b] M. A. L. Thathachar and V. V. Phansalkar. Learning the global maximum with parameterized learning automata. *IEEE Transactions on Neural Networks*, 6:398–406, 1995.

[TR82] M. A. L. Thathachar and K. R. Ramakrishnan. On-line optimization with a team of learning automata. In *Proc. IFAC Symposium of Theory and Application of Digital Control*, pages 8–13. New Delhi, India, 1982.

[TR84] M. A. L. Thathachar and K. R. Ramakrishnan. A cooperative game of a pair of automata. *Automatica*, 20:797–801, 1984.

[TS85] M. A. L. Thathachar and P. S. Sastry. A new approach to the design of reinforcement schemes for learning automata. *IEEE Transactions on Systems, Man and Cybernetics*, 15:168–175, January 1985.

[TS86a] M. A. L. Thathachar and P. S. Sastry. Estimator algorithms for learning automata. In *Proc. Platinum Jubilee Conf. Syst. Signal Processing*. Dept. Electrical Engineering, IISc, Bangalore, 1986.

[TS86b] M. A. L. Thathachar and P. S. Sastry. Relaxation labeling with learning automata. *IEEE Transactions on Pattern Analysis and Machine Intelligence*, 8(2):256–267, March 1986.

[TS87] M. A. L. Thathachar and P. S. Sastry. Learning optimal discriminant functions through a cooperative game of automata. *IEEE Transactions on Systems, Man and Cybernetics*, 17(1):73–85, 1987.

[TS91] M. A. L. Thathachar and P. S. Sastry. Learning automata in stochastic games with incomplete information. In *Systems and Signal Processing*, pages 137–140, New Delhi, 1991. Oxford and IBH.

[TS01] M. A. L. Thathachar and P. S. Sastry. Adaptive stochastic algorithms for pattern classification. In S. K. Pal and A. Pal, editors, *Pattern Recognition: From Classical to Modern Approaches*, pages 67–114, Singapore, 2001. World Scientific.

[TS02] M. A. L. Thathachar and P. S. Sastry. Varieties of learning automata: An overview. *IEEE Transactions on Systems, Man and Cybernetics: Part B*, 32:711–722, 2002.

[Tse62] M. L. Tsetlin. On the behaviour of finite automata in random media. *Automation and Remote Control*, 22:1210–1219, 1962.

[Tse73] M. L. Tsetlin. *Automata Theory and Modeling of Biological Systems*. Academic, New York, 1973.

[Tsy71] Ya. Z. Tsypkin. *Adaptation and learning in automatic systems*. Academic Press, New York, 1971.

[Tsy73] Ya. Z. Tsypkin. *Foundations of the theory of learning systems*. Academic Press, New York, 1973.

[UKB99] C. Unsal, P. Kachroo, and J. S. Bay. Multiple stochastic learning automata for vehicle path control in an automated highway system. *IEEE Transactions on Systems, Man and Cybernetics: Part A*, 29:120–128, 1999.

[Vap97] V. Vapnik. *Nature of Statistical Learning Theory*. Springer-Verlag, New York, 1997.

REFERENCES

[Vid78] M. Vidyasagar. *Nonlinear systems analysis*. Prentice-Hall, Englewood Cliffs, New Jersey, 1978.

[Vis72] R. Viswanathan. *Learning Automata: Models and Applications*. PhD thesis, Yale University, New Haven, CT, USA, 1972.

[Vis95] P. Viswanath. Stability and rule generation in fuzzy systems. ME Project Report, Dept. CSA, Indian Institute of Science, Bangalore, 1995.

[VN74] R. Viswanathan and K. S. Narendra. Games of stochastic automata. *IEEE Transactions on Systems, Man and Cybernetics*, 4:131–135, 1974.

[VN02] K. Verbeeck and A. Nowe. Colonies of learning automata. *IEEE Transactions on Systems, Man and Cybernetics: Part B*, 32:772–780, 2002.

[VV63] V. I. Varshavskii and I. P. Vorontsova. On the behaviour of stochastic automata with a variable structure. *Automation and Remote Control*, 24:327–333, 1963.

[Wal75] S. Walker. *Learning and Reinforcement*. Methuen, London, 1975.

[WF00] I. H. Witten and E. Frank. *Data Mining: Practical Machine Learning Tools and Techniques with JAVA Implementations*. Morgan Kaufmann, New York, 2000.

[WH85] E. Wong and B. Hajek. *Stochastic Processes in Engineering Systems*. Springer-Verlag, New York, 1985.

[Wil88] R.J. Williams. Toward a theory of reinforcement learning connectionist systems. Nu-ccs-88-3, Northeastern Univ., Boston, MA, July 1988.

[Wil92] R. J. Williams. Simple statistical gradient-following algorithms for connectionist reinforcement learning. *Machine Learning*, 8:229–256, 1992.

[WN86] R. M. Wheeler, Jr. and K. S. Narendra. Decentralized learning in finite Markov chains. *IEEE Transactions on Automatic Control*, 31(6):519–526, 1986.

[Wu95] Q. H. Wu. Learning coordinated control of power systems using interconnected learning automata. *International Journal of Electric Power and Energy Systems*, 17:91–99, 1995.

[ZBL99] J. Zhou, E. Billard, and S. Lakshmivarahan. Learning in multilevel games with incomplete information: Part II. *IEEE Transactions on Systems, Man and Cybernetics*, 29:340–349, 1999.

[ZZV00] X. Zeng, J. Zhou, and C. Vasseur. A strategy for controlling nonlinear systems using a learning automaton. *Automatica*, 36:1517–1524, 2000.

Index

ϵ-optimality, 14

absolute expediency, 16, 128
action, 3, 10
action probability distribution, 10
adaptive decision making, 8
adaptive signal processing, 84, 219
approximating ODE, 227, 229, 231
    L_{R-I}, 236
    CALA, 237
    GLA, 38
    LA network, 119
    PLA, 45
    stochastic game, 65
artificial neural networks, 2, 224
    weights, 2
associative reinforcement learning, 33, 113, 224
asymptotic properties, 227
asymptotic solutions, 62
asymptotically stable, 62, 125
automatic generation control, 220
average reinforcement, 27
average reward, 14

backpropagation, 7, 84
bang-bang control, 106
Bayes classifier, 143
Bayesian network, 207
broadcast networks, 215
Brownian motion, 132
bursty traffic, 215

circuit switched networks, 214
class conditional densities, 142, 143
class label, 164
classical conditioning, 2
classification noise, 148
CLearn algorithm, 96
cognitive learning, 2
common payoff game, 74

CALA, 83
FALA, 78
multimodal game matrix, 196
parallelization, 196
team of FALA, 148
unimodal game matrix, 196
common reinforcement, 53
communication network, 8
compatibility functions, 78
complex structures of LA, 137
concept learning, 53, 92
conjunctive concept, 94
conjunctive logic expression, 92
connectionist systems, 1
consistent labeling, 78
constraint satisfaction, 219
context vector, 33, 106, 107, 110
    control problem, 33
    pattern recognition, 33
continuous action learning automata, 24
control of synchronous machines, 220
Courant Fisher Minimax Theorem, 210

data networks, 214
datamining, 5, 170, 225
decision tree, 106, 164, 170
    axis-parallel or univariate, 165
    hyperplane, 164
    leaf node, 164
    learning algorithms, 166
    left child, 166, 169
    non-leaf node, 164
    oblique or multivariate, 165
    pruning, 176
    right child, 166, 169
diffusion, 44
discretized learning automata, 22
discriminant function, 143, 144, 148, 152, 158
    linear, 144

nonlinear, 152
parameter vector, 144
Donsker's Theorem, 250
drift, 119
dynamical systems, 89

empirical expectation, 6
equilibrium point, 122
estimator algorithms, 18
evolutionary algorithms, 224
expert systems, 1

feature grouping, 206
feature space, 106
finite action learning automaton, 10
fixed structure stochastic automata, 48
FONKT, 118
Fubini's theorem, 58
function learning, 171
    example, 172

G-environment, 107
game matrix, 55
game of automata, 52
Gaussian density, 27
generalization, 3, 6
generalized environment, 107
generalized learning automaton, 33
generate and test, 3
GLA
    context vectors, 34
    internal state, 34
    optimal mapping, 35
    probability generating function, 34, 35, 38
global convergence, 129
global maximum, 44, 51, 79, 129, 130, 150, 161
globally Lipschitz, 38
Goore game, 82
gradient ascent, 38, 145
gradient-free optimization, 76, 84
graph bisection, 209
graph partitioning, 207, 221
Gronwall lemma, 232

hierarchical organization, 137
hierarchy, 127
hierarchy of automata, 49
high-level vision, 206
hill climbing, 3
hybrid networks, 138
hyperplane, 155, 156, 164, 167

incomplete information, 52
induction, 3
infinitely oscillatory, 127
intelligent behavior, 1
internal state

GLA, 34, 39
PLA, 43
invariant probability measure, 130
inverted pendulum, 4
Iris data, 97, 160, 170
isolated maxima, 125

Kiefer–Wolfowitz scheme, 85
Kuhn–Tucker conditions, 118

Langevin equation, 130
Laplacian matrix, 210
LaSalle's Theorem, 126
learning agents, 190
learning automaton, 5, 8, 10
    accuracy, 177
    action, 10
    action probability distribution, 10
    CALA, 24
    CARLA, 32
    FALA, 10
    GLA, 33
    learning algorithm, 14
    module, 178
    module of CALA, 185
    optimal action, 13
    parallel operation of, 177
    PLA, 43
    reaction, 10
    speed of convergence, 177
learning control, 219
learning from examples, 2
learning machines, 2
learning with a critic, 3
linear attribute, 93
linear regressor model, 219
linear reward ϵ-penalty algorithm, 17
linear reward–inaction algorithm, 14
linear reward–penalty algorithm, 17
linear system identification, 90
LMS algorithm, 85, 88
local maximum, 51, 55, 78, 83, 117, 150, 151, 161
low-level vision, 206
Lyapunov function, 70

Markov process, 13
Markovian decision processes, 220
martingale, 229
martingale convergence theorem, 181
mean burst length, 218
memory neuron network, 219
mixed strategy, 52, 82
model tree, 170
    examples, 172
module of CALA, 185
module of FALA, 178
    fuser, 178

INDEX

modules of GLA, 201
modules of PLA, 199
multi-criteria optimization, 54
multi-machine power system, 220
multiagent systems, 223
multidimensional Gaussian, 58, 143
multistage decision process, 33
multiteacher environments, 204

Nash equilibrium, 55, 77
network of LA, 109
    parallelization, 197
network of learning units, 106
network of PLA, 161
networks of GLA, 136
neural networks, 84
nominal attribute, 92
nonlinear algorithms, 48

object partitioning, 221
object recognition, 206
ODE method, 16, 227
operant conditioning, 2
optimal action, 13
optimal action mapping, 108
optimization, 51

packet transmission, 216
parallel pursuit algorithm, 187
parameter optimization, 205
parameterized learning automaton, 43
parameterized model, 90
pathological cases, 127
pattern recognition, 3, 8, 53, 139
    class label, 140, 141
    classification, 140
    decision rule, 139, 141
    decision tree, 164
    examples of, 140
    feature extraction, 140
    feature space, 141
    feature vector, 140, 141
    loss function, 141, 142, 146
    network of PLA, 161
    noisy training samples, 143
    three layer network, 155
    training patterns, 140, 143, 144
payoff, 52
payoff matrix, 78
    mode, 78
perceptual organization, 206
performance index, 3, 5, 9
perturbation, 129
piece-wise linear function, 170
PLA
    internal state, 43
    probability generating function, 43

polyhedral sets, 171
posterior probability, 142, 146
prior probabilities, 142, 143
prisoner's dilemma game, 82
probabilistic teacher, 3, 8
probability generating function
    GLA, 34, 35, 38
    PLA, 43
probability of misclassification, 140, 142, 146
pure maximal point, 61
pure strategy, 82
pursuit algorithm, 18
    team of FALA, 79

Q-learning, 220
quantization, 51

random environment, 3, 10
    P–model, 12
    penalty, 12
    Q–model, 12
    reward, 12
    S–model, 12
    stationary, 12
random perturbation, 44
rate of learning, 177
reaction, 10
regression function, 84, 146
regression tree, 170
REINFORCE algorithm, 36, 132
reinforcement, 3, 10
reinforcement learning, 3, 4, 6
relaxation labeling, 77, 221
reward probability matrix, 150
Robbins–Munro algorithm, 85, 88, 145

saddle point, 82
scalar reinforcement signal, 107, 113
scene structure graph, 207
segmentation, 206
shared memory system, 191
Shubert function, 30
simulated annealing, 44, 129
simultaneous perturbation stochastic approximation, 86
soft computing, 225
speed of convergence, 177
SRS, 107, 113
star automaton, 48
statistical learning theory, 6, 144
stochastic approximation, 6, 85, 145
stochastic difference equation, 26
stochastic differential equation, 130
stochastic estimator algorithm, 23
stochastic game, 52
    approximating ODE, 65
    CALA, 54
    FALA, 54

incomplete information, 52
maximal point, 59, 68, 72, 75
mixed strategy, 52
modal point, 61, 83
optimal point, 54, 57, 77
parallel operation of, 192
payoff, 52
payoff function, 54
play, 52
players, 52
solution, 52
stochastic gradient following, 27, 36
stochastic hill climbing, 4
stochastic neural network, 107
stochastic optimization, 53, 84
strict maximal point, 62
strict modal point, 62
submartingale, 15, 181
subregular function, 181
supermartingale, 15
supervised learning, 3
symbolic AI, 2

system identification, 53, 89

TD(λ) algorithm, 204
TDMA protocols, 215
team of automata, 75
team of modules of FALA, 192
temporal difference learning, 204
tight, 248
time-optimal control, 106
training patterns, 143
training sample, 84
trial and error, 3

unstable, 62

variable structure stochastic automata, 48
vehicle suspension control, 219

weak convergence, 130, 247
wine recognition data, 98

zero-sum game, 82